Power Systems Handbook

Harmonic Generation Effects
Propagation and Control
Volume 3

Power Systems Handbook

Series Author
J.C. Das
Power System Studies, Inc., Snellville, Georgia, USA

Volume 1: Short-Circuits in AC and DC Systems:
ANSI, IEEE, and IEC Standards

Volume 2: Load Flow Optimization and Optimal Power Flow

Volume 3: Harmonic Generation Effects Propagation and Control

Volume 4: Power Systems Protective Relaying

Harmonic Generation Effects
Propagation and Control
Volume 3

J.C. Das

CRC Press
Taylor & Francis Group
Boca Raton London New York

CRC Press is an imprint of the
Taylor & Francis Group, an **informa** business

CRC Press
Taylor & Francis Group
6000 Broken Sound Parkway NW, Suite 300
Boca Raton, FL 33487-2742

First issued in paperback 2022

ISBN-13: 978-1-498-74546-8 (hbk)
ISBN-13: 978-1-03-233943-6 (pbk)
DOI: 10.1201/9781351228305

Visit the Taylor & Francis Web site at
http://www.taylorandfrancis.com

and the CRC Press Web site at
http://www.crcpress.com

Contents

Series Preface

This handbook on power systems consists of four volumes. These are carefully planned and designed to provide state-of-the-art material on the major aspects of electrical power systems, short-circuit currents, load flow, harmonics, and protective relaying.

An effort has been made to provide a comprehensive coverage, with practical applications, case studies, examples, problems, extensive references, and bibliography.

The material is organized with sound theoretical base and its practical applications. The objective of creating this series is to provide the reader with a comprehensive treatise that could serve as a reference and day-to-day application guide for solving the real-world problem. It is written for plasticizing engineers and academia at the level of upper-undergraduate and graduate degrees.

Though there are published texts on similar subjects, this series provides a unique approach to the practical problems that an application engineer or consultant may face in conducting system studies and applying it to varied system problems.

Some parts of the work are fairly advanced on a postgraduate level and get into higher mathematics. Yet the continuity of the thought process and basic conceptual base are maintained. A beginner and advanced reader will equally benefit from the material covered. An underground level of education is assumed, with a fundamental knowledge of electrical circuit theory, rotating machines, and matrices.

Currently, power systems, large or small, are analyzed on digital computers with appropriate software. However, it is necessary to understand the theory and basis of these calculations to debug and decipher the results.

A reader may be interested only in one aspect of power systems and may choose to purchase only one of the volumes. Many aspects of power systems are transparent between different types of studies and analyses—for example, knowledge of short-circuit currents and symmetrical component is required for protective relaying and fundamental frequency load flow is required for harmonic analysis. Though appropriate references are provided, the material is not repeated from one volume to another.

The series is a culmination of the vast experience of the author in solving real-world problems in the industrial and utility power systems for more than 40 years.

Another key point is that the solutions to the problems are provided in Appendix D. Readers should be able to independently solve these problems after perusing the contents of a chapter and then look back to the solutions provided as a secondary help. The problems are organized so these can be solved with manual manipulations, without the help of any digital computer power system software.

It is hoped the series will be a welcome addition to the current technical literature.

The author thanks CRC Press editor Nora Konopka for her help and cooperation throughout the publication effort.

—J.C. Das

Preface to Volume 3: Harmonic Generation Effects Propagation and Control

The power system harmonics is a subject of interest to many power system professionals engaged in harmonic analysis and mitigation. It is one of the major power quality concerns.

Volume 3 provides coverage of generation, effects, and control of harmonics, including interharmonics, measurements, estimation of harmonics, harmonic resonance, and harmonic limitations according to standards. The intention is that the book can serve as a practical guide to practicing engineers on harmonics.

A beginner should be able to form a clear base for understanding the subject of harmonics and an advanced reader's interest should be stimulated to explore further. In writing this book, an undergraduate level of knowledge is assumed. It has the potentiality of serving as advance undergraduate and graduate textbook. Surely, it can serve as continuing education textbook and supplementary reading material.

The effects of harmonics can be experienced at a distance, and the effect on power system components is a dynamic and evolving field. These interactions have been analyzed in terms of current thinking. The concepts of modeling filter designs and harmonic penetrations (propagations) in industrial systems, distribution, and transmission systems are amply covered with the application of SVCs and FACTS controllers. An introduction to the active filters, multilevel inverters, and active current shaping is provided.

A chapter is included on harmonic analysis in wind and solar generating plants. Many case studies and practical examples are included. The problems at the end of a chapter can be solved by hand without resort to any computer software packages. Appendix A contains Fourier analysis, pertinent to harmonic analysis, and Appendix B provides solutions to the problems.

—J.C. Das

Author

J.C. Das is an independent consultant, Power System Studies, Inc. Snellville, Georgia. Earlier, he headed the electrical power systems department at AMEC Foster Wheeler for 30 years. He has varied experience in the utility industry, industrial establishments, hydroelectric generation, and atomic energy. He is responsible for power system studies, including short circuit, load flow, harmonics, stability, arc flash hazard, grounding, switching transients, and protective relaying. He conducts courses for continuing education in power systems and is the author or coauthor of about 70 technical publications nationally and internationally. He is the author of the following books:

- *Arc Flash Hazard Analysis and Mitigation*, IEEE Press, 2012.
- *Power System Harmonics and Passive Filter Designs*, IEEE Press, 2015.
- *Transients in Electrical Systems: Analysis Recognition and Mitigation*, McGraw-Hill, 2010.
- *Power System Analysis: Short-Circuit Load Flow and Harmonics*, Second Edition, CRC Press 2011.
- *Understanding Symmetrical Components for Power System Modeling*, IEEE Press, 2017.

These books provide extensive converge, running into more than 3000 pages, and are well received in the technical circles. His interests include power system transients, EMTP simulations, harmonics, passive filter designs, power quality, protection, and relaying. He has published more than 200 electrical power system study reports for his clients.

He has published more than 200 study reports of power systems analysis addressing one problem or the other.

Das is a Life Fellow of the Institute of Electrical and Electronics Engineers, IEEE, (USA), Member of the IEEE Industry Applications and IEEE Power Engineering societies, a Fellow of the Institution of Engineering Technology (UK), a Life Fellow of the Institution of Engineers (India), a Member of the Federation of European Engineers (France), a Member of CIGRE (France), etc. He is registered Professional Engineer in the states of Georgia and Oklahoma, a Chartered Engineer (CEng) in the UK, and a European Engineer (EurIng) in Europe. He received a meritorious award in engineering, IEEE Pulp and Paper Industry in 2005.

He earned a PhD in electrical engineering at Atlantic International University, Honolulu, an MSEE at Tulsa University, Tulsa, Oklahoma, and a BA in advanced mathematics and a BEE at Panjab University, India.

1

Harmonics Generation

Harmonics in power systems can be studied under distinct sections:

- Generation of characteristic and noncharacteristic harmonics
- Interharmonics and flicker
- Resonance, secondary resonance, and harmonic resonance
- Effects of harmonics
- Limitations of harmonics according to the IEEE and IEC standards
- Measurements of harmonics
- Harmonic propagation, modeling, and analysis
- Mitigation of harmonics, passive and active filters

Harmonics cause distortions of the voltage and current waveforms, which have adverse effects on electrical equipment. Harmonics are one of the major power quality concerns. The estimation of harmonics from nonlinear loads is the first step in a harmonic analysis and this may not be straightforward. There is an interaction between the harmonic producing equipment, which can have varied topologies, and the electrical system. Over the course of recent years, much attention has been focused on the analysis and control of harmonics, and standards have been established for permissible harmonic current and voltage distortions.

In this chapter, we will discuss the nature of harmonics and their generation by electrical equipment. Harmonic emission can have varied amplitudes and frequencies. The most common harmonics in power systems are sinusoidal components of a periodic waveform, which have frequencies that can be resolved into some multiples of the fundamental frequency. Fourier analysis is the mathematical tool employed for such analysis, and Appendix A provides an overview. It is recommended that the reader becomes familiarized with Fourier analysis before proceeding with the subject of harmonics. Power systems also have harmonics that are noninteger multiples of the fundamental frequency and have aperiodic waveforms, see Chapter 2. The generation of harmonics in power system occurs from two distinct types of loads as follows:

1. Linear time-invariant loads are characterized so that an application of a sinusoidal voltage results in a sinusoidal flow of current. These loads display constant steady-state impedance during the applied sinusoidal voltage. If the voltage is increased, the current also increases in direct proportion. Incandescent lighting is an example of such a load. Transformers and rotating machines, under normal loading conditions, approximately meet this definition, though the flux wave in the air gap of a rotating machine is not sinusoidal. Tooth ripples and slotting may produce forward and reverse rotating harmonics. Magnetic circuits

can saturate and generate harmonics. As an example, saturation in a transformer on abnormally high voltage produces harmonics, as the relationship between magnetic flux density B and the magnetic field intensity H in the transformer core is not linear. The inrush current of a transformer contains odd and even harmonics, including a dc component. Yet, under normal operating conditions, these effects are small. Synchronous generators in power systems produce sinusoidal voltages and the loads draw nearly sinusoidal currents. For the sinusoidal input voltages, the harmonic pollution produced due to these load types of loads is small.

2. The second category of loads is described as nonlinear. In a nonlinear device, the application of a sinusoidal voltage does not result in a sinusoidal flow of current. These loads do not exhibit constant impedance during the entire cycle of applied sinusoidal voltage. *Nonlinearity is not the same as the frequency dependence of impedance*, i.e., the impedance of a reactor changes in proportion to the applied frequency, but it is linear at each applied frequency. On the other hand, nonlinear loads draw a current that may even be discontinuous, or flow in pulses for a part of the sinusoidal voltage cycle. Some examples of nonlinear loads are as follows:

   - Adjustable drive systems
   - Cycloconverters
   - Arc furnaces and rolling mills
   - Switching mode power supplies (SMPSs)
   - Computers, copy machines, and television sets
   - Static var compensators (SVCs)
   - HVDC transmission
   - Electric traction
   - Switching mode power supplies
   - Wind and solar power generation
   - Pulse burst modulation (PBM)
   - Battery charging and fuel cells
   - Slip recovery schemes of induction motors
   - Fluorescent lighting and electronic ballasts
   - Silicon-controlled rectifier (SCR) heating, induction heating, and arc welding

The distortion produced by nonlinear loads can be resolved into a number of categories:

- A distorted waveform having a Fourier series with fundamental frequency equal to power system frequency, and a periodic steady state exists. This is the most common case in harmonic studies.
- A distorted waveform having a submultiple of power system frequency, and a periodic steady state exists. Certain types of pulsed loads and integral cycle controllers produce these types of waveforms.
- The waveform is aperiodic, but perhaps almost periodic. A trigonometric series expansion may still exist. Examples are arcing devices, e.g., arc furnaces,

fluorescent, mercury, and sodium vapor lighting. The process is not periodic in nature, and a periodic waveform is obtained if the conditions of operation are kept constant for a length of time.

The components in a Fourier series that are not an integral multiple of the power frequency are called noninteger harmonics, see Chapter 2.

The arc furnace loads are highly polluting; cause phase unbalance, flicker, impact loading, harmonics, and resonance; and may give rise to torsional vibrations in rotating equipment.

1.1 Sequence Components of Harmonics

In a three-phase balanced system under nonsinusoidal conditions, the hth-order harmonic voltage (or current) can be expressed as follows:

$$V_{ah} = V_h \sin(h\omega_0 t + \theta_h) \tag{1.1}$$

$$V_{bh} = V_h \sin\left(h\omega_0 t - \frac{2h\pi}{3} + \theta_h\right) \tag{1.2}$$

$$V_{ch} = V_h \sin\left(h\omega_0 t + \frac{2h\pi}{3} + \theta_h\right) \tag{1.3}$$

Based on Equations 1.1 through 1.3 and counterclockwise rotation of the fundamental phasors, we can write

$$V_a = V_1 \sin\omega t + V_2 \sin 2\omega t + V_3 \sin 3\omega t + V_4 \sin 4\omega t + V_5 \sin 5\omega t + \cdots$$

$$V_b = V_1 \sin(\omega t - 120°) + V_2 \sin(2\omega t - 240°) + V_3 \sin(3\omega t - 360°) + V_4 \sin(4\omega t - 480°)$$

$$+ V_5 \sin(5\omega t - 600°) + \cdots$$

$$= V_1 \sin(\omega t - 120°) + V_2 \sin(2\omega t + 120°) + V_3 \sin 3\omega t + V_4 \sin(4\omega t - 120°)$$

$$+ V_5 \sin(5\omega t + 120°) + \cdots \tag{1.4}$$

$$V_c = V_1 \sin(\omega t + 120°) + V_2 \sin(2\omega t + 240°) + V_3 \sin(3\omega t + 360°) + V_4 \sin(4\omega t + 480°)$$

$$+ V_5 \sin(5\omega t + 600°) + \cdots$$

$$= V_1 \sin(\omega t + 120°) + V_2 \sin(2\omega t - 120°) + V_3 \sin 3\omega t + V_4 \sin(4\omega t + 120°)$$

$$+ V_5 \sin(5\omega t - 120°) + \cdots$$

Under balanced conditions, the hth harmonic (frequency of harmonic = h times the fundamental frequency) of phase b lags h times 120° behind that of the same harmonic in phase a. The hth harmonic of phase c lags h times 240° behind that of the same harmonic in phase a. In the case of triplen harmonics, shifting the phase angles by three times 120° or

TABLE 1.1
Sequence of Harmonics

Harmonic Order	Sequence of the Harmonic
1	+
2	−
3	0
4	+
5	−
6	0
7	+
8	−
9	0
10, 11, 12	+, −0

three times 240° results in cophasial vectors. Table 1.1 shows the sequence of harmonics, and the pattern is clearly positive–negative–zero. We can write

$$\text{Harmonics of the order } 3h+1 \text{ have positive sequence} \qquad (1.5)$$

$$\text{Harmonics of the order } 3h+2 \text{ have negative sequence} \qquad (1.6)$$

and

$$\text{Harmonics of the order } 3h \text{ are of zero sequence} \qquad (1.7)$$

All triplen harmonics generated by nonlinear loads are zero sequence phasors. These add up in the neutral. In a three-phase four-wire system, with perfectly balanced single-phase loads between the phase and the neutral, all positive and negative sequence harmonics will cancel out, leaving only the zero sequence harmonics. In an unbalanced single-phase load, the neutral carries zero sequence and the residual unbalance of positive and negative sequence currents. Even harmonics are absent in the line because of phase symmetry (Appendix A) and unsymmetrical waveforms will add even harmonics to the phase conductors.

1.2 Increases in Nonlinear Loads

Nonlinear loads are continually on the increase. It is estimated that, during the next 10 years, 60% of the loads on utility systems will be nonlinear. Concerns for harmonics originate from meeting a certain power quality, which leads to the related issues of (1) effects on the operation of electrical equipment, (2) harmonic analysis, and (3) harmonic control. A growing number of consumer loads are sensitive to poor power quality and it is estimated that power-quality problems cost US industry tens of billions of dollars per year. While the expanded use of consumer automation equipment and power electronic controls is leading to higher productivity, these very loads are a source of electrical noise, and harmonics are less tolerant to poor power quality. For example, adjustable speed drives (ASDs) are less tolerant to voltage sags and swells, and a voltage dip of 10% of certain duration may precipitate a shutdown.

1.3 Harmonic Factor

An index of merit has been defined as a harmonic distortion factor [1] (harmonic factor). It is the ratio of the root mean square (RMS) of the harmonic content to the RMS value of the fundamental quantity, expressed as a percentage of the fundamental:

$$DF = \sqrt{\frac{\sum \text{of squares of amplitudes of all harmonics}}{\text{square of the amplitude of the fundamental}}} \times 100\% \tag{1.8}$$

Voltage and current harmonic distortion indices, defined in Chapter 5, are the most commonly used indices. Total harmonic distortion (THD) in common use is the same as DF.

1.3.1 Equations for Common Harmonic Indices

We can write the following equations.
RMS voltage in the presence of harmonics can be written as follows:

$$V_{\text{rms}} = \sqrt{\sum_{h=1}^{h=\infty} V_{h,\text{rms}}^2} \tag{1.9}$$

And similarly the expression for the current is

$$I_{\text{rms}} = \sqrt{\sum_{h=1}^{h=\infty} I_{h,\text{rms}}^2} \tag{1.10}$$

The total distortion factor for the voltage is

$$\text{THD}_V = \frac{\sqrt{\sum_{h=2}^{h=\infty} V_{h,\text{rms}}^2}}{V_{f,\text{rms}}} \tag{1.11}$$

where $V_{f,\text{rms}}$ is the fundamental frequency voltage. This can be written as follows:

$$\text{THD}_V = \sqrt{\left(\frac{V_{\text{rms}}}{V_{f,\text{rms}}}\right)^2 - 1} \tag{1.12}$$

or

$$V_{\text{rms}} = V_{f,\text{rms}} \sqrt{1 + \text{THD}_V^2} \tag{1.13}$$

Similarly

$$\text{THD}_I = \frac{\sqrt{\sum\limits_{h=2}^{h=\infty} I_{h,\text{rms}}^2}}{I_{f,\text{rms}}} = \sqrt{\left(\frac{I_{\text{rms}}}{I_{f,\text{rms}}}\right)^2 - 1} \qquad (1.14)$$

$$I_{\text{rms}} = I_{f,\text{rms}} \sqrt{1 + \text{THD}_I^2} \qquad (1.15)$$

where $I_{f,\text{rms}}$ is the fundamental frequency current.

The total demand distortion (TDD) is defined as follows:

$$\text{TDD} = \frac{\sqrt{\sum\limits_{h=2}^{h=\infty} I_h^2}}{I_L} \qquad (1.16)$$

where I_L is the load demand current.

The partial weighted harmonic distortion (PWHD) of current is defined as follows:

$$\text{PWHD}_I = \frac{\sqrt{\sum\limits_{h=14}^{h=40} h I_h^2}}{I_{f,\text{rms}}} \qquad (1.17)$$

Similar expression is applicable for the voltage. The PWHD evaluates influence of current or voltage harmonics of higher order. The sum parameters are calculated with single harmonic current components I_h.

1.4 Three-Phase Windings in Electrical Machines

The armature windings of a machine consist of phase coils that span approximately a pole pitch. A phase winding consists of a number of coils connected in series, and the EMF generated in these coils is time displaced in phase by a certain angle. The air gap is bounded on either side by iron surfaces and provided with slots and duct openings and is skewed. Simple methods of estimating the reluctance of the gap to carry a certain flux across the gap are not applicable and the flux density in the air gap is not sinusoidal. Figure 1.1 shows that armature reaction varies between a pointed and flat-topped trapezium for a phase spread of $\pi/3$. Fourier analysis of the pointed waveform in Figure 1.1 gives

$$F = \frac{4}{\pi} F_m \cos \omega t \left[\sum_{h=1}^{h=\infty} \frac{1}{h} k_{mn} \sin hx \right] \qquad (1.18)$$

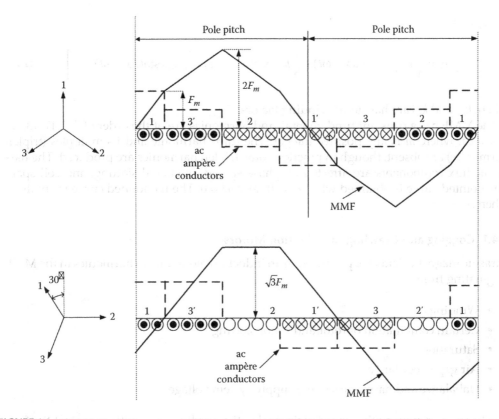

FIGURE 1.1
Armature reaction of a three-phase winding spanning a pole pitch.

where k_{mn} is a winding distribution factor for the hth harmonic:

$$k_{mn} = \frac{\sin(1/2)h\sigma}{g'\sin(1/2)\,(h\sigma/g')} \tag{1.19}$$

in which g' is the number of slots per pole per phase and σ is the phase spread, h is order of harmonic.

The MMFs of three phases will be given by considering the time displacement of currents and space displacement of axes as follows:

$$F_t = \frac{4}{\pi}F_m\cos\omega t\left[\sum_{h=1}^{h=\infty}\frac{1}{n}k_{mn}\sin hx\right] + \frac{4}{\pi}F_m\cos\left(\omega t - \frac{2}{3}\pi\right)$$

$$\times\left[\sum_{h=1}^{h=\infty}\frac{1}{h}k_{mn}\sin h\left(x - \frac{2}{3}\pi\right)\right] \tag{1.20}$$

$$+\frac{4}{\pi}F_m\cos\left(\omega t - \frac{4}{3}\pi\right)\left[\sum_{h=1}^{h=\infty}\frac{1}{h}k_{mn}\sin h\left(x - \frac{4}{3}\pi\right)\right]$$

This gives

$$F_t = \frac{6}{\pi} F_m \left[F_{mi} \sin(x - \omega t) + \frac{1}{5} k_{m5} \sin(5x - \omega t) - \frac{1}{7} k_{m7} \sin(7x - \omega t) + \cdots \right] \qquad (1.21)$$

where k_{m5} and k_{m7} are harmonic winding factors.

The MMF has a constant fundamental, and harmonics are of the order of 5, 7, 11, 13, ..., or $6m \pm 1$, where m is any positive integer. The third harmonic and its multiples (triplen harmonics) are absent, though in practice, some triplen harmonics are produced. The harmonic flux components are affected by phase spread, fractional slotting, and coil span. The pointed curve is obtained when $\sigma = 60°$ and $\omega t = 0$. The flat topped curve is obtained when $\omega t = \pi/6$.

1.4.1 Cogging and Crawling of Induction Motors

Parasitic magnetic fields are produced in an induction motor due to harmonics in the MMF originating from

- Windings
- Certain combination of rotor and stator slotting
- Saturation
- Air gap irregularity
- Unbalance and harmonics in the supply system voltage

The harmonics move with a speed reciprocal to their order, either with or against the fundamental. Harmonics of the order of $6m + 1$ move in the same direction as the fundamental magnetic field while those of $6m - 1$ move in the opposite direction.

1.4.2 Harmonic Induction Torques

The harmonics can be considered to produce, by an additional set of rotating poles, rotor EMF's, currents, and harmonic torques akin to the fundamental frequency at synchronous speeds depending upon the order of the harmonics. Then, the resultant speed–torque curve will be a combination of the fundamental and harmonic torques. This produces a saddle in the torque speed characteristics and the motor can crawl at the lower speed of 1/7th of the fundamental, see Figure 1.2a. This torque speed curve is called the *harmonic induction torque curve*.

This harmonic torque can be augmented by stator and rotor slotting. In n-phase winding, with g' slots per pole per phase, EMF distribution factors of the harmonics are

$$h = 6Ag' \pm 1 \qquad (1.22)$$

where A is any integer, 0, 1, 2, 3, ...

The harmonics of the order $6Ag' + 1$ rotate in the same direction as the fundamental, while those of order $6Ag' - 1$ rotate in the opposite direction.

A four-pole motor with 36 slots, $g' = 3$ slots per pole per phase, will give rise to 17th and 19th harmonic torque saddles, observable at +1/19 and −1/17 speed, similar to the saddles shown in Figure 1.2a.

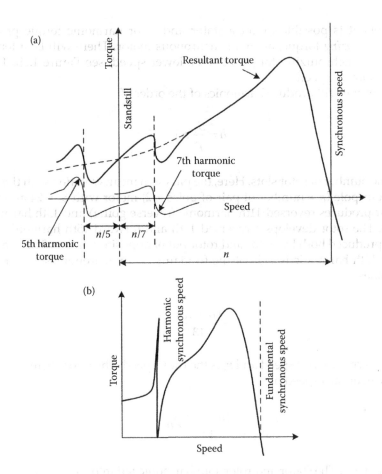

FIGURE 1.2
(a) Harmonic induction torques and (b) synchronous torques in an induction motor.

Consider 24 slots in the stator of a four-pole machine. Then $g' = 2$ and 11th and 13th harmonics will be produced strongly. The harmonic induction torque thus produced can be augmented by the rotor slotting. For a rotor with 44 slots, 11th harmonic has 44 half waves each corresponding to a rotor bar in a squirrel cage induction motor. This will accentuate 11th harmonic torque and produce strong vibrations.

If the numbers of stator slots are equal to the number of rotor slots, the motor may not start at all, a phenomenon called *cogging*.

The phenomena will be more pronounced in squirrel cage induction motors as compared to wound rotor motors, as the effect of harmonics can be reduced by coil pitch, see Section 3.4. In the cage induction motor design, S_2 (number of slots in the rotor) should not exceed S_1 (number of slots in the stator) by more than 50%–60%; otherwise, there will be some tendency toward saddle harmonic torques.

1.4.3 Harmonic Synchronous Torques

Consider that the fifth and seventh harmonics are present in the gap of a three-phase induction motor. With this harmonic content and with certain combination of the stator and rotor slots, it is possible to get a stator and rotor harmonic torque producing a *harmonic synchronizing* torque, as in a synchronous motor. There will be a tendency to develop sharp synchronizing torque at some lower speed, see Figure 1.2b. The motor may crawl at a lower speed.

The rotor slotting will produce harmonics of the order of

$$h = \frac{S_2}{p} \pm 1 \tag{1.23}$$

where S_2 is the number of rotor slots. Here, the plus sign means rotation with the machine. Consider a four-pole (p = number of pair of poles = 2) motor with $S_1 = 24$ and with $S_2 = 28$. The stator produces reversed 11th harmonic (reverse going) and 13th harmonic (forward going). The rotor develops a reversed 13th and forward 15th harmonic. The 13th harmonic is produced both by stator and rotor but of opposite rotation. The synchronous speed of the 13th harmonic is 1/13 of the fundamental synchronous speed. Relative to rotor, it becomes

$$-\frac{(n_s - n_r)}{13} \tag{1.24}$$

where n_s is the synchronous speed and n_r is the rotor speed. The rotor, therefore, rotates its own 13th harmonic at a speed of

$$-\frac{(n_s - n_r)}{13} + n_r \tag{1.25}$$

relative to the stator. The stator and rotor 13th harmonic fall into step when

$$+\frac{n_s}{13} = -\frac{(n_s - n_r)}{13} + n_r \tag{1.26}$$

This gives $n_r = n_s/7$, i.e., torque discontinuity is produced not by 7th harmonic but by 13th harmonic in the stator and rotor rotating in opposite directions. The torque–speed curve is shown in Figure 1.3.

- The synchronous torque at 1800/7 = 257 rpm
- Induction torque due 13th stator harmonic = 138 rpm
- Induction torque due to reversed 11th harmonic = 164 rpm

Typical synchronous torques in four-pole cage induction motors are listed in Table 1.2. If $S_1 = S_2$, the same order harmonics will be strongly produced, and each pair of harmonics will produce a synchronizing torque, and the rotor may remain at standstill (cogging), unless the fundamental frequency torque is large enough to start the motor.

The harmonic torques are avoided in the design of induction machines by proper selection of the rotor and stator slotting and winding designs.

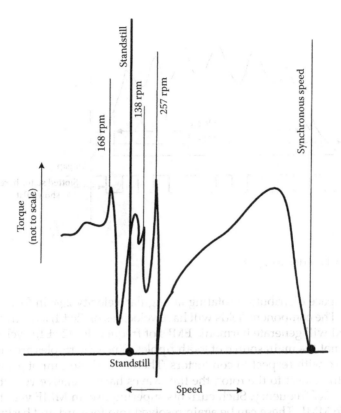

FIGURE 1.3
Torque speed curve of a four-pole, 60 Hz motor, considering harmonic synchronous torques.

TABLE 1.2

Typical Synchronous Torques Four-Pole Cage Induction Motors

Stator Slots	Rotor Slots	Stator Harmonics		Rotor Harmonics	
S_1	S_2	Negative	Positive	Negative	Positive
24	20	−11	+13	−9	+11
24	28	−11	+13	−13	+15
36	32	−17	+19	−15	+17
36	40	−17	+19	−19	+21
48	44	−23	+25	−21	+23

1.5 Tooth Ripples in Electrical Machines

Tooth ripples in electrical machinery are produced by slotting as these affect air-gap permeance. Figure 1.4 shows ripples in the air-gap flux distribution (exaggerated) because of variation in gap permeance. The frequency of flux pulsations corresponds to the rate at which slots cross the pole face, i.e., it is given by $2gf$, where g is the number of slots per pole and f is the system frequency. This stationary pulsation may be regarded as two waves

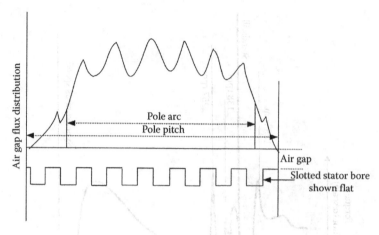

FIGURE 1.4
Gap flux distribution due to tooth ripples.

of fundamental space distribution rotating at angular velocity $2g\omega$ in forward and backward directions. The component fields will have velocities of $(2g \pm 1)\omega$ relative to the armature winding and will generate harmonic EMFs of frequencies $(2g \pm 1)f$ cycles per second. However, this is not the main source of tooth ripples. Since the ripples are due to slotting, these do not move with respect to conductors. Therefore, these cannot generate an EMF of pulsation. With respect to the rotor, the flux waves have a relative velocity of $2g\omega$ and generate EMFs of $2gf$ frequency. Such currents superimpose an MMF variation of $2gf$ on the resultant pole MMF. These can be again resolved into forward and backward moving components with respect to the *rotor*, and $(2g \pm 1)\omega$ with respect to the *stator*. Thus, stator EMFs at frequencies $(2g \pm 1)f$ are generated, which are the principle tooth ripples.

1.6 Synchronous Generators Waveforms

The terminal voltage wave of synchronous generators must meet the requirements of NEMA, which states that the deviation factor of the open line-to-line terminal voltage of the generator shall not exceed 0.1.

Figure 1.5 shows a plot of a hypothetical generated wave, superimposed on a sinusoid, and the deviation factor is defined as follows:

$$F_{\text{DEV}} = \frac{\Delta E}{E_{\text{OM}}} \tag{1.27}$$

where E_{OM} is calculated from a number of samples of instantaneous values:

$$E_{\text{OM}} = \sqrt{\frac{2}{J} \sum_{j=1}^{J} E_j^2} \tag{1.28}$$

The deviation from a sinusoid is very small.

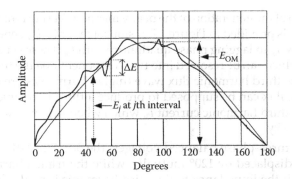

FIGURE 1.5
Measurements of deviation factor of a generator voltage.

Generator neutrals have predominant third harmonic voltages. In a wye-connected generator, with the neutral grounded through high impedance, the third harmonic voltage for a ground fault increases toward the neutral, while the fundamental frequency voltage decreases. The third harmonic voltages at line and neutral can vary considerably with load.

1.7 Transformers: Harmonics

Harmonics in transformers originate as a result of saturation, switching, high-flux densities, and winding connections. The following summarizes the main factors with respect to harmonic generation:

1. For economy in design and manufacture, transformers are operated close to the knee point of saturation characteristics of magnetic materials. Figure 1.6 shows a *B–H* curve and the magnetizing current waveform. A sinusoidal flux wave, required by sinusoidal applied voltage, demands a magnetizing current with a harmonic content. Conversely, with a sinusoidal magnetizing current, the induced EMF is peaky and the flux is flat topped.

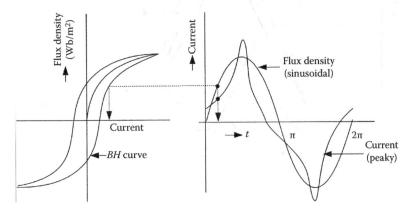

FIGURE 1.6
B–H curve of magnetic material and peaky transformer magnetizing current.

2. An explanation of the generation of the peaky magnetizing current considering the third harmonic is provided in Figure 1.7. A sinusoidal EMF, E_a, generates a sinusoidal current flow, I_a, in lagging phase quadrature with E_a. These set up a flat-topped flux wave, ϕ_1, which can be resolved into two components: ϕa the fundamental flux wave and ϕ_3 the third harmonic flux wave (higher harmonics are neglected). The third harmonic flux can be supposed to produce a third harmonic EMF E_3 and a corresponding third harmonic current I_3, which when summed with I_a makes the total current peaky.

3. In a system of three-phase balanced voltages, the 5th, 7th, 11th, and so on produce voltages displaced by 120° mutually, while the triplen harmonic voltages are cophasial. If the impedance to the third harmonic is negligible, only a very small third harmonic EMF is required to circulate a magnetizing current additive to the fundamental frequency, so as to maintain a sinusoidal flux. This is true if the transformer windings are delta connected. In wye–wye connected transformers with isolated neutrals, as all the triplen harmonics are either directed inward or outward, these cancel between the lines, no third harmonic currents flow, and the flux wave in the transformer is flat topped. The effect on a wye-connected point is to make it oscillate at three times the fundamental frequency, giving rise to distortion of the phase voltages (Figure 1.8). Tertiary delta-connected windings are included in wye–wye connected transformers for neutral stabilization.

4. Three-phase core-type transformers have magnetically interlinked phases, and the return paths of triplen harmonic fluxes lie outside the core, through the tank and transformer fluid, which have high reluctance. In five-limb transformers, the end limbs provide return paths for triplen harmonics.

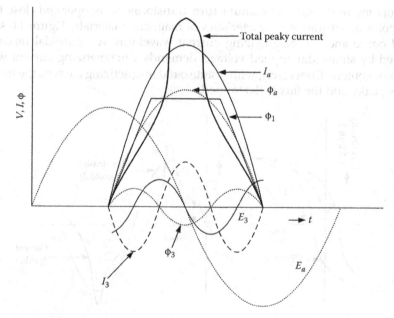

FIGURE 1.7
Origin of flat-topped flux wave in a transformer, third-harmonic current, and overall peaky magnetizing current.

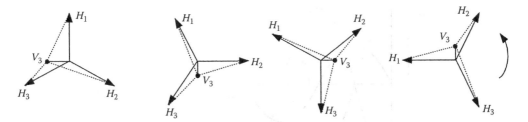

FIGURE 1.8
Phenomena of neutral oscillation in a wye–wye connected transformer, due to third-harmonic voltages.

5. It can be said that power transformers generate very low levels of harmonic currents in steady-state operation, and the harmonics are controlled by design and transformer winding connections. The higher order harmonics, i.e., the fifth and seventh, may be <0.1% of the transformer full-load current.

6. Energizing a power transformer does generate a high order of harmonics, including a dc component. Figure 1.9 shows three conditions of energizing of a power transformer: (1) the switch closed at the peak value of the voltage, (2) the switch closed at the zero value of the voltage, and (3) energizing with some residual trapped flux in the magnetic core due to retentivity of the magnetic materials. Figure 1.9d shows the spectrum of magnetizing inrush current, which resembles a rectified current and its peak value may reach 8–15 times the transformer full-load current, mainly depending on the transformer size. The asymmetrical loss due to conductor and core heating rapidly reduces the flux wave to symmetry about the time axis and typically the inrush currents last for a short duration (0.1 s).

Typical harmonics generated by the transformer inrush current are shown in Figure 1.10. Overexcitation of transformers in steady-state operation can produce harmonics. The generated fundamental frequency EMF is given by

$$V = 4.44 f T_{ph} B_m A_c \tag{1.29}$$

where T_{ph} is the number of turns in a phase, B_m is the flux density (consisting of fundamental and higher order harmonics), and A_c is the area of core. Thus, the factor V/f is a measure of the overexcitation, though these currents do not normally cause a wave distortion of any significance. Exciting currents increase rapidly with voltage, and transformer standards specify application of 110% voltage without overheating the transformer. Under certain system upset conditions, the transformers may be subjected to even higher voltages and overexcitation. ANSI protective device number 24, volts per hertz relay is used for overexcitation protection.

1.8 Harmonics due to Saturation of Current Transformers

Saturation of current transformers under fault conditions produces harmonics in the secondary circuits. Accuracy classification of current transformers is designated by one letter, C or T, depending on current transformer construction [2]. Classification C covers bushing-type transformers with uniformly distributed windings, and the leakage

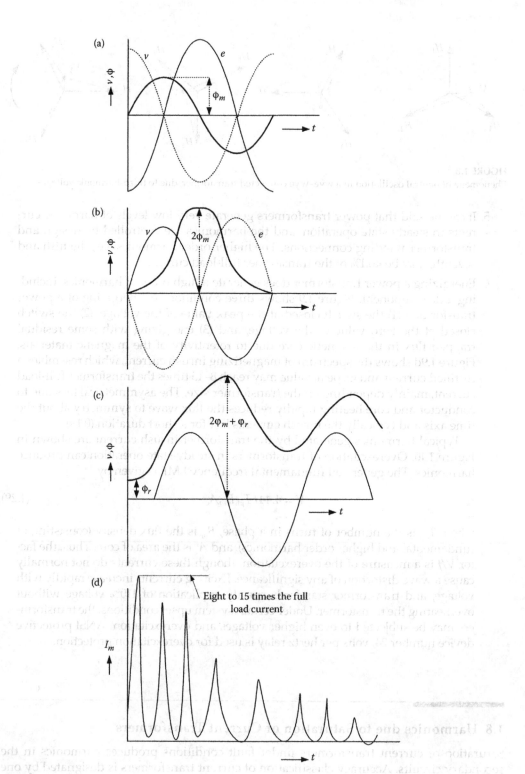

FIGURE 1.9
(a through d) Switching inrush current transients in a transformer (see the text).

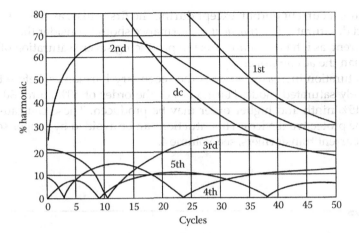

FIGURE 1.10
Harmonic components of the inrush current of a transformer.

flux has a negligible effect on the ratio within the defined limits. A transformer with relaying accuracy class C200 means that the percentage ratio correction will not exceed 10% at any current from 1 to 20 times the rated secondary current at a standard burden of 2.0 Ω, which will generate 200 V. The secondary voltage as given by maximum fault current reflected on the secondary side multiplied by connected burden ($R + jX$) should not exceed the assigned C accuracy class. When current transformers are improperly applied, saturation can occur, as shown in Figure 1.11 [3]. A completely saturated CT

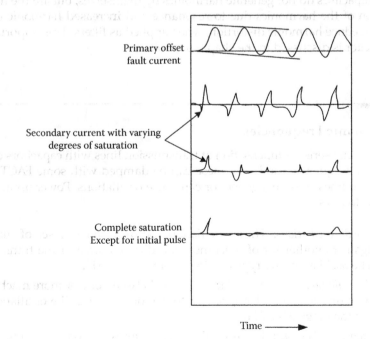

FIGURE 1.11
Saturation of a current transformer on asymmetrical fault current and origin of harmonics.

does not produce a current output, except during the first pulse, as there is a finite time to saturate and desaturate. The *transient* performance should consider the dc component of the fault current, as it has far more effect in producing severe saturation of the current transformer than the ac component.

As the CT saturation increases, so does the secondary harmonics, before the CT goes into a completely saturated mode. Harmonics of the order of 50% third, 30% fifth, 18% seventh, and 15% ninth and higher order may be produced. These can cause improper operation of the protective devices. This situation can be avoided by proper selection and application of current transformers, see Volume 4.

1.9 Switching of Shunt Capacitor Banks

High frequencies of inrush currents on switching of shunt capacitors occur, which are discussed in Volume 1. The frequency of the system transient is typically <1 kHz for an isolated capacitor bank and <5 kHz for back-to-back switching. Series filter reactors and switching inrush current limiting reactors reduce these frequencies. The fast front of the switching surge voltage can cause a part-winding resonance and harmonic generation in a transformer, if the frequency coincides with the transformer's natural frequency which is of the order of 10–100 kHz, the first resonance occurring in the range 7–15 kHz. The likelihood of exciting a part-winding resonance on switching is remote, but switching overvoltages are of major concern, see Volume 1.

The power capacitors do not generate harmonics by themselves, but are the main cause of amplification of the harmonics due to resonance and increased harmonic distortion. These can also reduce harmonic distortion, when applied as filters. This important aspect of power capacitor is discussed in the chapters to follow.

1.10 Subharmonic Frequencies

Volume 2 shows that series compensation of transmission lines with capacitors can generate subharmonic frequencies and how these can be damped with some FACTS devices. Switching of long lines close to a generator can cause oscillations. Power oscillations can be described as follows:

- *Interarea mode oscillations*: These oscillations occur when one set of machines swings against another set of machines in a different area of the transmission system. The oscillations are typically in the range 0.2–0.5 Hz.
- *Local mode oscillations*: These oscillations occur between one or more machines in a plant swinging against a large power source or network. The oscillations are typically in the range 0.7–2.0 Hz.
- *Interunit mode oscillations*: These oscillations occur when one machine swings against another machine in the same area in the same power plant. The oscillations are typically in the range 1.5–3.0 Hz.

Power system stabilizers in the excitation systems of the machines are used to stabilize the oscillations.

It can, generally, be said that the harmonics in the power systems from sources other than nonlinear loads are comparatively small, though these cannot always be ignored. Major sources of harmonies are nonlinear loads [4].

1.11 Static Power Converters

The primary sources of harmonics in the power system are power converters, rectifiers, inverters, and ASDs. The *characteristic* harmonics are those produced by the power electronic converters during normal operation and these harmonics are integer multiples of the fundamental frequency of the power system. *Noncharacteristic* harmonics are usually produced by sources other than power electronic equipment and may be at frequencies other than the integer multiple of the fundamental power frequency. The converters do produce some noncharacteristic harmonics, as ideal conditions of commutation and control are not achieved in practice. The ignition delay angles may not be uniform, and there may be an unbalance in the supply voltages and the bridge circuits.

1.11.1 Single-Phase Bridge Circuit

The single-phase rectifier full-bridge circuit of Figure 1.12 is first considered. It is assumed that there is no voltage drop or leakage current, the switching is instantaneous, the voltage source is sinusoidal, and the load is resistive. For full-wave conduction, the waveforms of input and output currents are then as shown in Figure 1.12b and c. The *average* dc current is

$$I_{dc} = \frac{1}{2\pi} \int_0^{2\pi} \frac{E_m}{R} \sin \omega t \, d\omega t = \frac{2E_m}{\pi R} \tag{1.30}$$

and the rms value or the effective value of the output current, including all harmonics, is

$$I_{rms} = \sqrt{\frac{1}{2\pi} \int_0^{\pi} \left(\frac{E_m}{R}\right)^2 \sin^2 \omega t \, d\omega t} = \frac{E_m}{\sqrt{2}R} \tag{1.31}$$

The *input current has no harmonics*. The average dc voltage is given by

$$E_{dc} = \frac{2E_m}{\pi} \tag{1.32}$$

The output ac power is defined as follows:

$$P_{ac} = E_{rms} I_{rms} = \frac{(0.707 E_{rms})^2}{R} \tag{1.33}$$

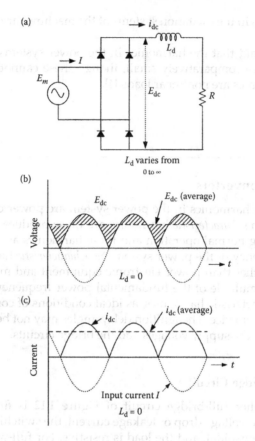

FIGURE 1.12
(a) A single-phase full rectifier bridge circuit, with resistive load; and (b and c) waveforms with zero dc reactor.

where E_{rms} considers the effect of harmonics on the output. The dc output power is

$$P_{dc} = E_{dc}I_{dc} = \left(\frac{2E_m}{\pi}\right)\left(\frac{2E_m}{\pi R}\right) = \frac{0.405E_m^2}{R} \tag{1.34}$$

The efficiency of rectification is given by P_{dc}/P_{ac} (81%). The *form factor* is a measure of the shape of the output voltage or current and it is defined as follows:

$$FF = \frac{I_{rms}}{I_{dc}} = 1.11 \tag{1.35}$$

The *ripple factor*, which is a measure of the ripple content of the output current or voltage, is defined as the rms value of output voltage or current, including all harmonics, divided by the average value:

$$RF = \sqrt{\left(\frac{I_{rms}}{I_{dc}}\right)^2 - 1} = \sqrt{FF^2 - 1} \tag{1.36}$$

For the single-phase bridge circuit with resistive load, the ripple factor is

$$\mathrm{RF} = \sqrt{\left(\frac{I_{\mathrm{rms}}}{I_{\mathrm{dc}}}\right)^2 - 1} = 0.48 \qquad (1.37)$$

This shows that the ripple content of the dc output voltage is high, see Figure 1.12b. This is not acceptable even for the simplest of applications. Let a series reactor be added in the dc circuit. The load current is no longer a sine wave, but the average current is still equal to $2E_m/\pi R$. The ac line current is no longer sinusoidal, but approximates a poorly defined square wave with superimposed ripples, see Figure 1.13a and b. The inductance has reduced the harmonic content of the load current by increasing the harmonic content of the ac line current. When the inductance is large, the ripple across the load is insignificant and can be assumed constant, and the ac current wave is now a square wave, see Figure 1.13c and d.

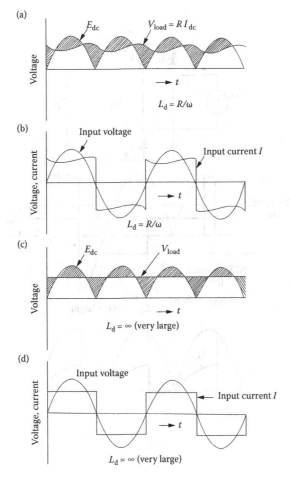

FIGURE 1.13
(a) Waveforms of a single-phase full rectifier bridge with small dc output reactor and (b) with large dc output reactor.

1.11.1.1 Phase Control

An SCR can be turned on by applying a short pulse to its gate and turned off due to natural or line commutation. The term thyristor pertains to the family of semiconducting devices for power control. The angle by which the conduction is delayed after the input voltage starts to go positive until the thyristor is fired is called the delay angle. Figure 1.14b shows waveforms with a large dc reactor, and Figure 1.14c shows waveform with no dc reactor but identical firing angle. Thyristors 1–4 are fired in pairs as shown in Figure 1.14b. Even when the polarity of the voltage is reversed, the current keeps flowing in thyristors 1 and 2 until thyristors 3 and 4 are fired, see Figure 1.14a. Firing of thyristors 3 and 4 reverse biases thyristors 1 and 2 and turns them off. (This is referred to as class F-type forced commutation or line commutation.) The average dc voltage is

$$E_{dc} = \frac{2}{2\pi} \int_{\alpha}^{\pi+\alpha} E_m \sin \omega t \, d\omega(\omega t) = \frac{2E_m}{\pi} \cos \alpha \tag{1.38}$$

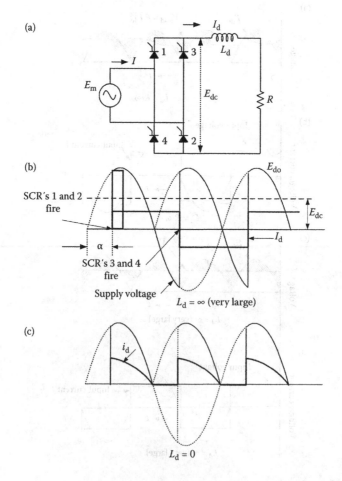

FIGURE 1.14
(a) Circuit of a single-phase fully controlled bridge; and (b and c) waveforms with large dc reactor and with zero dc reactor.

and the Fourier analysis of the rectangular current wave in Figure 1.14b gives

$$a_h = -\frac{4I_a}{h\pi}\sin\, h\alpha, \quad h = 1,3,5,\ldots$$

$$= 0 \qquad\qquad h = 2,4,6,\ldots \qquad\qquad (1.39)$$

$$b_h = \frac{4I_a}{h\pi}\cos\, h\alpha \quad h = 1,3,5,\ldots$$

$$= 0 \qquad\qquad h = 2,4,6,\ldots \qquad\qquad (1.40)$$

Since

$$I = \sum_{h=1,2,\ldots}^{\infty} \left[a_h \cos(h\omega t) + b_h \sin(h\omega t)\right] \qquad\qquad (1.41)$$

The rms input current is given by

$$I = \frac{4}{\pi}I_d\left[\sin(\omega t - \alpha) + \frac{1}{3}\sin 3(\omega t - \alpha) + \frac{1}{5}\sin 5(\omega t - \alpha) + \cdots\right] \qquad (1.42)$$

Triplen harmonics are present. Figure 1.15 shows harmonics as a function of the delay angle for a resistive load. The overlap angle (defined further) decreases the magnitude of the harmonics. When the output reactor is small, the current goes to zero, the input current wave is no longer rectangular, and the line harmonics increase.

1.11.1.2 Power Factor, Distortion Factor, and Total Power Factor

For sinusoidal voltages and currents, the power factor is defined as kW/kVA and the power factor angle ϕ is

$$\phi = \cos^{-1}\frac{kW}{kVA} = \tan^{-1}\frac{kvar}{kW} \qquad\qquad (1.43)$$

The power factor of a converter is made up of two components: displacement and distortion. The effect of the two is combined in total power factor. The displacement component is the ratio of active power of the fundamental wave in watts to apparent power of fundamental wave in voltampères. This is the power factor as seen by the watt-hour and var-hour meters. The distortion component is that part associated with harmonic voltages and currents.

$$PF_t = PF_f \times PF_{distortion} \qquad\qquad (1.44)$$

At fundamental frequency, the displacement power factor will be equal to the total power factor, as the displacement power factor does not include kVA due to harmonics, while the total power factor does include it. For harmonic generating loads, the total power factor will always be less than the displacement power factor. The discussion is continued in Chapter 7.

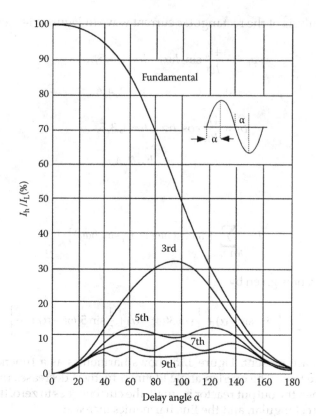

FIGURE 1.15
Harmonic generation as a function of phase-angle control, of delay angle, and of resistive load.

The fundamental input power factor angle is equal to the firing angle α. For the single-phase bridge circuit, the input active and reactive power is

$$\text{Active power} = \frac{4}{2\pi} I_d E_m \cos\alpha \tag{1.45}$$

$$\text{Reactive power} = \frac{4}{2\pi} I_d E_m \sin\alpha \tag{1.46}$$

The power factor becomes depressed for large firing angles. This is the case whenever large phase control is used in the converter circuits. The maximum reactive power input for a half-controlled bridge will be one-half of that of a fully controlled bridge. The reactive power requirements of converters become important in many installations and this can be limited by limiting amount of phase control, reducing reactance of converter transformers, which limits μ and sequential control of converters, which has been popular in HVDC transmission systems. In the sequential control, two or more converter sections can operate in series, with one section fully phased on and the other sections adding or subtracting from the voltage of the first section, see Figure 1.16.

Where controlled reactive power supply is not required, the reactive power consumed by the converters can be supplied by shunt capacitors and filters. In these

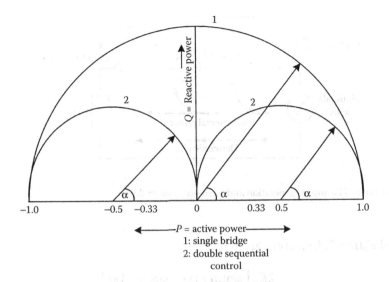

FIGURE 1.16
Reduction in reactive power with sequential control of converters.

cases, line-commutated converters are economical and for HVDC transmission, line-commutated converters have been used exclusively [5]. The thyristor-based converters become economical in large power handling capabilities as per device basis; the thyristors can handle two to three times more power than gate turn-off thyristors (GTOs), integrated gate bipolar transistors (IGCTs), and MOS turn-off thyristors (MTOs). There are variations in the current source converters such as resonant converters, hybrid converters, and artificial commutation converters, which are not discussed. New converter topologies are being developed. With GTOs, the forced commutation can improve the power factor and reduce the input harmonic levels. The forced commutation techniques are as follows:

- Extinction angle control
- Pulse-width modulation (PWM) and sinusoidal PWM, see section 1.17.
- Symmetrical angle control

A PWM converter with sinusoidal ac currents and minimum filter requirements is described in Reference [6]. A full-range four-quadrant operation is described, with control of input power factor. Experimental test results on a three-phase bipolar transistor-controlled current PWM modulator with leading power factor are described in Reference [7]. See Chapter 8 for some near to unity power factor converter topologies [8].

1.11.1.3 Harmonics on Output Side

The output waveform (on the dc side) contains even harmonics of the input frequency and the Fourier expansion is

$$E_{d0} = E_{dc} + e_2 \sin 2\omega t + e_2' \cos 2\omega t + e_4 \sin 4\omega t + e_4' \cos 4\omega t + \cdots \qquad (1.47)$$

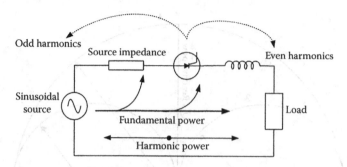

FIGURE 1.17
Converter as a source of harmonic generation and harmonic power flow.

where e_m and $e'_m(m = 2, 4, 6, \ldots)$ are given by

$$e_m = \frac{2E_m}{\pi}\left[\frac{\sin(m+1)\alpha}{m+1} - \frac{\sin(m-1)\alpha}{m-1}\right] \tag{1.48}$$

$$e'_m = \frac{2E_m}{\pi}\left[\frac{\cos(m+1)\alpha}{m+1} - \frac{\cos(m-1)\alpha}{m-1}\right] \tag{1.49}$$

The converter can be considered as a harmonic current source; the even harmonics go into the load and the odd harmonics into the supply source. The harmonics fed into the supply system propagate into the power system. These can be either magnified or attenuated and are the subject of study in this book. The harmonics in the load circuit have an adverse effect on the loads, but are, generally, localized to the loads to which these connect. There is harmonic power associated with harmonic currents, which is a function of relative system impedances and load-side impedance. Figure 1.17 shows this action of the converter.

- The effect of overlap angle γ (see Section 1.11.2.2) is
  - For small values of γ, harmonic magnitude increases with an increase in α
  - For constant α, harmonics decrease and reach a first minimum at approximately γ = π/n
  - For γ = 2π/n, there is a maximum and then there is a further minimum at γ = 3π/n

1.11.2 Three-Phase Bridge Circuit

A three-phase bridge has two forms: (1) half-controlled and (2) fully controlled. The three-phase fully controlled bridge is described, as it is most commonly used.

Figure 1.18a shows a three-phase fully controlled bridge circuit, and Figure 1.18b shows its current and voltage waveforms. The firing sequence of thyristors is shown in Table 1.3. At any time two thyristors are conducting. The firing frequency is six times the fundamental frequency and the firing angle can be measured from point O shown in Figure 1.18b. With a large output reactor, the output dc current is continuous and the input current is a rectangular pulse of $2\pi/3$ duration and amplitude i_d. The average dc voltage is

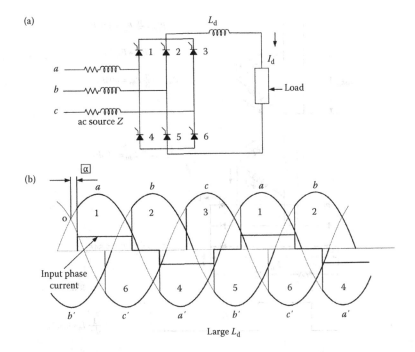

FIGURE 1.18
(a) Circuit of a three-phase fully controlled bridge, with large dc output reactor and (b) voltage and current waveforms for a certain delay angle α.

TABLE 1.3

Firing Sequence of Thyristors in Six-Pulse Full Converter

Conducting thyristors	5, 3	1, 5	6, 1	2,6	4, 2	3,4
Thyristor to be fired	1	6	2	4	3	5
Thyristor turning off	3	5	1	6	2	4

$$E_d = 2\left[\frac{3}{2\pi}\int_{-n/3+\alpha}^{\pi/3+\alpha} E_m \cos\omega t\, d(\omega t)\right] = \frac{3\sqrt{3}}{\pi}E_m \cos\alpha \tag{1.50}$$

where E_m is the peak value of line-to-neutral voltage. For firing angles $> \pi/2$, the circuit can work as an inverter, i.e., dc power is fed back into the ac system. This requires that a dc source with opposite polarity is connected at the output. The power factor is lagging for rectifier operation and leading for inverter operation.

Figure 1.19 shows a connection diagram and waveforms for a three-phase fully controlled bridge, with delta–delta connection of the rectifier transformer and firing angle α = 0; the input current is rectangular and its Fourier analysis gives

$$i_a = \frac{2\sqrt{3}}{\pi}I_d\left[\cos\omega t - \frac{1}{5}\cos 5\omega t + \frac{1}{7}\cos 7\omega t - \frac{1}{11}\cos 11\omega t + \frac{1}{13}\cos 13\omega t\right] \tag{1.51}$$

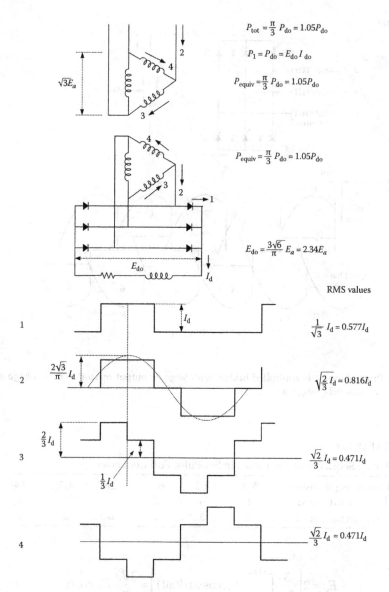

FIGURE 1.19
A six-pulse bridge circuit, zero delay angle, large output reactor, and delta–delta input transformer. Voltage and current relations.

Thus, the maximum fundamental frequency current is

$$\frac{2\sqrt{3}}{\pi} I_d \text{ peak} = \frac{\sqrt{6}}{\pi} I_d \text{ rms} \qquad (1.52)$$

Figure 1.20 shows a similar connection diagram and waveforms for delta–wye rectifier transformer connections. The input current is stepped and the resulting Fourier series for the current waveform is

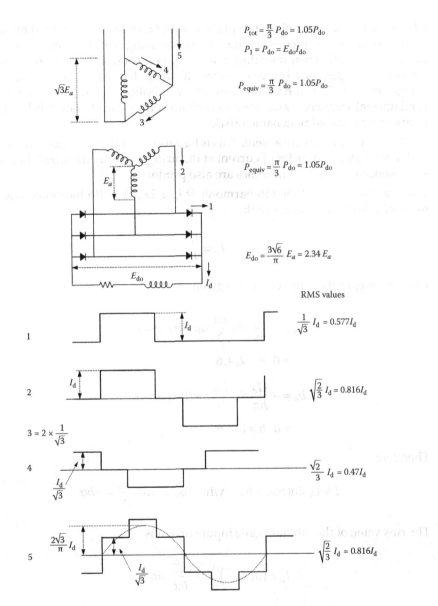

FIGURE 1.20
A six-pulse bridge circuit, zero delay angle, large output reactor, and delta–wye input transformer. Voltage and current relations.

$$i_a = \frac{2\sqrt{3}}{\pi} I_d \left(\cos \omega t + \frac{1}{5} \cos 5\omega t - \frac{1}{7} \cos 7\omega t - \frac{1}{11} \cos 11\omega t + \frac{1}{13} \cos 13\omega t -,... \right) \quad (1.53)$$

From these equations, the following observations can be made:

1. The line harmonics are of the order:

$$h = pm \pm 1, \quad m = 1, 2, ... \quad (1.54)$$

where p is the pulse number. The pulse number is defined as the total number of successive nonsimultaneous commutations occurring within the converter circuit during each cycle when operating without phase control. This relationship also holds for a single-phase bridge converter, as the pulse number for a single-phase bridge circuit is 2. The harmonics given by Equation 1.54 are an integer of the fundamental frequency and are called characteristic harmonics, while all other harmonics are called noncharacteristic.

2. The triplen harmonics are absent. This is because an ideal rectangular wave shape and instantaneous transfer of current at the firing angle are assumed. In practice, some noncharacteristic harmonics are also produced.

3. The rms magnitude of the nth harmonic is I_f/h, i.e., the fifth harmonic is a maximum of 20% of the fundamental:

$$I_h = \frac{I_f}{h} \tag{1.55}$$

4. Fourier series of the input current is given by

$$a_n = -\frac{4I_d}{h\pi}\sin\frac{n\pi}{3}\sin(h\alpha)\quad h = 1,3,5$$

$$= 0 \quad h = 2,4,6$$

$$b_n = -\frac{4I_d}{h\pi}\sin\frac{h\pi}{3}\cos(h\alpha)\quad h = 1,3,5 \tag{1.56}$$

$$= 0 \quad h = 2,4,6.$$

Therefore

$$I = I_h\sin(h\omega t + \phi_h), \quad \text{where } \phi_h = \tan^{-1}\frac{a_h}{b_h} = -h\alpha \tag{1.57}$$

The rms value of the nth harmonic input current is

$$I_h = \left(a_h^2 + b_h^2\right)^{1/2} = \frac{2\sqrt{2}}{h\pi}\sin\frac{h\pi}{3} \tag{1.58}$$

The rms value of the fundamental current is

$$I_1 = \frac{\sqrt{6}}{\pi}I_d = 0.779I_d \tag{1.59}$$

The rms input current (including harmonics) is

$$\left[\frac{2}{\pi}\int_{(-\pi/3)+\alpha}^{(\pi/3)+\alpha} I_d^2\, d(\omega t)\right]^{1/2} = I_d\sqrt{\frac{2}{3}} = 0.8165I_d \tag{1.60}$$

7. The ripple factor of a six-pulse converter, with zero firing angle, is 0.076 and the lowest harmonic in the output of the converter is the sixth. As the pulse number of the converter increases, the ripple in the dc output voltage and the harmonic content in the input current are reduced. Also, for a given voltage and firing angle, the average dc voltage increases with the pulse number.

8. From Equation 1.60, the rms current including harmonics is $0.8165I_d$, and from Equation 1.53, the fundamental current is $0.7797I_d$. Then, the total harmonic current is

$$I_h = \left[(0.8165I_d)^2 - (0.7797I_d)^2 \right]^{1/2} = 0.24I_d \tag{1.61}$$

In the above analysis, we assumed that the commutation is instantaneous. It may take some time before the current is commutated, through the inductive circuit of the ac system. This is discussed in the next section.

1.11.2.1 Cancellation of Harmonics due to Phase Multiplication

Equations 1.51 and 1.53 show that the harmonics 5th, 7th, and 17th are of opposite sign. We know that there is a 30° phase shift between the primary and secondary voltage vectors of a delta–wye transformer, while for a delta–delta or wye–wye connected transformer, this phase shift is 0°. If the load is equally divided on two transformers, one with delta–delta connections and the other with wye–delta or delta–wye connections, harmonics of the order of 5th, 7th, 17th, and so on are eliminated, and the system behaves like a 12-pulse circuit. This is called phase multiplication. The circuit is shown in Figure 1.21a and the waveform in the time domain in Figure 1.21b. Extending this concept, 24-pulse operation can be achieved with four transformers with 15° mutual phase shifts. As the magnitude of the harmonic is inversely proportional to the pulse number, the troublesome lower order harmonics of larger magnitude are eliminated. This cancellation of harmonics, though, is not 100% as the ideal conditions of operation are rarely met in practice. The transformers should have exactly the same ratios and same impedances, the loads should be equally divided and converters should have exactly the same delay angle. Approximately 75% cancellation may be achieved in practice, and in harmonic analysis studies, 25% residual harmonics are modeled.

1.11.2.2 Effect of Source Impedance

The commutation of current from one SCR to another will take place instantaneously if the source impedance is zero. The commutation is delayed by an angle μ due to source inductance, and during this period, a short circuit occurs through the conducting devices, the ac circulating current being limited by the source impedance; μ is called the overlap angle. When α is zero, the short-circuit conditions are those corresponding to maximum asymmetry and μ is large, i.e., slow initial rise. At $\alpha = 90°$, the conditions are of zero asymmetry with its fast rate of rise of current. Commutation produces two primary notches per cycle and four secondary notches of lesser amplitude, which are due to notch reflection from the other legs of the bridge (Figure 1.22). For a purely inductive source impedance, the output average dc voltage is reduced, which is given by

$$E_d = E_{do} - \frac{3\omega L_s}{\pi} I_d \tag{1.62}$$

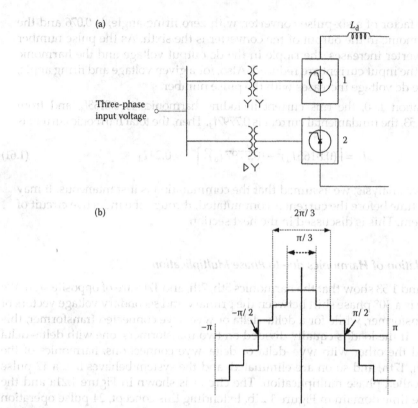

FIGURE 1.21
(a) Harmonic elimination with phase multiplication: circuit diagram and (b) input current waveform.

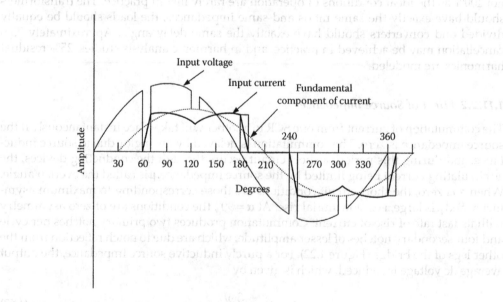

FIGURE 1.22
Voltage notching due to commutation in a six-pulse fully controlled bridge with dc output reactor.

where L_s is the source inductance, and for a six-pulse fully controlled bridge E_{do} is given by Equation 1.50 and it is called the internal voltage of the rectifier.

Figure 1.23 shows that the overlap helps in reducing the harmonic content in the input current wave, which is rounded off and is more close to a sinusoid. AC harmonics at overlap are given by the following [1]:

$$I_h = I_{dc}\left[\sqrt{\frac{6}{\pi}}\frac{\sqrt{A^2+B^2-2AB\cos(2\alpha+\mu)}}{h[\cos\alpha-\cos(\alpha+\mu)]}\right] \tag{1.63}$$

where

$$A = \frac{\sin[(h-1)\mu/2]}{h-1} \tag{1.64}$$

$$B = \frac{\sin[(h+1)\mu/2]}{h+1} \tag{1.65}$$

The depth of the voltage notch is calculated by the IZ drop and is a function of the impedance. The width of the notch is the commutation angle:

$$\mu = \cos^{-1}[\cos\alpha - (X_s + X_t)I_d] - \alpha \tag{1.66}$$

$$\cos\mu = 1 - \frac{2E_x}{E_{do}} \tag{1.67}$$

where X_s is the system reactance in per unit on converter base, X_t is the transformer reactance in per unit on converter base, I_d is the dc current in per unit on converter base, and E_x is the dc voltage drop caused by commutating reactance. Notches cause electromagnetic interference (EMI) problems and misoperation of electronic devices which sense the true zero crossing of the voltage wave.

As a six-pulse converter is most frequently used in industry, Chapter 3 provides graphical/analytical methods for estimation of harmonics, with varying overlap angles. The assumption of a flat-topped wave is not correct. The actual wave and its effect on line harmonics

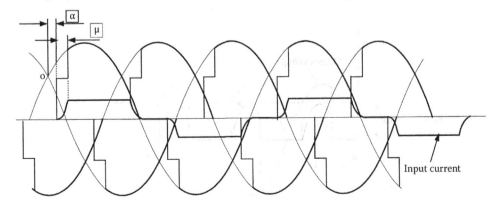

FIGURE 1.23
Effect of overlap angle on the current waveform.

TABLE 1.4

Theoretical and Typical Harmonic Spectrum of Six-Pulse Full Converters with Large DC Reactor

h	5	7	11	13	17	19	23	25
$1/sh$	0.200	0.143	0.091	0.077	0.059	0.053	0.043	0.040
Typical	0.175	0.111	0.045	0.029	0.015	0.010	0.009	0.008

are discussed in this chapter. Table 1.4 shows the theoretical magnitude of harmonics given by Equation 1.54 and a typical harmonic spectrum, assuming ripple-free dc output current and instantaneous commutation.

1.11.2.3 Effect of Output Reactor

The foregoing treatment of six-pulse converters assumes large dc inductance, so that the output dc current is continuous. At large phase-control angles and low output value of the reactor, the current will be discontinuous, giving rise to increased line harmonics, see Chapter 5 for further discussions.

1.11.2.4 Effect of Load with Back EMF

Harmonic magnitude will be largely affected if the output dc load is active, e.g., a battery charge. A dc motor has low inductance as well as a back EMF. The value of inductance at which the load current becomes discontinuous can be calculated by writing a differential equation of the following form:

$$L_d \frac{di_d}{dt} + i_d R_d = E_m \cos(\omega t + \alpha) \tag{1.68}$$

and solving for i_d, equating it to zero, and evaluating L_d.

When the load has a back EMF, the load current waveform is decided not only by the firing angle, but also by the opposing voltage of the load. Figure 1.24 shows the waveforms of output voltage and the load current for a single-phase fully controlled circuit feeding a battery charger, neglecting source impedance; β is called the conduction angle. The

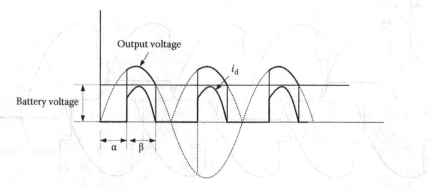

FIGURE 1.24

Current waveform with delay angle and a load with back EMF.

harmonics are increased. At large control angles and with dc motor loads, having a back EMF, the discontinuous nature of the dc voltage and ac current, for a six-pulse converter, may give rise to higher harmonics and the fifth harmonic can reach peak levels up to three times that of the rectangular wave.

1.11.2.5 Inverter Operation

From Equation 1.50, the dc voltage is zero for a 90° delay angle. As the delay angle is further increased, the average dc voltage becomes negative. (This can be examined by drawing output voltage waveform with advancing delay angle α in Figure 1.23.) If instantaneous commutation is assumed then at 180°, the negative voltage is as large as that for a rectifier at zero delay angle. However, operation at 180° is not possible, as some delay, denoted by γ, is needed for commutation of current and some additional time, angle δ, is needed for the outgoing thyristors to turn off before voltage across it reverses. Otherwise, the outgoing thyristor will commutate the current back, and lead to commutation failure. In HVDC transmission, two modes of operation are recognized. In mode 1, the firing angle is varied and current remains constant. The duration for which the thyristor is reverse biased will reach a minimum value at some level of internal voltage and any further increase will result in commutation failure. Thus, in the second mode, the inverter operates at a constant margin angle or extinction angle, see Reference [9] for further details.

1.11.3 Diode Bridge Converter

Converters with an output dc reactor and front-end thyristors follow the dc link voltage. A full converter controls the amount of dc power from zero to full dc output. The voltage and current waveforms of this type of converter are discussed in Section 1.11.2. The harmonic injection into the supply system may be represented by a Norton equivalent. This type of converter is used at the front end of current source inverters.

The full-wave diode bridge with capacitor load, as shown in Figure 1.25, is the second type of converter. It converts from ac to dc and does not control the amount of dc power. This type of converter does not cause line notching, but the current drawn is more like a pulse current rather than the approximate square-wave current of the full converter. The voltage and current waveforms are shown in Figure 1.25b. This circuit is better represented by a Thévenin equivalent and the source impedance has a greater impact.

Typical current harmonics for comparison are shown in Table 1.5. In the diode converter with dc link capacitor, the fifth harmonic is higher by a factor of 3–4 times and the seventh harmonic by a factor of 3. This type of converter with dc link capacitor is used in voltage source inverters (VSIs). Sometimes, a controlled bridge may replace the diode bridge preceding the dc link capacitor.

1.12 Switch Mode Power Supplies

Single-phase rectifiers are used for power supplies in copiers, computers, TV sets, and household appliances. In these applications, the rectifiers use a dc filter capacitor and draw impulsive current from the ac supply. The harmonic current is worse than that given by Equation 1.42. Figure 1.26a and b shows conventional power supplies and SMPSs. In the

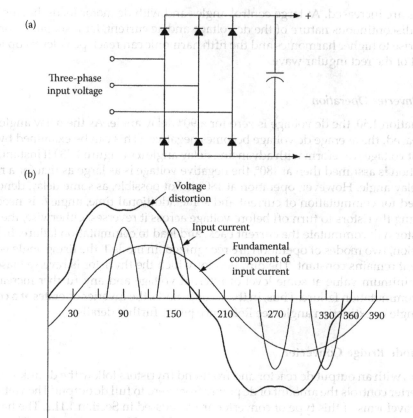

FIGURE 1.25
(a) Circuit of a diode bridge with dc link bus capacitor and (b) input current waveform.

TABLE 1.5

Typical Current Harmonics for a Six-Pulse
Converter and Diode Bridge Converter, as a
Percentage of Fundamental Frequency Current

Current Harmonic	Six-Pulse Converter	Diode Bridge Converter
5	17.94	64.5
7	11.5	34.6
11	4.48	5.25
13	2.95	5.89

conventional power supply system, the main ripple frequency is 120 Hz, and the current drawn is relatively linear. Capacitors C_1 and C_2 and inductor act as a passive filter. In the SMPSs, the incoming voltage is rectified at line voltage and the high dc voltage is stored in capacitor C_1. The transistorized switcher and controls switch the dc voltage from C_1 at a high rate (10–100 kHz). These high-frequency pulses are stepped down in a transformer and rectified. The switcher eliminates the series regulator and its losses in conventional

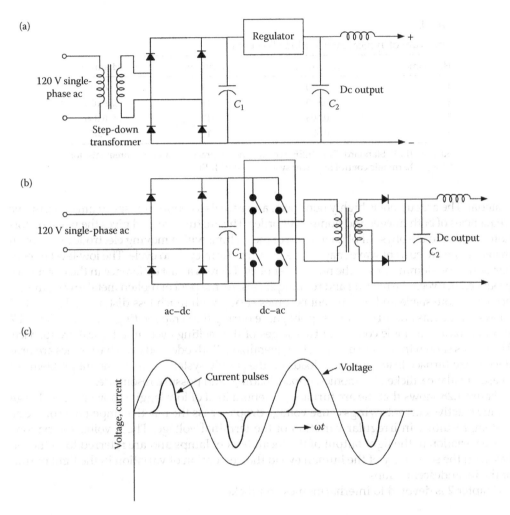

FIGURE 1.26
(a) Conventional power supply circuit; (b) SMPS circuit; and (c) input current pulse waveform.

power supplies. There are four common configurations used with the switched mode operation of the dc to ac conversion stage, and these are fly back, push–pull, half-bridge, and full bridge. The input current wave for such an SMPS is highly nonlinear, flowing in pulses for part of the sinusoidal ac voltage cycle, see Figure 1.26c. The spectrum of an SMPS is given in Table 1.6 and shows the high magnitude of the third and fifth harmonics.

1.13 Arc Furnaces

Arc furnaces may range from small units of a few ton capacities, power rating 2–3 MVA, to larger units having 400 ton capacity and a power requirement of 100 MVA. The harmonics produced by electric arc furnaces are not definitely predicted due to variation of the arc feed

TABLE 1.6

Spectrum of Typical Switch Mode Power Supply

Harmonic	Magnitude	Harmonic	Magnitude
1	1.000	9	0.157
3	0.810	11	0.024
5	0.606	13	0.063
7	0.370	15	0.07

Source: IEEE Standard 519, IEEE recommended practices and requirements for harmonic control in power systems © 1992 IEEE.

material. The arc current is highly nonlinear, and reveals a continuous spectrum of harmonic frequencies of both integer and noninteger order. The arc furnace load gives the worst distortion, and due to the physical phenomenon of the melting with a moving electrode and molten material, the arc current wave may not be the same from cycle to cycle. The low-level integer harmonics predominate over the noninteger ones. There is a vast difference in the harmonics produced between the melting and refining stages. As the pool of molten metal grows, the arc becomes more stable and the current becomes steady with much less distortion. Figure 1.27 shows erratic rms arc current in a supply phase during the scrap melting cycle, and Table 1.7 shows typical harmonic content of two stages of the melting cycle in a typical arc furnace. The values shown in this table cannot be generalized. Both odd and even harmonics are produced. Arc furnace loads are harsh loads on the supply system, with attendant problems of phase unbalance, flicker, harmonics, impact loading, and possible resonance.

Figure 1.28 shows that the arc furnace presents a load of low lagging power factor. Large erratic reactive current swings cause voltage drops across the reactive impedance of the ac system, resulting in irregular variation of the terminal voltage. These voltage variations cause variation in the light output of the incandescent lamps and are referred to as flicker, based on the sensitivity of the human eye to the perception of variation in the light output of the incandescent lamps.

Chapter 2 is devoted to interharmonics and flicker.

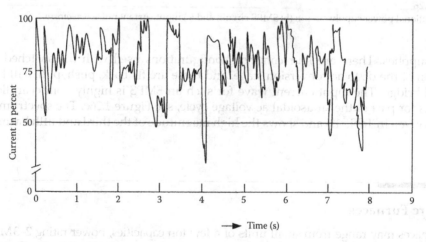

FIGURE 1.27
Erratic melting current in one-phase supply circuit of an arc furnace.

TABLE 1.7

Harmonic Content of Arc Furnace Current as a Percentage of Fundamental

h	Initial Melting	Refining
2	7.7	0.0
3	5.8	2.0
4	2.5	0.0
5	4.2	2.1
7	3.1	

Source: IEEE Standard 519, IEEE recommended practices and requirements for harmonic control in power systems © 1992 IEEE.

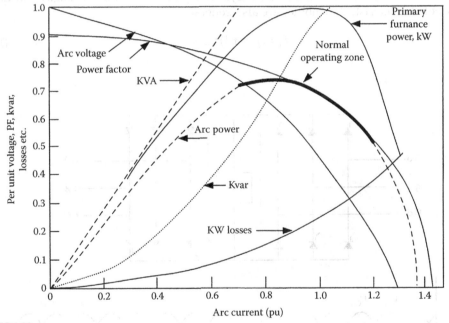

FIGURE 1.28

Typical performance curves for an arc furnace, showing the normal operating zone in thick line.

1.14 Cycloconverters

Cycloconverters are used in a wide spectrum of applications from ball mill and linear motor drives to static var generators. The range of application for synchronous or induction motors varies from 1000 to 50,000 hp and the speed control in the ratio of 50:1. Figure 1.29a shows the circuit of a three-phase single-phase cycloconverter, which synthesizes a 12 Hz output, and Figure 1.29c shows the output voltage waveform with resistive load. The positive converter operates for half the period of the output frequency and the negative converter operates for the other half. The output voltage is made of segments of input voltages (Figure 1.29b), and the average value of a segment depends on the delay angle for that segment; α_p is the delay angle of the positive converter and $\pi - \alpha_p$ is the delay angle of the negative converter. The output voltage contains harmonics and the input power factor is poor. For three-phase, three systems Figure 1.29a, i.e., a total of 36 thyristors are required.

The voltage of a segment depends on the delay angle. If the delay angles of the segments are varied so that the average value of the segment corresponds as closely as possible to the variations in the desired sinusoidal output voltage, the harmonics in the output are minimized. Such delay angles can be generated by comparing a cosine signal at source frequency with an output sinusoidal voltage.

Cycloconverters have a characteristic harmonic frequency of

$$f_h = (pm \pm 1)f \pm 6nf_0 \qquad (1.69)$$

where f_0 is the output frequency of the cycloconverter, m, $n = 1, 2, 3, \ldots$, and p is the pulse number. Because of load unbalance and asymmetry between phase voltages and firing angles, noncharacteristic harmonics are also generated:

$$f_h = (pm \pm 1)f \pm 2nf_0 \qquad (1.70)$$

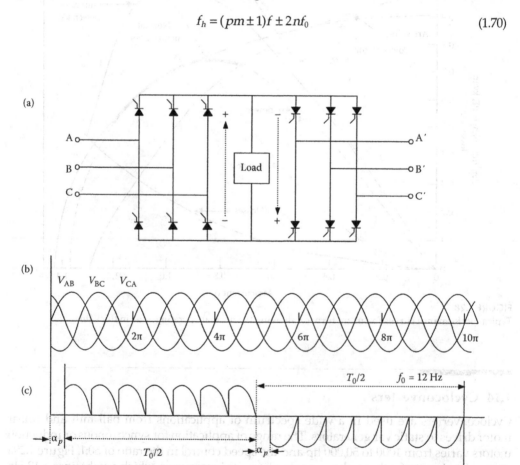

FIGURE 1.29
To illustrate the principle of a cycloconverter.

As the output frequency varies, so does the spectrum of harmonics. Therefore, control of harmonics with single tuned filters becomes ineffective (Chapter 8). The work of Pelly [10] is entirely devoted to cycloconverters.

1.15 Thyristor-Controlled Reactor

Consider a thyristor controlled reactor (TCR), controlled by two thyristors in an antiparallel circuit as shown in Figure 1.30a. If both thyristors are gated at maximum voltage, there are no harmonics and the reactor is connected directly across the voltage, producing a

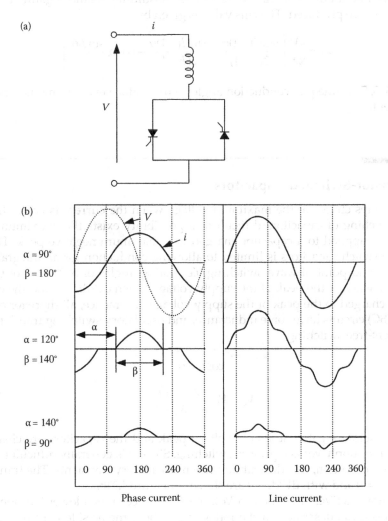

FIGURE 1.30
(a) Circuit of a thyristor-controller reactor and (b) current waveforms due to varying firing and conduction angles.

90° lagging current, ignoring the losses. If the gating is delayed, the waveforms as shown in Figure 1.30b result. The instantaneous current through the reactor is

$$i = \sqrt{2}\frac{V}{X}(\cos\alpha - \cos\omega t) \text{ for } \alpha < \omega t < \alpha + \beta \tag{1.71}$$

$$= 0 \text{ for } \alpha + \beta < \omega t < \alpha + \pi \tag{1.72}$$

where V is the line-to-line fundamental rms voltage, α is the gating angle, and β is the conduction angle. The fundamental component can be written as follows:

$$I_f = \frac{\beta - \sin\beta}{\pi X}V \tag{1.73}$$

Considerable amount of harmonics is generated. Assuming balanced gating angles only odd harmonics are produced. The rms value is given by

$$I_h = \frac{4V}{\pi X}\left[\frac{\sin(h+1)\alpha}{2(h+1)} + \frac{\sin(h-1)\alpha}{2(h-1)} - \cos\alpha\frac{\sin h\alpha}{h}\right] \tag{1.74}$$

where $h = 3, 5, 7, \ldots$ unequal conduction angles will produce even harmonics including a dc component.

1.16 Thyristor-Switched Capacitors

A capacitor traps charge at the maximum voltage when the current is zero. This makes thyristor switching of capacitors difficult as a possibility exists that maximum ac peak voltage can be applied to a capacitor charged to a maximum negative peak. The use of thyristors to switch capacitors is limited to allowing conduction for an integral number of half-cycles and point of wave switching, i.e., gating angles of >90° are not used. The thyristors are gated at the peak of the supply voltage, when $dV/dt = 0$, and the capacitors are already charged to the peak of the supply voltage. In practice, all thyristor controlled capacitor (TSC) circuits have some inductance, and oscillatory switching transients result. For a transient-free switching

$$\cos\alpha = 0 \tag{1.75}$$

$$V_c = \pm V\frac{(X_c/X_L)}{X_c/X_L - 1} \tag{1.76}$$

i.e., the capacitors are gated at supply voltage peak and the capacitors are charged to a higher than the supply voltage prior to switching. Since it is generally difficult to guarantee the second condition, it is difficult to prevent oscillatory transients. The transients on gating with $V = 0$ and with $dV/dt = 0$ are shown in Figure 1.31b.

The synchronous SVC, described in Volume 2, produces much lower harmonics. With pulse-width control, described in the following section, the ac-side harmonics are controlled. Thyristor-controlled series compensation produces harmonics in series with the line (Volume 2).

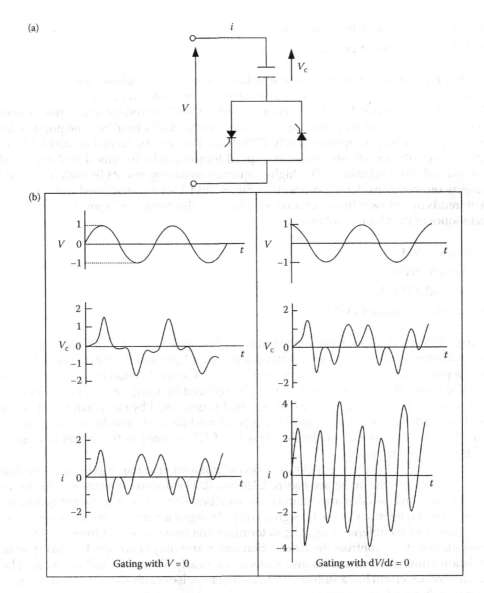

FIGURE 1.31
(a) Circuit of a thyristor-switched capacitor and (b) switching transients with $V = 0$ and with $dV/dt = 0$.

1.17 Pulse-Width Modulation

Over the years, the pulse switching time of the power devices has been drastically reduced:

- SCR (fast thyristor): $4\,\mu s$
- GTO (gate turn-off thyristor): $1.0\,\mu s$
- GTR (giant transistor): $0.8\,\mu s$

- IGBT (insulated gate bipolar): 0.2 μs
- IGBT (power plate type): 0.1 μs

The VSIs using IGBTs operate from a dc link bus. The inverter synthesizes a variable voltage, of variable-frequency waveform (V/f = constant), by switching the dc bus voltage at high frequencies (10–20 kHz). The inverter output line-to-line voltage is a series of voltage pulses with constant amplitude and varying widths. IGBTs have become popular for power output levels up to approximately 250 kW, as these can be turned on and off from simple low-cost driver circuits. Motor low-speed torque can be increased and improved low-speed stability is obtained. The high-frequency switching results in high dV/dt and the effects on motor insulation, connecting cables, and EMI are discussed in Chapter 6. Recent trends in soft-switching technology reduce the rise time, see Figure 1.32f and g.

Techniques of PWM are as follows:

- Single PWM
- Multiple PWM
- Sinusoidal PWM
- Modified sinusoidal PWM

In a single-PWM technique, there is one pulse per half-cycle and the width of the pulse is varied to control the inverter output voltage, see Figure 1.32a and b. By varying A_r from 0 to A_c, the pulse width δ can be varied from 0° to 180°. The modulation index is defined as A_r/A_c. The harmonic content is high, but can be reduced by using several pulses in each half-cycle of the output voltage. The gating signal is generated by comparing a reference signal with a triangular carrier wave. This type of modulation is also known as uniform PWM. The number of pulses per half-cycle is $N = f_c/2f_0$, where f_c is the carrier frequency and f_0 is the output frequency.

In sinusoidal PWM, the pulse width is varied in proportion to the amplitude of the sine wave at the center of the pulse, see Figure 1.32c and d. The distortion factor and the lower order harmonic magnitudes are reduced considerably. The gating signals are generated by comparing a reference sinusoidal signal with a triangular carrier wave of frequency f_c. The frequency of the reference signal f_r determines the inverter output frequency f_0, and its peak amplitude A_r controls the modulation index and output voltage V_0. This type of modulation eliminates harmonics and generates a nearly sinusoidal voltage wave. The input current waveform has a sinusoidal shape due to pulse-width shaping; however, harmonics at switching frequency are superimposed, see Figure 1.32e.

1.18 Adjustable Speed Drives

ASDs account for the largest percentage of nonlinear loads in the industry. A comparison of electronic drive systems with type of motor, horse power rating, and drive system topology is shown in Table 1.8. Most drive systems require that the incoming ac power supply be converted into dc. The dc power is then inverted back to ac at a frequency demanded by the speed reference of the ac variable-frequency drive or the dc feeds directly to dc drive systems through two or four quadrant converters. The fully controlled bridge circuit with

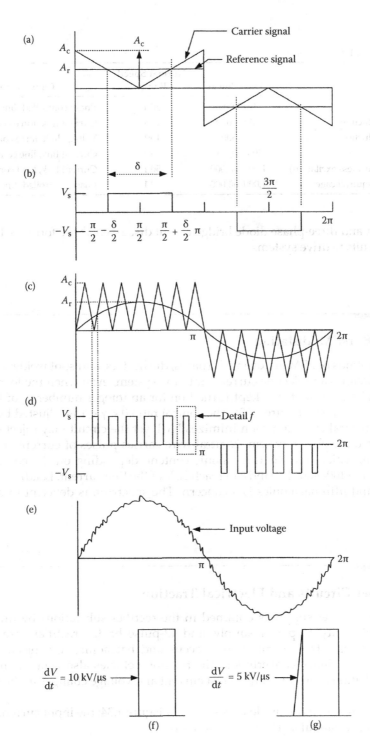

FIGURE 1.32
(a,b) Single PWM; (c,d) sinusoidal PWM; (e) reflection of switching transients in the input current wave; (f) high dV/dt due to high-frequency switching; and (g) reduced dV/dt due to soft switching.

TABLE 1.8

Adjustable Speed Drive Systems

Drive Motor	Horse Power	Normal Speed Range	Converter Type
DC	1–10,000	50:1	Phase controlled, line commutated
Squirrel-cage induction	100–4,000	10:1	Current link, force commutated
Squirrel-cage induction	1–1,500	10:1	Voltage link, force commutated
Wound rotor	500–20,000	3:1	Current link, line commutated
Synchronous (brushless excitation)	1,000–60,000	50:1	Current link, load commutated
Synchronous or squirrel cage	1,000–60,000	50:1	Phase controlled, line commutated

output reactor and three-phase diode bridge circuit discussed above form the basic front-end input circuits to drive systems.

1.19 Pulse Burst Modulation

Typical applications of PBM are ovens, furnaces, die heaters, and spot welders [4]. Three-phase PBM circuits can inject dc currents into the system, even when the load is purely resistive. A solid-state switch is kept turned on for an integer number γ_n of half-cycles out of a total of n cycles (Figure 1.33). The control ratio $0 < \gamma < 1$ is adjusted by feedback control. The integral cycle control minimizes EMI, yet the circuit may inject significant dc currents into the power system. Neutral wire carries pulses of current at switch off and switch on, which have high harmonic content, depending on the control ratio γ. Harmonics in the 100–400 Hz band can reach 20% of the line current. Loading of neutrals with triplen and fifth harmonics is a concern. The spectrum is deficient in high-order harmonics.

1.20 Chopper Circuits and Electrical Traction

The dc traction power supply is obtained in the rectifier substations by unsmoothed rectification of utility ac power supply, and 12-pulse bridge rectifiers are common. Switching transients from commutation occur and harmonics are injected into the supply system. Auxiliary converters in the traction vehicles also generate harmonics, while EMI radiation is produced from fast current and voltage changes in the switching equipment.

A chopper with high inductive load is shown in Figure 1.34; the input current is pulsed and assumed as rectangular. The Fourier series is

$$i_c(t) = kI_d + \frac{I_a}{n\pi} \sum_{h=1}^{\infty} \sin\ 2h\pi k \cos 2h\pi f_c t + \frac{I_d}{h\pi} \sum_{h=1}^{\infty} (1 - \cos 2h\pi k)\sin 2h\pi f_c t \qquad (1.77)$$

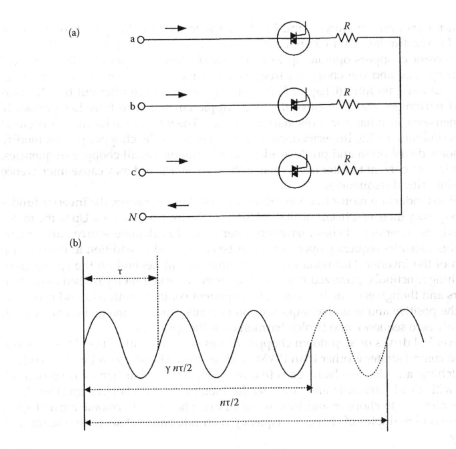

FIGURE 1.33
(a) Circuit for PBM and (b) PBM control, current waveform.

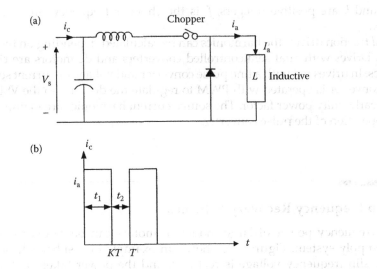

FIGURE 1.34
(a) A chopper circuit with input filter and (b) current waveform.

where f_c is the chopping frequency and k is the mark-period ratio (duty cycle of the chopper $= t_1/T$). The fundamental component is given for $h = 1$. In railway dc-fed traction drives, thyristor choppers operate up to about 400 Hz. The chopper circuit is operated at fixed frequency and the chopping frequency is superimposed on the line harmonics. An input low-pass filter (Chapter 8) is normally connected to filter out the chopper-generated harmonics and to control the large ripple current. The filter has physically large dimensions, as it has a low resonant frequency. The worst-case harmonics occur at a mark-period ratio of 0.5. Imperfections and unbalances in the chopper phases modify the harmonic distribution and produce additional harmonics at all chopper frequencies. Transient inrush current to filter occurs when the train starts and may cause interference if it contains critical frequencies.

In a VSI-fed induction motor traction drive from a *dc traction system*, the inverter fundamental frequency increases from 0 to about 120 Hz as the train accelerates. Up to the motor base speed, the inverter switches many times per cycle. To calculate source current harmonics, this variable-frequency operation must be considered in addition to three-phase operation of the inverter. Harmonics in the dc link current depend on the spectrum of the switching function. Optimized PWM with quarter-wave symmetry is used in traction converters and though each dc link current component contains both odd and even harmonics, the positive and negative sequence components cancel in the dc link waveform, leaving only zero sequence and triplen harmonics in the spectrum.

Multilevel VSI drives or step-down chopper drives are used with GTOs. VSIs may use a different control strategy, other than PWM, such as torque band control with asynchronous switching, and the disadvantage is that relatively high $6h$ harmonics are produced, together with third harmonic and some subharmonics. In the chopper inverter drive, the harmonics due to chopper and inverter combine. The input harmonic current spectrum consists of multiples of chopper frequency with side bands at six times the inverter frequency:

$$f_h = kf_c \pm 6hf_i \tag{1.78}$$

where k and h are positive integers, f_c is the chopper frequency, and f_i is the inverter frequency.

In *ac-fed* traction drives, the harmonics can be calculated depending on the drive system topology. Drives with dual semicontrolled converters and dc motors are rich sources of harmonics. In drives fed with a line pulse converter and voltage or current source inverter, the line converter is operated with PWM to regulate the demand to the VSI, while maintaining nearly unity power factor. The source current harmonics are mainly derived from the line operation of the pulse converter.

1.21 Slip Frequency Recovery Schemes

The slip frequency power of large induction motors can be recovered and fed back into the supply system. Figure 1.35 shows an example of a subsynchronous cascade. The rotor slip frequency voltage is rectified, and the power taken by the rotor is fed into the supply system through a line-commutated inverter. The speed of the induction motor can be adjusted as desired throughout the subsynchronous range, without

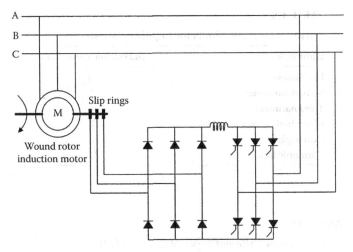

FIGURE 1.35
A slip recovery scheme for a wound rotor induction motor.

losses, though the reactive power consumption of the motor cannot be corrected in the arrangement shown in Figure 1.35.

Such a system can cause subharmonics in the ac system. For six-pulse rectification, the power returned to the system pulsates at six times the rotor slip frequency. Torsional oscillations can be excited if the first or second natural torsional frequency of the mechanical system is excited, resulting in shaft stresses [11]. The ac harmonics for this type of load cannot be reduced by phase multiplication as dc current ripple is independent of the rectifier ripple.

1.22 Lighting Ballasts

Lighting ballasts may produce large harmonic distortions and third harmonic currents in the neutral. The newer rapid start ballast has a much lower harmonic distortion. The current harmonic limits for lighting ballasts are given in Tables 1.9 and 1.10. Table 1.9 shows that the limits for the newer ballasts are much lower as compared to earlier ballasts (Table 1.10). This also compares distortion produced by the lighting ballasts with other office equipment [12,13].

This chapter shows that it may not be always possible to estimate clearly harmonic emission. The topologies of harmonic producing equipment are changing fast, and manufacturer's data can be used, wherever practical. The system impedance plays a major role. Consider that load currents are highly distorted but the system impedance is low. Thus, the voltages will not be much distorted. Chapter 5 shows ANSI limits for harmonic currents that a user can inject into utility system, and these take the system impedance into cognizance. The limits are specified on the basis of the I_s/I_r ratio, where I_s is the short-circuit current and I_r is the load demand current. Further discussions continue in Chapter 5.

TABLE 1.9

Current Harmonic Limits for Lighting Ballasts

Harmonic	Maximum Value (%)
Fundamental	100
Second harmonic	5
Third harmonic	30
Individual harmonics > 11th	7
Odd triplens	30
Harmonic factor	32

TABLE 1.10

THD Ranges for Different Types of Lighting Ballasts

Device Type	THD (%)
Older rapid start magnetic ballast	10–29
Electronic IC-based ballast	4–10
Electronic discrete based ballast	18–30
Newer rapid start electronic ballast	<10
Newer instant start electronic ballast	15–27
High intensity discharge ballast	15–27
Office equipment	50–150

1.23 Home Appliances

A number of households are served from the same utility transformer. The harmonics can be measured at the metering point of each residential house. The nonlinear loads in households are increasing because of replacement of incandescent lighting with fluorescent lighting, and use of variable-speed air conditioning systems for energy efficiency. There are three main considerations with respect to harmonic emissions from home appliances:

- The harmonic emissions from various load types will not arithmetically sum up. Generally, there will be a reduction of harmonics when measured at a common bus. This situation is not unique to residential loads and can occur, in general, in all power systems, see Chapter 5.

- In the harmonic analysis, it is assumed that the supply source is sinusoidal, ignoring the effect of harmonic generation. The waveforms and emissions may considerably change with distorted supply inputs, see Chapter 5.

- Referring to Equation 1.55, the characteristic harmonics are given by the inverse of the harmonic order. This is because the odd harmonic waveform is approximated as a square wave. For home appliances, this type of characterization has not been established. The following equation is proposed:

$$I_h = \frac{I_1}{h^{\alpha}} \tag{1.79}$$

where α is a parameter that determines the decline rate of the current spectrum and is estimated by performing curve fitting on the normalized spectra of home appliances.

Table 1.11 from Reference [14] shows the residential loads and harmonic characteristics. The wide variations are noteworthy.

The work in Reference [15] is mainly devoted to harmonics in household loads.

1.24 Voltage Source Converters

The FACTS use voltage source bridges. The current source converters have been extensively used for HVDC transmission. In the 1990s, HVDC transmission using voltage source

TABLE 1.11

Residential Loads and Whole House Harmonic Current Characteristics

Type of Load	RMS Load Current	THDi (%)	h_3	h_5	h_7	h_9
Clothes dryer	25.3	4.6	3.9	2.3	0.3	0.3
Stovetop	24.3	3.6	3.0	1.8	0.9	0.2
Refrigerator #1	2.7	13.4	9.2	8.9	1.2	0.6
Refrigerator #2	3.2	10.4	9.6	3.7	0.8	0.2
Desktop computer/printer	1.1	140.0	91.0	75.2	58.2	39.0
Conventional heat pump #1	23.8	10.6	8.0	6.8	0.5	0.6
Conventional heat pump #2	25.7	13.2	12.7	3.2	0.7	0.2
ASD heat pump #1	14.4	123.0	84.6	68.3	47.8	27.7
ASD heat pump #2	27.7	16.1	15.0	4.2	2.3	1.9
ASD heat pump #3	13.0	53.6	48.8	6.3	17.0	10.1
Color television	0.7	120.8	85.0	60.6	34.6	14.6
Microwave #1	11.7	18.2	15.7	5.1	3.2	2.1
Microwave #2	11.7	26.4	23.3	9.6	2.2	1.6
Vehicle Battery charger	0.5	51.7	43.2	26.9	2.6	4.2
Light Dimmer	1.6	49.7	41.2	16.0	12.1	10.0
Electric Dryer	25.3	4.6	3.9	2.3	0.3	0.3
Fluorescent ceiling light	2.5	39.5	36.9	13.8	2.3	1.4
Fluorescent desk lamp	0.6	17.6	17.1	3.4	2.0	0.6
Vacuum	6	25.9	25.7	2.7	1.8	0.4
House #1	72.0	4.8	3.6	3.2	0.2	
House #2	51.3	7.7	7.2	2.6	0.8	0.2
House #3	41.6	10.9	8.6	6.5	1.5	0.2
House #4	19.9	6.4	5.5	2.7	1.1	1.2
House #5	6.6	16.2	11.7	10.2	4.1	1.2
House #6	60.8	8.5	6.9	4.9	0.6	0.2
House#7	30.4	11.8	10.7	5.0	0.3	0.3
House #8	62.6	31.6	29.5	6.9	6.8	5.0

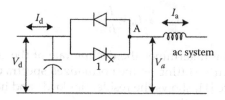

FIGURE 1.36
Principle of a voltage source converter.

converters with PWM was introduced, which was commercially called HVDC *light* [5]. The requirements of FACT controllers are as follows:

- The converter should be able to act as an inverter or rectifier with leading or lagging reactive power, i.e., four-quadrant operation is required, as compared to current source line-commutated converter which has two quadrant operations.
- The active and reactive power should be independently controllable with control of phase angle.

The principle is illustrated with respect to single valve operation, see Figure 1.36. Consider that dc voltage remains constant and the turn-off device is turned on by gate control. Then, the positive of dc voltage is applied to terminal A, and the current flows from $+V_d$ to A, i.e., inverter action. If the current flows from A to $+V_d$, *even when device 1 is turned on*, it will flow through the parallel diode, rectifier action. Thus, the power can flow in either direction. A valve with a combination of turn-off device and diode can handle power in either direction.

1.24.1 Three-Level Converter

Figure 1.37 shows the circuit of a three-level converter and associated waveforms. In Figure 1.37a, each half of the phase leg is split into two series-connected circuits, midpoint connected through diodes, which ensure better voltage sharing between the two sections. Waveforms in Figure 1.37b through e are obtained corresponding to one three-phase leg. Waveform (b) is obtained with 180° conduction of the devices. Waveform (c) is obtained if 1 is turned off and 2A is turned on at an angle α *earlier* than for 180° conduction. The ac voltage V_a is clamped to zero with respect to midpoint N of the two capacitors. This occurs because devices 1A and 2A conduct and in combination with diodes clamp the voltage to zero. This continues for a period of 2α, till 1A is turned off and 2 is turned on and the voltage is now $-V_d/2$, with both the 2 and 2A turned off and 1 and 1A turned on. The angle α is variable and the output voltage V_a is square waves $\sigma = 180° - 2\alpha°$. The converter is called a three-level converter as dc voltage has three levels, $V_d/2$, 0, and $-V_d/2$. The magnitude of the ac voltage can be varied without changing the magnitude of the dc voltage by varying angle α. Figure 1.37d shows the voltage V_b and Figure 1.37e, the phase-to-phase voltage V_{ab}.

The harmonic and fundamental rms voltages are given by

$$V_h = \frac{2\sqrt{2}}{\pi}\left(\frac{V_d}{2}\right)\frac{1}{2}\sin\frac{h\alpha}{2}$$

$$V_f = \frac{2\sqrt{2}}{\pi}\left(\frac{V_d}{2}\right)\sin\frac{\alpha}{2}$$

(1.80)

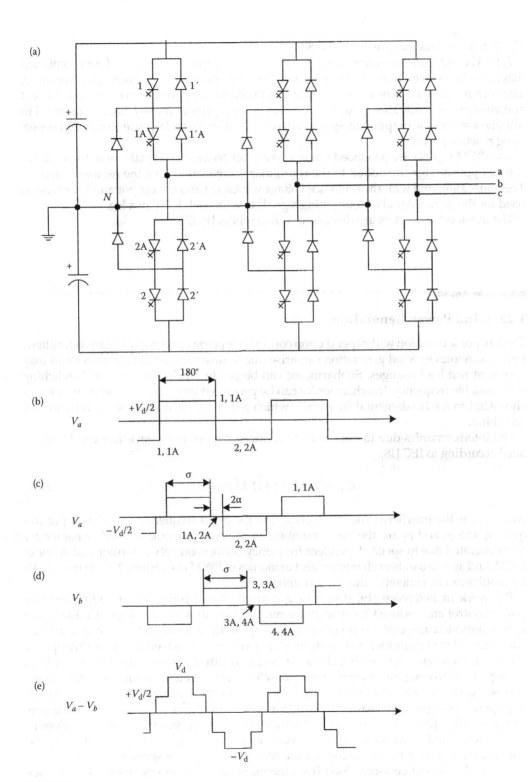

FIGURE 1.37
(a) A three-level three-phase voltage source converter. (b through e) Operational waveforms, see the text.

$V_h = 0$ at $\alpha = 0°$ and maximum at $\alpha = 180°$.

A STATCOM may use many six-pulse converters, output phase shifted and combined magnetically to give a pulse number of 24 or 48 for the transmission systems, see Volume 2. The output waveform is nearly sinusoidal and the harmonics present in the output current and voltage are small. This ensures waveform quality without passive filters. Figure 9.5 in Volume 2 shows the output voltage and current waveform of a 48-pulse STATCOM generating reactive power.

The PWM signals are produced using two carrier triangular signals; two different offsets of opposite sign are added to the triangular waveform to give the required carriers. These are compared with the reference voltage waveform and the output and its inverse is used for the gating signals—four gating signals one for each IGBT in a leg.

The five-level converters are discussed in References [16,17].

1.25 Wind Power Generation

The harmonic emission will depend upon converter topology and applied harmonic filters. Even harmonics in wind generation can arise due to unsymmetrical half waves and may appear at fast load changes. Subharmonic can be produced due to periodical switching with variable frequency. Interharmonics can be generated when the frequency is not synchronized to the fundamental frequency, which may happen at low and high frequency switching.

The interharmonics due to back-to-back configuration of two converters can be calculated according to IEC [18]:

$$f_{n,m} = \left[\left(p_1 k_1 \right) \pm 1 \right] f \pm \left(p_2 k_2 \right) F \tag{1.81}$$

where $f_{n,m}$ is the interharmonics frequency, f_1 is the input frequency, F is the output frequency, and p_1 and p_2 are the pulse numbers of the two converters. Interharmonics are also generated due to speed-dependent frequency conversion between rotor and stator of DFIM, and as side bands of characteristic harmonics of PWM converters. Noncharacteristic harmonics can be generated due to grid unbalance.

The work in Reference [19] describes a current source converter providing reactive power control and reduced harmonics for multimegawatt wind turbines. It utilizes two series-connected three-phase inverters that employ fully controllable switches and an interconnection transformer on the utility side. Two converters feed into two, three-phase wye-connected windings, with a delta-connected winding on the utility source side. A three-phase active harmonic current compensator is used on the output of the converter.

This chapter is indicative of the wide variations in the harmonic emissions, depending upon topologies. Thousands of such topologies exist, while every year some new ones are added [20]. For the purpose of harmonic load flow and harmonic filter designs in Chapters 7 and 8, the harmonic emission from loads should be first estimated. This is not an easy task due to varied topologies. Mostly, appropriate spectrums and the harmonic angles should be ascertained from the manufacturers. Furthermore, the emission is also a function of load as well as system source impedance. Figure 1.38 shows the harmonic current spectrum calculated based upon specified system impedance at the point of

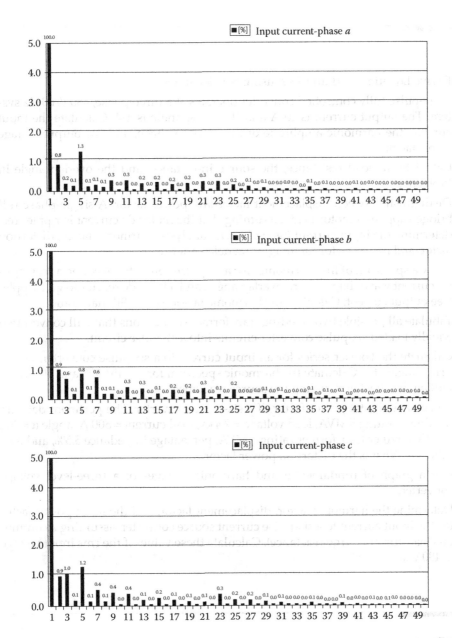

FIGURE 1.38
Harmonic emission, phases *a*, *b*, and *c*; 24-pulse PWM converter for a medium-voltage drive system.

interconnection. The drive system is PWM, 18-pulse. Note the difference in the harmonic emission in phases *a*, *b*, and *c*. Generally, for harmonic load flow, three-phase models are not used, though it may be necessary to do so where harmonic phase unbalance exists. The emission is small and meets the requirements of IEEE 519 [1], without additional filters.

Note that *all* sources of harmonic generation are not described; e.g., ferroresonance gives rise to overvoltages and harmonics, See Appendix C, Volume 1.

Problems

1.1 Derive Equations 1.41 and 1.42, using Fourier series.

1.2 A six-pulse fully controlled converter operates at a three-phase, 480 V, 60 Hz system. The output current is 10 A and the firing angle is $\pi/4$. Calculate the input current, the harmonic amplitude of the output voltage, and the output voltage ripple factor.

1.3 Calculate the load resistance, the source inductance, and the overlap angle in Problem 2.

1.4 Distinguish between displacement factor and power factor. A single-phase full bridge supplies a motor load. Assuming that the motor dc current is ripple free, determine the input current (using Fourier analysis), harmonic factor, distortion factor, and power factor for an ignition delay angle of α.

1.5 Draw a spectrum of line harmonics for a 12-pulse converter. Assume a rectangular current wave shape, a zero overlap angle, and a large dc reactor to give ripple-free output current. Calculate the harmonic factor to the 29th harmonic.

1.6 Tabulate all possible two-winding transformer connections that will convert two equally loaded six-pulse converter circuits into a 12-pulse circuit.

1.7 Calculate the Fourier series for an input current to a six-pulse converter, with a firing angle of α. Calculate the harmonic spectrum for $\alpha = 15°$, $45°$, and $60°$. What will be the effect of doubling the source reactance?

1.8 Calculate the total power factor of a six-pulse converter. The pertinent data are maximum rating 5 MVA, load voltage = 2.4 kV, load current = 500 A, angle $\alpha = 30°$, 13.8 kV rectifier transformer rating 5 MVA, percentage impedance 5.5%, and X/R ratio = 8. What is the distortion power factor?

1.9 Plot a graph of fundamental and harmonic voltages of a three-level voltage converter.

1.10 Determine the harmonic factor, displacement factor, and the input power factor for the input current to a six-pulse current source converter, assuming a continuous load current (large reactance). Calculate these values if the rms input voltage is 480 V and $\alpha = \pi/3$.

References

1. IEEE Standard 519, IEEE recommended practices and requirements for harmonic control in power systems, 1992.
2. ANSI/IEEE Standard C57.13, Requirements for instrument transformers, 2005.
3. J.C. Das, J.R. Linders, Power system relaying, *Encyclopedia of Electronics and Electrical Engineers*, Vol. 17, John Wiley & Sons, New York, 2002.
4. IEEE Working Group on Power System Harmonics, Power system harmonics: An overview, *IEEE Trans Power Apparatus Syst PAS*, 102, 2455–2459, 1983.
5. J. Arrillaga, *High Voltage Direct Current Transmission*, 2nd Ed., IEEE Press, Piscataway, NJ, 1998.

6. L. Malesani, P. Tenti, Three-phase AC/DC PWM converter with sinusoidal AC currents and minimum filter requirements, *IEEE Trans Ind Appl*, 23(1), 71–78, 1987.
7. B.T. Ooi, J.C. Salmon, J.W. Dixon, A. Kulkarni, A three-phase controlled-current PWM converter with leading power factor, *IEEE Trans Ind Appl*, 23(1), 78–84, 1987.
8. J. Cardosa, T. Lipo, Current stiff converter topologies with resonant snubbers, *IEEE Industry Application Society Annual Meeting*, New Orleans, LA, 1322–1329, 1997.
9. M.H. Rashid, *Power Electronics*, Prentice Hall, Englewood Cliffs, NJ, 1988.
10. B.R. Pelly, *Thyristor Phase-Controlled Converters and Cycloconverters, Operation Control and Performation*, John Wiley & Sons, New York, 1971.
11. H. Flick, Excitation of sub-synchronous torsional oscillations in turbine generator sets by a current-source converter, *Siemens Power Eng IV*, 4(2), 83–86, 1982.
12. R. Arthur, R.A. Sanhan, Neutral currents in three-phase wye systems. Square D Company, WI, 0104ED9501R8/96, August 1996.
13. National Electrical Code, NFPA 70, 2017.
14. A.E. Emanuel, J.A. Orr, D. Cyganski, E.M. Gulachenski, A survey of harmonic voltages and currents at the consumer's bus, *IEEE Trans Power Deliv*, 8(1), 411–421, 1993.
15. A. Nasif, *Harmonics in Power Systems: Modeling, Measurement and Mitigation of Power System Harmonics*, CRC Press, Boca Raton, FL, 2010.
16. N. Hatti, Y. Kondo, H. Akagi, Five level diode-clamped PWM converters connected back-to-back for motor drives, *IEEE Trans Ind Appl*, 44(4), 1268–1276, 2008.
17. H. Akagi, H. Fujita, S. Yonetani, Y. Kondo, A 6.6 kV transformer-less STATCOM based on a five level diode-clamped PWM converter: System design and experimentation of 200-V, 10 kVA laboratory model, *IEEE Trans Ind Appl*, 44(2), 672–680, 2008.
18. CEI/IEC 1000-2-1:1990, Electromagnetic compatibility, Part 2: Environment, Section 1: Description of the environment—Electromagnetic environment for low-frequency conducted disturbances and signaling in public power supply systems, 1st Edition, 1990–2005, IEC Standard.
19. P. Tenca, A.A. Rockhill, T.A. Lipo, Wind turbine current-source converter providing reactive power control and reduced harmonics, *IEEE Trans Ind Appl*, 43(4), 1050–1060, 2007.
20. F.L. Luo, H. Ye, *Power Electronics*, CRC Press, Boca Raton, FL, 2010.

Popular Books on Harmonics

E. Acha, M. Madrigal, *Power System Harmonics: Computer Modeling and Analysis*, John Wiley & Sons, 2001.
J. Arrillaga, N.R. Watson, *Power System Harmonics*, 2nd Edition, John Wiley & Sons, 2003.
J. Arrillaga, B.C. Smith, N.R. Watson, A.R. Wood, *Power System Harmonic Analysis*, John Wiley & Sons, 2000.
J.C. Das, *Power System Harmonics and Passive Filter Designs*, IEEE Press, NJ, 2015.
F.C. De La Rosa, *Harmonics Power System*, CRC Press, 2006.
A. Fadnis, *Harmonics in Power Systems: Effects of Power Switching Devices in Power Systems*, Lambert Academic Publishing, 2012.
S.M. Ismail, S.F. Mekhamer, A.Y. Abdelaziz, *Power System Harmonics in Industrial Electrical Systems-Techno-Economical Assessment*, Lambert Academic Publishing, 2013.
A. Nassif, *Harmonics in Power System-Modeling, Measurement and Mitigation*, CRC Press 2010.
G.J. Wakileh, *Power System Harmonics: Fundamentals, Analysis and Filter Design*, Springer, 2001.

2

Interharmonics and Flicker

2.1 Interharmonics

IEC [1] defines interharmonics as "Between the harmonics of the power frequency voltage and current, further frequencies can be observed, which are not an integer of the fundamental." They appear as discrete frequencies or as a wide-band spectrum. A more recent definition is any frequency which is not an integer multiple of the fundamental frequency.

Noninteger harmonics are the same as interharmonics. Also see IEEE task force on harmonic modeling and simulation [2]. IEEE Standard 519d address interharmonics related to flicker.

2.2 Generation of Interharmonics

Cycloconverters are a major source of interharmonics. Large mill motor drives using cycloconverters ranging up to 8 MVA appeared in 1970s. In 1992, rolling mill drives of 26 MVA were installed. These are also used in 22 Hz railroad traction power applications. The cycloconverters can be thought of as frequency converters. The dc voltage is modulated by the output frequency of the converter and interharmonic currents appear in the input, as discussed further.

Electrical arc furnaces (EAFs) are another major source. Arcing devices give rise to interharmonics. These include arc welders. These loads are associated with low-frequency voltage fluctuations giving rise to flicker. These also exhibit higher frequency interharmonic components. The interharmonic limits must be accounted for in filter designs. Other sources are as follows:

- Induction furnaces
- Integral cycle control
- Low-frequency power line carrier; ripple control
- HVDC
- Traction drives
- Adjustable speed drives (ASDs)
- Slip frequency recovery schemes

2.2.1 Imperfect System Conditions

Practically, the ideal conditions of operation are not obtained, which give rise to noncharacteristics harmonics and interharmonics. Consider the following:

- The ac system three-phase voltages are not perfectly balanced. The utilities and industrial power systems have some single-phase loads also, which give rise to unbalance. A 1% lower voltage in one phase will give approximately 7% third and 4.3% second harmonics in a standard 12-pulse converter.

- The impedances in the three phases are not exactly equal, especially unequal commutation reactances or unequal phase impedances of the converter transformer.

- With a commutating reactance of 0.20 pu and a variation of 7.2% in each phase, a firing angle 12° will give a second harmonic of 33% of fundamental.

- DC modulation and cross modulation. Addition of a harmonic h on the dc side will transfer to ac side; the harmonic will be of different order but of same phase sequence. The back-to-back frequency conversion represents the worst case of interharmonic generation. Consider that two ac systems are interconnected through a dc link, and operate at different frequencies. An equivalent circuit is shown in Figure 2.1. The source of ripple in this figure is voltage from the remote end of the dc link and the distortion caused by the converter itself, the ripple current depending on the dc link reactor. The system can represent HVDC link, the ac power systems operating at different frequencies.

A conventional six-pulse three-phase bridge circuit can be conceived as a combination of *a switching function and a modulating function*. The switching function can be termed as follows:

$$s(t) = k\left[\cos\omega_1 t - \frac{1}{5}\cos 5\omega_1 t + \frac{1}{7}\cos 7\omega_1 t - \cdots\right] \tag{2.1}$$

and the modulating function as the sum of dc current and superimposed ripple content:

$$i(t) = I_d \sum_{z=1}^{\infty} a_z \sin(\omega_z t + \phi) \tag{2.2}$$

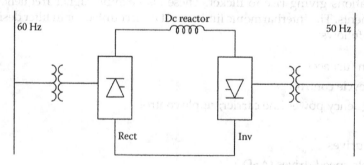

FIGURE 2.1
Two ac systems connected through a dc link, equivalent circuit.

where a_z is the peak magnitude of sinusoidal components and ω_z can be any value and not an integer multiple of ω_1. Then, the harmonics on the ac side are as follows:

$$i_{ac}(t) = i(t)s(t) \tag{2.3}$$

For a 12-pulse operation, the following equations can be written for the harmonics on the ac side:

$$i_{ac} = ki_d(\text{12th, 13th, 23rd}, \cdots)$$

$$+ \frac{kb}{2}[\sin(\omega_1 t + 12\omega_2 t + \phi_{12}) - \sin(\omega_1 t - 12\omega_2 t - \phi_{12})]$$

$$- \frac{kb}{22}[\sin(11\omega_1 t + 12\omega_2 t + \phi_{12}) - \sin(11\omega_1 t - 12\omega_2 t - \phi_{12})]$$

$$+ \frac{kb}{26}[\sin(13\omega_1 t + 12\omega_2 t + \phi_{12}) - \sin(13\omega_1 t - 12\omega_2 t - \phi_{12})] \tag{2.4}$$

$$- \frac{kc}{46}[\sin(23\omega_1 t + 24\omega_2 t + \phi_{24}) - \sin(23\omega_1 t - 24\omega_2 t - \phi_{24})]$$

$$+ \frac{kc}{50}[\sin(25\omega_1 t + 24\omega_2 t + \phi_{24}) - \sin(25\omega_1 t - 24\omega_2 t - \phi_{24})]$$

$$- \frac{kd}{70}[\sin(35\omega_1 t + 36\omega_2 t + \phi_{36}) - \sin(35\omega_1 t - 36\omega_2 t - \phi_{36})]$$

etc.

where $k = 2\sqrt{3}\pi$, b, c, and d are the magnitudes of the 12th, 24th, and 36th harmonic currents on the dc side.

In a rectifier–dc link-converter system linking two isolated ac systems, if the frequency on two sides differs by Δf_0, then for a 12-pulse system, the dc side voltage at a frequency of $12n(f_0 + \Delta f_0)$ will be modulated by another converter:

$$(12m \pm 1)f_0 + 12n(f_0 + \Delta f_0) \tag{2.5}$$

The ac side, among other frequencies, includes the frequency

$$f_0 + 12n\Delta f_0 \tag{2.6}$$

This will beat with the fundamental component at a frequency of $12n\Delta f_0$, which will allow flicker producing currents to flow.

The control systems and gate control of the electronic switching devices are not perfectly symmetrical. These system conditions will give rise to noncharacteristics and interharmonics, which would have been absent if the systems were perfectly symmetrical.

2.3 Interharmonics from ASDs

The interharmonics can originate from the converters by interaction of a harmonic from the dc link into the power source. Consider an ASD, with the motor running at 44 Hz; this frequency will be present at the dc link as a ripple of 44 Hz times the pulse number of the inverter. The current on the dc link contains both the 60 and 44 Hz ripples. The 44 Hz ripples will pass on to the supply side and present themselves as interharmonics because a 44× pulse number is not an integer of 60 Hz.

If m is the mth motor harmonic and nth is the PWM inverter harmonic and ω is the inverter operating frequency, then significant components of inverter input current will exist at frequencies $(n \pm m)\omega$ for n with a significant switching frequency component and m with a significant motor harmonic current component.

The unbalance causes low-order harmonics particularly the 2nd and 12th [3]. The 2nd and 12th inverter current harmonics in the dc link cause interharmonics when reflected to the ac side of the rectifier:

$$f_h = \left| \mu f_I \pm k f_s \right| \tag{2.7}$$

where,

f_h is the frequency of the interharmonic, μ is the order of current harmonic, typically 2 or 12
f_I is the inverter operating frequency, $k = 1, 2, 7, \ldots$
f_s is the source frequency = 60 Hz.

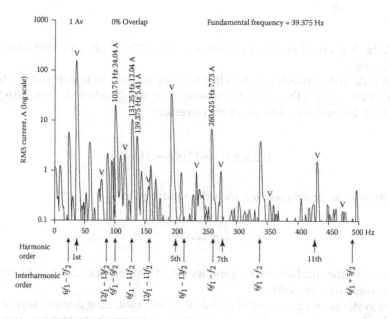

FIGURE 2.2
Harmonic spectrum from an actual ASD. (R. Yacamini, Power system harmonics—Part 4, interharmonics, *IEEE Power Eng J* © 1996 IEEE.)

The most significant values of interharmonics will occur with $\mu = 2$ and $k = 1$. For inverter frequencies of 22, 37.2, and 48 Hz and source frequency of 60 Hz, the side band pairs are 10 and 110, 115 and 132 Hz, and 36 and 126 Hz, respectively. The frequency modulation rate m_f was chosen so that switching frequency is in the range 1.8–2 kHz.

With overmodulation, $m_a > 1$, where $m_a = V_{AN}/(0.5V_d)$, defined as amplitude modulation ratio. Lower order harmonics appear on the dc link in balanced case, most dominant being the sixth harmonic. When it is reflected to the ac side, interharmonics at 228 and 348 Hz occur with inverter operating frequency of 48 Hz. This assumes that the dc reactance is substantial and source inductance is negligible. As L_d is reduced, the rectifier current harmonics rise till the rectifier output current becomes discontinuous. The effect of the source inductance will be to change apparent dc link inductance and therefore the tuning of the dc link components [3].

An example of harmonic spectrum from an actual operating system is shown in Figure 2.2 [4]. The motor is fed at 39.4 Hz (20 Hz source frequency). If the harmonics or interharmonics coincide with the natural frequency of the motor/shaft/load mechanical system, then shaft damage is possible.

2.3.1 Interharmonics from HVDC Systems

HVDC systems are another possible source of interharmonics. Interharmonics of the order of 0.1% of the rated current can be expected in HVDC systems, when two ends are working at even slightly different frequencies [5, 6].

The modulation theory has been used in harmonic interactions in HVDC systems. When ac networks operate at different frequencies, interharmonics will be produced.

2.3.1.1 DC Side

The voltage harmonics will contain frequency groups $6n\omega_1$ and $(6n\omega_1 + \omega_m)$, where ω_m is the mth harmonic frequency which may be an integer harmonic of either of the two ac systems—call it a disturbing frequency.

The characteristic harmonics in dc voltage ($=6n\omega_1$) appear in the dc current. Harmonic frequencies will be $\omega_m = 6m\omega_2$. This gives a new group of harmonics:

$$6n\omega_1 \pm 6m\omega_1 = 6(n \pm m)\omega_1 = 6k\omega_1 \qquad (2.8)$$

where n, m, and k are integers.

All characteristic harmonics from the inverter will appear in the dc current and the frequency will be $6m\omega_2$. Thus, the second group of harmonics on the dc side is

$$6n\omega_1 \pm 6m\omega_2 = 6(n\omega_1 \pm m\omega_2) \qquad (2.9)$$

The third set of frequencies from Equation (2.9) will also appear in the dc current, so that $\omega_m = 6(n\omega_1 \pm p\omega_2)$; and, therefore, the frequencies

$$6n\omega_1 \pm 6(m\omega_1 \pm p\omega_2) = 6[(n \pm m)\omega_1 \pm p\omega_2] = 6(k\omega_1 \pm p\omega_2) \qquad (2.10)$$

where n, m, p, and k are integers.

2.3.1.2 AC Side

The harmonics on the ac side will be as follows:

Those caused by dc characteristic harmonics, $\omega_m = 6m\omega_1$. The harmonics transferred to ac side are as follows:

$$6m\omega_1 \pm (6n \pm 1)\omega_1 = (6k \pm 1)\omega_1 \tag{2.11}$$

Those caused by characteristic dc voltage harmonics generated at far end, $\omega_m = 6m\omega_2$:

$$6m\omega_2 \pm (6n \pm 1)\omega_1 \tag{2.12}$$

Those caused by

$$\omega_m = 6(m\omega_1 \pm p\omega_2) \tag{2.13}$$

The harmonics transferred to the ac side will be

$$6(m\omega_1 \pm p\omega_2) \pm (6n \pm 1)\omega_1 \tag{2.14}$$

This assumes low dc side impedance, also see References [7, 8]. Practically, dc link reactors and ac and dc filters are used to mitigate the harmonics.

The interharmonics due to electric traction are discussed in Chapter 1.

2.3.2 Cycloconverters

Cycloconverters are discussed in Chapter 1. Certain relationship exists between the converter pulse numbers, the harmonic frequencies present in the output voltage, and the input current. The harmonic frequencies in *the output voltage* are integer multiple of pulse number and input frequency, $(np)f$, to which are added and subtracted integer multiples of output frequency, i.e.,

$$h_{\text{output voltage}} = (np)f \pm mf_0 \tag{2.15}$$

Here, n is any integer and not the order of the harmonic, and m is also an integer as described in the following.

For cycloconverter with *single-phase output*, the harmonic frequencies present in the input current are related to those in the output voltage. There are two families of input harmonics:

$$h_{\text{input current}} = \left| \left[(np) - 1 \right] f \pm (m-1)f_0 \right|$$

$$h_{\text{input current}} = \left| \left[(np) + 1 \right] f \pm (m-1)f_0 \right| \tag{2.16}$$

where m is odd for (np) even and m is even for (np) odd.

In addition, the characteristic family of harmonics, independent of pulse number, is given by the following:

$$\left| f \pm 2mf_0 \right| \quad m \geq 1 \tag{2.17}$$

For a cycloconverter with a balanced three-phase output, for each family of output voltage harmonics, $(np)f \pm mf_0$, there are two families of input current harmonics:

$$h_{\text{input current}} = \left| \left[(np) - 1 \right] f \pm 3(m-1)f_0 \right|$$

$$h_{\text{input current}} = \left| \left[(hp) + 1 \right] f \pm 3(m-1)f_0 \right|$$

(2.18)

where m is odd for (np) even and m is even for (np) odd.

In addition, the characteristic family of harmonics, independent of pulse number, is given by

$$\left| f \pm 6mf_0 \right| \quad m \geq 1$$

(2.19)

For input current waveforms with higher pulse numbers, certain harmonic families are eliminated as shown [9].

The magnitude of the harmonic is a function of the output voltage and the load–displacement angle, but is independent of the frequency of the component. Thus, for a given output voltage ratio and load–displacement angle, those harmonic components that are present always have the same relative magnitude independent of pulse number or the number of output phases.

2.4 Arc Furnaces

A schematic diagram of an EAF installation is shown in Figure 2.3. The furnace is generally operated with static var compensation systems and passive shunt harmonic filters. The installations can compensate rapidly changing reactive power demand, arrest voltage fluctuations, reduce flicker and harmonics, and simultaneously improve the power factor to

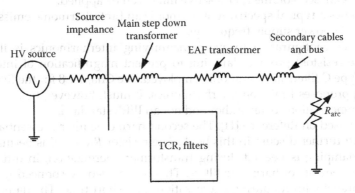

FIGURE 2.3
Schematic diagram of an EAF installation.

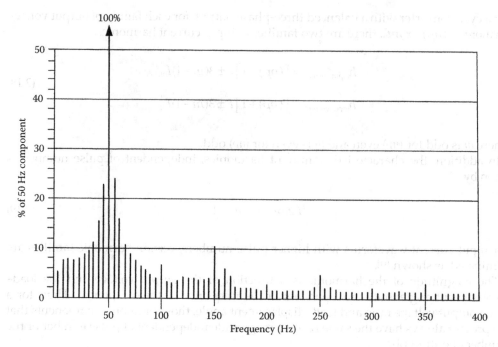

FIGURE 2.4
Typical spectrum of harmonic and interharmonic emission from an EAF.

unity. Typical harmonic emissions from IEEE standard 519 are shown in Table 1.6. In practice, a large variation in harmonics is noted. For example, maximum to minimum limits of *voltage distortions* at second, third, and fourth harmonics may vary from 17% to 2%, 29% to 20%, and 7.2% to 3%, respectively. The tap-to-tap time (the time for one cycle operation, melting, refining, tipping, and recharging) may vary between 20 and 60 min depending upon the processes, and the furnace transformer is de-energized and then re-energized during this operation. This gives rise to additional harmonics during switching; saturation of transformer due to dc and second harmonic components, dynamic stresses, and can bring about resonant conditions with improperly designed passive filters. New technologies like STATCOM, see Volume 2, and active filters can be applied.

Figure 2.4 shows a typical spectrum of harmonic and interharmonic emission from an arc furnace (50 Hz power supply frequency).

A typical filter configuration to avoid magnifying interharmonics is illustrated in Figure 2.5. The resistors provide damping to prevent magnification of interharmonics components. Type C filters are commonly employed. See Chapter 8 for the filter types.

Gunther [10] proposes limits on interharmonics. It must, however, be recognized that these are recommendations of an author and not an IEEE standard.

Figure 2.6 is based on Reference [11]. The second harmonic filter is essentially a type C filter, as we will further discuss in this book. The resistor R_D remains permanently connected. High damping is needed during transformer energization, in order to reduce stresses on the elements of harmonic filters. This is achieved by connecting a low resistance R_{TS} in parallel with R_D during energization for a short time. The damping of transients becomes an important consideration.

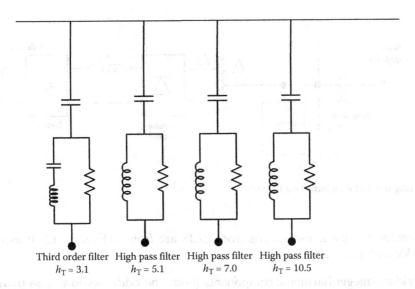

FIGURE 2.5
A typical harmonic filter configuration to avoid magnification of interharmonics in EAF.

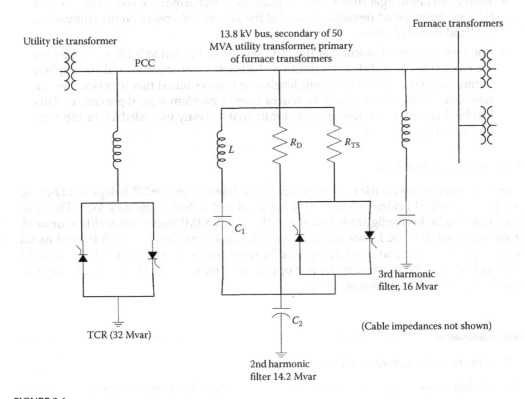

FIGURE 2.6
A configuration for harmonic emission control with type C filter and damping of transformer inrush current harmonics. (C.O. Gercek, M. Ermis, A. Ertas, K.N. Kose, O. Unsar, Design implementation and operation of a new C-type second-order harmonic filter for electric arc and ladle furnaces, *IEEE Trans Ind Appl,* © 2011 IEEE.)

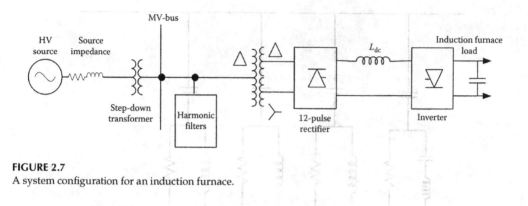

FIGURE 2.7
A system configuration for an induction furnace.

The following harmonic restrictions from EAFs are from Reference [12]. It assumes a PCC < 161 kV and SCR < 20:

- Individual integer harmonic components (even and odd) should be less than 2% of the specified demand current for the facility, 92% point on the cumulative probability distribution.

- Individual noninteger distortion components (interharmonics) should not exceed 0.2% of the specified demand current of the facility, 92% point on the cumulative probability distribution.

- The total demand distortion at the PCC should be limited to 2.2%, 92% point on the cumulative probability distribution. Additional restrictions should be applied to limit shorter duration harmonic levels, if it is determined that they could excite resonance in the power supply system or cause a problem at local generators. This can be done by specifying separate limits that are only exceeded 1% of the time (usually not necessary).

2.4.1 Induction Furnaces

A system configuration of the induction furnace is shown in Figure 2.7. It depicts a 12-pulse rectification with H-bridge inverter (see Chapter 8) and induction furnace load. The measurement data in this configuration are for of 22-t, 12 MVA IMF (induction melting furnace), from Reference [13]. The harmonics and interharmonics may interact with the industrial loads or passive filters and may be amplified by resonance with the passive filters. A model is generated, and the variable frequency operation is represented by a time-varying R–L circuit in parallel with a current source.

2.5 Effects of Interharmonics

The interharmonic frequency components, greater than the power frequency, produce heating effects, similar to those produced by harmonics. The low-frequency voltage interharmonics cause significant additional loss in induction motor stator windings.

The impact on light flicker is important. Modulation of power system voltage with interharmonic voltage introduces variations in system voltage rms value.

The IEC flicker meter is used to measure the light flicker indirectly by simulating the response of an incandescent lamp and human-eye-brain response to visual stimuli, IEC Standard [14]. The other impacts of concern are as follows:

- Excitation of subsynchronous conditions in turbo-generator shafts.
- Interharmonic voltage distortions similar to other harmonics.
- Interference with low-frequency power line carrier control signals.
- Overloading of conventional tuned filters: See Chapter 8 for the limitations of the passive filters, the tuning frequencies, the displaced frequencies, and possibility of a resonance with the series tuned frequency. As the interharmonics vary with the operating frequency of a cycloconverter, a resonance can be brought where none existed before making the design of single-tuned filters impractical.

The distortion indices for the interharmonics can be described similar to indices for harmonics. The total interharmonic distortion (THID) factor (voltage) is

$$\text{THID} = \frac{\sqrt{\sum_{i=1}^{n} V_i^2}}{V_1} \qquad (2.20)$$

where i is the total number of interharmonics being considered including subharmonics and n is the total number of frequency bins. A factor exclusively for subharmonics can be defined as total subharmonic distortion factor:

$$\text{TSHD} = \frac{\sqrt{\sum_{s=1}^{S} V_s^2}}{V_1} \qquad (2.21)$$

An important consideration is the torsional interaction may develop at the nearby generating facilities, see Section 2.10. In this case, it is necessary to impose severe restrictions on interharmonic components. In other cases, interharmonics need not be treated any different from integer harmonics.

2.6 Reduction of Interharmonics

The interharmonics can be controlled by the following:

- Higher pulse numbers
- DC filters, active or passive, to reduce the ripple content
- Size of the dc link reactor
- Pulse width modulated drives

2.7 Flicker

Voltage flicker occurs due to operation of rapidly varying loads, such as arc furnaces which affect the system voltage. This can cause annoyance by causing visible light flicker on tungsten filament lamps. The human eye is most sensitive to light variations in the frequency range 2–10 Hz and voltage variations of less than 0.2% in this frequency can cause annoying flicker from tungsten lamp lighting.

2.7.1 Perceptible Limits

The percentage pulsation of voltage related to frequency, at which it is most perceptible, from various references, is included in Figure 2.8 from Reference [15]. In this figure, the solid lines are composite curves of voltage flicker by General Electric Company (*General Electric Review*, 1922); Kansas Power and Light Company, *Electrical World* May 19, 1934; T7D Committee EEI, October 14, 1934, Chicago; Detroit Edison Company; West Pennsylvania Power Company; and Public Service Company of North Illinois. Dotted lines show voltage flicker allowed by two utilities, reference *Electric World* November 3, 1928 and June 1961. The flicker depends on the whole chain of "voltage fluctuations-luminance-eyes-brain."

Though this figure has been in use for a long time, it was superseded in IEEE Std. 1423 [16] with Figure 2.9. The solid-state compensators and loads may produce modulation

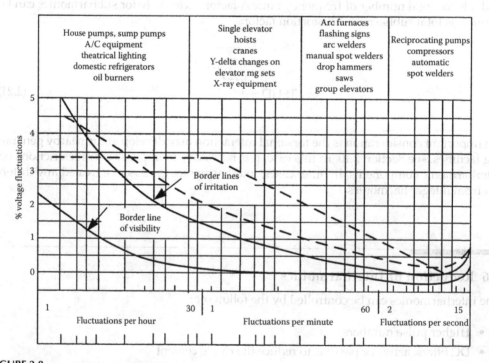

FIGURE 2.8
Maximum permissible voltage fluctuations see explanations in text. (IEEE Standard 519, IEEE recommended practices and requirements for harmonic control in power systems © 1992 IEEE, see also Revision of 2014.)

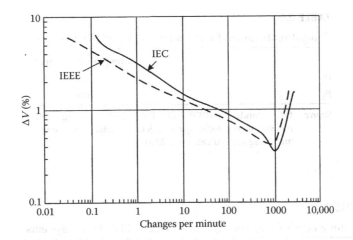

FIGURE 2.9
Comparison of IEC and IEEE standards with respect to flicker tolerance. (IEEE Standard 1453, Recommended practice for analysis of fluctuating installations © 2015 IEEE.)

of the voltage magnitude that is more complex than what was envisaged in the original flicker curves. This standard adopts IEC standard 61000-3-3 [17] in total. Define

$$P_{lt} = \sqrt[3]{\frac{1}{12} \times \sum_{j=1}^{12} P_{st\,j}^{\,3}} \qquad (2.22)$$

where P_{lt} is a measure of long-term perception of flicker obtained for a 2h period. This is made up of 12 consecutive P_{st} values, where P_{st} is a measure of short-term perception of flicker for 10 min intervals. This value is the standard output of the IEC flicker meter. Further qualification is that the IEC flicker meter is suited to events that occur once per hour or more often. The curves shown in Figure 2.8 are still useful for infrequent events like a motor start, once per day or even as frequent as some residential air conditioning equipment. Figure 2.9 depicts the comparison of IEEE and IEC for flicker irritation.

The short-term flicker severity is suitable for accessing the disturbances caused by individual sources with a short duty cycle. Where the combined effect of several disturbing loads operating randomly is required, it is necessary to provide a criterion for long-term flicker severity, P_{lt}. For this purpose, the P_{lt} is derived from short-term severity values over an appropriate period related to the duty cycle of the load, over which an observer may react to flicker.

For acceptance of flicker causing loads to utility systems, IEC standards [17–19] are recommended. The application of shape factors allows the effect of loads with voltages fluctuations other than the rectangular to be evaluated in terms of P_{st} values. Further research is needed in issues related to the effect of interharmonics on flicker and flicker transfer coefficients from HV to LV electrical power systems [20,21].

2.7.2 Planning and Compatibility Levels

Two levels, planning level and compatibility levels, are defined. Compatibility level is the specified disturbance level in a specified environment for coordination in setting the emission and immunity limits. Planning level, in a particular environment, adopted as a reference value for limits to be set for the emissions from large loads and installations, in

TABLE 2.1

Compatibility Levels for P_{st} and P_{lt} in LV and MV Systems

	Compatibility Level
P_{st}	1.0
P_{lt}	0.8

Source: IEC Standard 61000-4-12. Part 4–12. Testing and measurement techniques—Flicker meter—Functional and design specifications © 2010.

TABLE 2.2

Planning Levels for P_{st} and P_{lt} in MV, HV, and EHV Power Systems

	Planning Levels	
	MV	HV-EHV
P_{st}	0.9	0.8
P_{lt}	0.7	0.6

Source: IEC Standard 61000-4-12. Part 4–12. Testing and measurement techniques—Flicker meter—Functional and design specifications © 2010.

order to coordinate those limits with all the limits adopted for equipment intended to be connected to the power supply system.

As an example planning levels for P_{st} and P_{lt} in MV (voltages >1 kV and <32 kV), HV (voltages >32 kV and <230 kV) and EHV (voltages >230 kV) are shown in Table 2.1 and the compatibility levels for LV and MV power systems are shown in Table 2.2.

2.7.3 Flicker Caused by Arcing Loads

Arc furnaces cause flicker because the current drawn during melting and refining periods is erratic and fluctuates widely and the power factor is low, Chapter 1. An EAF current profile during melting and refining is depicted in Figure 2.10.

There are certain other loads that can also generate flicker; e.g., large spot welding machines often operate close to the flicker perception limits. Industrial processes may comprise a number of motors having rapidly varying loads or starting at regular intervals and even domestic appliances such as cookers and washing machines can cause flicker on weak systems. However, the harshest load for flicker is an arc furnace. During the melting cycle of a furnace, the reactive power demand is high. Figure 2.10 shows that an arc furnace current is random and no periodicity can be assigned, yet some harmonic spectrums have been established, Table 1.7, from IEEE 519. Note that even harmonics are produced during melting stage. The high reactive power demand and poor power factor cause cyclic voltage drops in the supply system. Reactive power flow in an inductive element requires voltage differential between sending end and receiving ends and there is reactive power loss in the element itself. When the reactive power demand is erratic, it causes corresponding swings in the voltage dips, much depending upon the stiffness of the system behind the application of the erratic load. This voltage drop is proportional to the short-circuit MVA of the supply system and the arc furnace load.

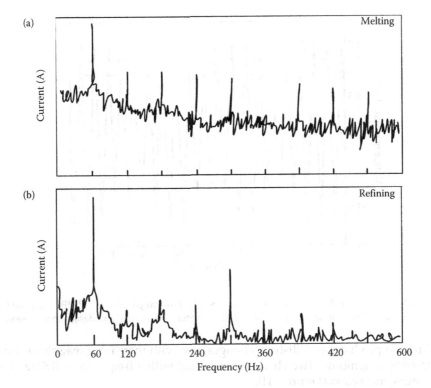

FIGURE 2.10
Erratic current spectrum of an EAF during melting and refining.

For a furnace installation, the short-circuit voltage depression (SCVD) is defined as follows:

$$SCVD = \frac{2MW_{furnace}}{MVA_{SC}} \tag{2.23}$$

where the installed load of the furnace in MW is $MW_{furnace}$ and MVA_{SC} is the short-circuit level of the utility's supply system. This gives an idea whether potential problems with flicker can be expected. An SCVD of 0.02–0.022 may be in the acceptable zone, between 0.03 and 0.032 in the borderline zone, and above 0.032 objectionable [22]. When there are multiple furnaces, these can be grouped into one equivalent MW. A study in Chapter 8 describes the use of tuned filters to compensate for the reactive power requirements of an arc furnace installation. The worst flicker occurs during the first 2–10 min of each heating cycle and decreases as the ratio of the solid to liquid metal decreases.

The significance of $\Delta V/V$ and number of voltage changes are illustrated with reference to Figure 2.11 from IEC [14]. This shows a 20 Hz waveform, having a 1.0 average voltage with a relative voltage change $\Delta v / \bar{v} = 40\%$ and with 8.8 Hz rectangular modulation. It can be written as follows:

$$v(t) = 1 \times \sin(2\pi \times 50t) \times \left\{ 1 + \frac{40}{100} \times \frac{1}{2} \times signum\left[2\pi \times 8.8 \times t\right] \right\} \tag{2.24}$$

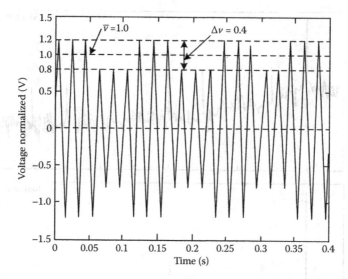

FIGURE 2.11
Modulation with rectangular voltage change $\Delta V/V = 40\%$, 8.8 Hz, 17.6 changes per second. (IEC Standard 61000-4-12. Part 4–12. Testing and measurement techniques—Flicker meter—Functional and design specifications, 2010.)

Each full period produces two distinct changes, one with increasing magnitude and one with decreasing magnitude. Two changes per period with a frequency of 8.8 Hz give rise to 17.6 changes per second (Figure 2.11).

2.8 Flicker Testing

The European test of flicker is designed for 230 V, 20 Hz power and the limits specified in IEC are based on the subjective severity of flicker from 230 V/60 W coiled-coil filament lamps, and fluctuations of the supply voltage. In the United States, the lighting circuits are connected at 112–120 V. For a three-phase system, a reference impedance of $0.4 + j0.22\ \Omega$, line to neutral, is recommended, and IEC standard 61000-3-3 is for equipments with currents $\leq$16 A. The corresponding values in the United States could be 32 A; statistical data of the impedance up to the point of common coupling with other consumers are required. As for a 112 V system, the current doubles, the authors in Reference [23] recommend a reference impedance of $0.2 + j0.12\ \Omega$.

An explanation of some terms used in IEC terminology is as follows:

$\Delta U(t)$: Voltage change characteristics. The time function of change in rms voltage between periods when the voltage is in a steady-state condition for at least 1 s

$U(t)$: Voltage shape, rms. The time function of the rms voltage evaluated stepwise over successive half periods of the fundamental voltage

ΔU_{max}: Maximum voltage change. The difference between maximum and minimum rms voltage change characteristics

ΔU_c: Steady-state voltage change. The difference between maximum and minimum rms values of voltage change characteristics

$d(t), d_{max}, d_c$: Ratios of the magnitudes of $\Delta U(t), \Delta U_{max}, \Delta U_c$ to the phase voltage

See Figure 2.12.

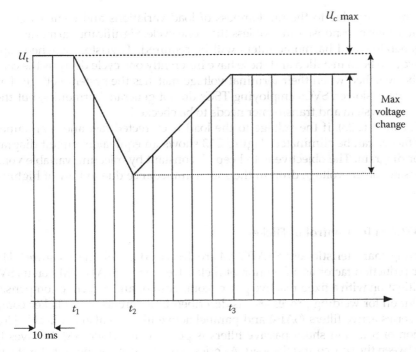

FIGURE 2.12
Explanation of the IEC terminology for calculation of P_{st}.

The following expression is used for P_{st} based on an observation period $T_{st} = 10\,\mathrm{min}$:

$$P_{st} = \sqrt{0.0314 P_{0.1} + 0.0525 P_{1s} + 0.0657 P_{3s} + 0.28 P_{10s} + 0.08 P_{50s}} \tag{2.25}$$

where percentiles $P_{0.1}, P_{1s}, P_{3s}, P_{10s}$, and P_{50s} are the flicker levels exceeded for 0.1%, 1%, 3%, 10%, and 20% of the time during the observation period. The suffix "s" indicates that the smoothed value should be used, obtained as follows:

$$P_{50s} = (P_{30} + P_{50} + P_{80})/3$$

$$P_{10s} = (P_6 + P_8 + P_{10} + P_{13} + P_{17})/5$$

$$P_{3s} = (P_{2.2} + P_3 + P_4)/3 \tag{2.26}$$

$$P_{1s} = (P_{0.7} + P_1 + P_{1.5})/3$$

2.9 Control of Flicker

The response of the passive compensating devices is slow. When it is essential to compensate load fluctuations within a few milliseconds, SVCs are required. Large TCR flicker compensators of 200 MW have been installed for arc furnace installations. Closed-loop control is necessary due to the randomness of load variations and complex circuitry is required to achieve response times of less than one cycle. Significant harmonic distortion may be generated, and harmonic filters will be required. Thyristor switched capacitors (TSCs) have also been installed and these have inherently one cycle delay as the capacitors can only be switched when their terminal voltage matches the system voltage. Thus, the response time is slower. SVCs employing TSCs do not generate harmonics, but the resonance with the system and transformer needs to be checked.

From Equation (2.23), if the voltage to the load is corrected fast and maintained constant, the flicker can be eliminated. Figure 2.13 shows an equivalent circuit diagram and the phasor diagram. The objective is to keep V_L constant by injecting variable voltage V_c, to compensate for the voltage drop in the source impedance due to flow of highly erratic current I_s.

2.9.1 STATCOM for Control of Flicker

The operating characteristics of a STATCOM are described in Volume 2. Figure 2.14 shows the flicker reduction factor as a function of flicker frequency, STATCOM versus SVC [24]. Flicker mitigation with a fixed reactive power compensator and an active compensator—a hybrid solution for welding processes—is described in Reference [25]. Flicker compensation with series active filters (SAFs) and parallel active filters is also applicable [26,27]. A combination of SAF and shunt passive filters is possible, in which SAF behaves like an isolator between the source and the load. A series capacitor can compensate for the voltage drop due to system impedance and fluctuating load demand, thus, stabilizing the system voltage and suppress flicker and noise.

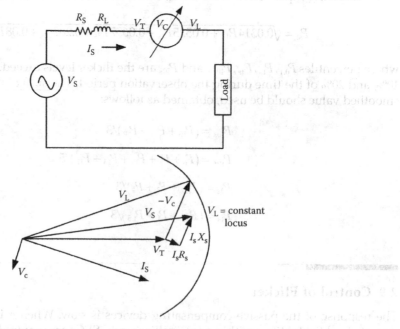

FIGURE 2.13
(a) Equivalent circuit diagram for compensation of load voltage and (b) phasor diagram.

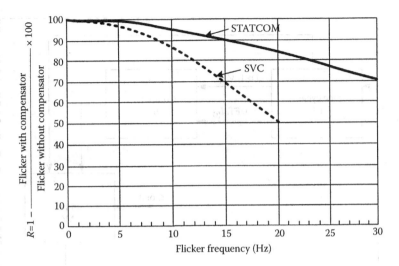

FIGURE 2.14
Flicker factor *R* for a STATCOM and SVC.

2.10 Tracing of Flicker and Interharmonics

Studies [28–32] provide flicker source identification methods. The active power- and impedance-based methods focus on major source of flicker, yet, how much each source contributes to flicker level at PCC is of much interest. The reactive load current component method is based on the concept that variation of the fundamental component of the voltage waveform in time causes the amplitude modulation effect which causes flicker. The background flicker contributed by the supply source adds to the flicker caused by loads at the PCC.

The system impedance and its angle are not constant quantities, and the values can be based on measurements.

Also source voltage cannot be assumed constant. Then, a time-varying relation can be written as follows:

$$e_s = iR_s + L_s \frac{di}{dt} + v_{pcc} \tag{2.27}$$

where v_{pcc} is the voltage at PCC.

This can be approximated as follows:

$$E_s \approx v_{pcc} + X_s I \sin \theta \tag{2.28}$$

where Θ is the impedance angle.

A block circuit diagram of this method is shown in Figure 2.15. Perera et al. [30] illustrate flicker contributions measurements of some sample plants, such as steel plants and EAFs by the proposed flicker contribution meter.

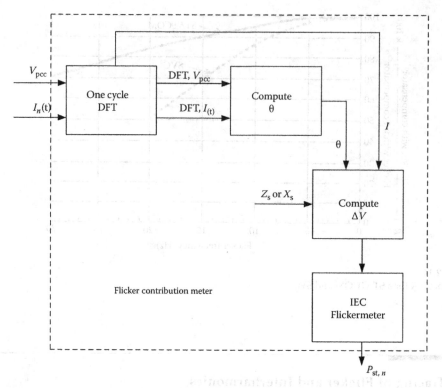

FIGURE 2.15
A block circuit diagram of individual flicker contribution meter. (S. Perera, D. Robinson, S. Elphick, D. Geddy, N. Browne, V. Smith, V. Gosbell, Synchronized flicker measurements for flicker transfer evaluation in power systems, *IEEE Trans Power Delivery* © 2006 IEEE.)

A reverse power flow procedure can be used to identify the source of harmonics. Line and bus data at several points in the network are used with a least square estimator to calculate the injection spectrum at buses expected to be harmonic sources. When energy at harmonic frequencies is found to be injected into the network, that bus is identified as a harmonic source.

2.10.1 IEEE 519, 2014 Revision

IEEE 519 specifies interharmonic voltage limits based on flicker. Figure 2.16 shows the interharmonic voltage limits based on flicker for frequencies up to 120 Hz for 60 Hz systems and Table 2.3 shows the limits corresponding to this figure. Weekly, 95th percentile short time harmonic voltages (see Chapter 5 for explanation and definition of short time measurements) are limited to the values shown in this figure, up to 120 Hz for 60 Hz systems. Depending upon the voltage level, the integer harmonic limits may be more restrictive and should be used. *This is applicable for voltages at PCC less than 1 kV.* The effects of interharmonics on other equipment and systems such as generator mechanical systems, motors, transformers, signaling, and communication systems, and filters are not included [12].

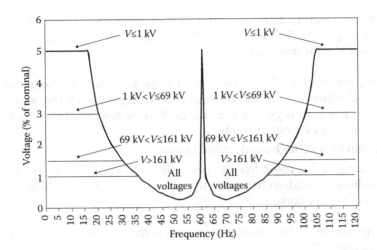

FIGURE 2.16
Interharmonic voltage limits based on flicker for frequencies up to 120 Hz for 60 Hz systems. (IEEE P219.1, Draft guide for applying harmonic limits on power systems © 2004 IEEE, see also 2014 Revision.)

TABLE 2.3

Voltage Interharmonic Limits, Corresponding to Figure 2.16 for PCC Voltage <1 kV

Frequency (Hz)	Magnitude (%)	Frequency (Hz)	Magnitude (%)	Frequency (Hz)	Magnitude (%)	Frequency (Hz)	Magnitude (%)
16	5.00	27	1.78	38	0.81	49	0.28
17	4.50	28	1.64	39	0.78	50	0.25
18	3.90	29	1.54	40	0.71	51	0.23
19	3.45	30	1.43	41	0.64	52	0.25
20	3.00	31	1.33	42	0.57	53	0.27
21	2.77	32	1.26	43	0.50	54	0.29
22	2.53	43	1.20	44	0.48	55	0.35
23	2.30	34	1.13	45	0.43	56	0.40
24	2.15	35	1.05	46	0.38	57	0.58
25	2.03	36	0.95	47	0.34	58	0.77
26	1.90	37	0.85	48	0.31	59	0.95

Source: IEEE P219.1, Draft guide for applying harmonic limits on power systems © 2004 IEEE, see also 2014 Revision. For the frequency range above 60 Hz but less than 120 Hz, say 61 Hz, subtract it from 120, it is equal to 59 Hz, and read the value from the table, = 0.95. The frequency resolution is 1 Hz.

2.11 Subsynchronous Resonance

We have defined the SSR in the opening paragraph of this chapter. The exchange of the energy with a turbine generator takes place at one or more of the natural frequencies of the combined system and these frequencies are below the synchronous frequency of the system [33]

The turbine generator shaft has natural modes of oscillations, which can be at subsynchronous frequencies. If the induced subsynchronous torque coincides with one of the shaft natural modes of oscillation, the shaft will oscillate at this natural frequency, sometimes with high amplitude. This may cause shaft fatigue and possible failure. The interactions can be caused by the following:

1. *Induction generator effect:* The resistance of rotor to subsynchronous currents is negative and network presents a resistance which is positive. If the negative resistance of the generator is greater than the positive system resistance, there will be sustained subsynchronous currents.
2. Torsional interaction has been described above.
3. Transient torques that result from a system disturbance cause changes in the network, resulting in sudden changes in the current that will oscillate at the natural frequency of the network.

The series compensation of the transmission lines is the most common cause of subsynchronous resonance (SSR).

2.11.1 Series Compensation of Transmission Lines

2.11.1.1 Subsynchronous Interharmonics (Subharmonics)

Groups of harmonics which are characterized by a frequency lower than the system fundamental frequency, i.e., having periods larger than the fundamental frequency, have been commonly called subsynchronous frequency components or subsynchronous interharmonics. In earlier documents, these were called subharmonics. The term subharmonic is popular in the engineering community, but it has no official definition. Series compensation of HV transmission lines is discussed in Volume 2. With a brief repetition, a series capacitor has a natural resonant frequency given by

$$f_n = \frac{1}{2\pi\sqrt{LC}} \qquad\qquad (2.29)$$

f_n is usually less than the power system frequency. At this frequency, the electrical system may reinforce one of the frequencies of the mechanical resonance, causing SSR. If f_r is the SSR frequency of the compensated line, then at resonance

$$2\pi f_r L = \frac{1}{2\pi f_r C}$$
$$f_r = f\sqrt{K_{sc}} \qquad\qquad (2.30)$$

This shows that the SSR occurs at frequency f_r which is equal to normal frequency multiplied by the square root of the degree of compensation, it is typically between 12 and 30 Hz. As the compensation is in 22%–72% range, f_r is lower than f. The transient currents at subharmonic frequency are superimposed upon power frequency component and may be damped out within a few cycles by the resistance of the line. Under certain conditions, subharmonic currents can have a destabilizing effect on rotating machines. If the electrical circuit oscillates, then the subharmonic component of the current results in a corresponding subharmonic field in the generator. This field rotates backward with respect to the main field and produces an alternating torque on the rotor at the difference frequency $f-f_r$. If the mechanical resonance frequency of the shaft of the generator coincides with this frequency, damage to the generator shaft can occur. A dramatic voltage rise can occur if the generator quadrature axis reactance and the system capacitive reactance are in resonance. There is no field winding or voltage regulator to control quadrature axis flux in a generator. Magnetic circuits of transformers can be driven to saturation and surge arresters can fail. The inherent dominant subsynchronous frequency characteristics of the series capacitor can be modified by a parallel connected TCR.

If the series capacitor is thyristor or GTO controlled, then the whole operation changes. It can be modulated to damp out any subsynchronous as well as low-frequency oscillations. Thyristor-controlled series capacitors have been employed for many HVDC projects [33].

2.11.2 SSR Drive Systems

In Section 2.2.2, the interharmonics due to drive systems are discussed. In this section, we stated that for inverter frequencies of 22, 37.2, and 48 Hz and source frequency of 60 Hz, the side band pairs are 10 and 110 Hz, 12 and 132 Hz, and 36 and 126 Hz, respectively. These can create SSR, though a number of conditions and parameters must coincide for such an event, see References [33–38] for further reading.

Problems

2.1 A 1 MW arc furnace load is connected to a 13.8 kV, 60 Hz, three-phase system having a short circuit of 20 kA at the point of connection of the arc furnace load. Is a flicker problem expected?

2.2 Two ac systems are interlinked through a dc link, 12-pulse converters on either side, and the frequency on inverter side is 43 Hz. Write the harmonics expected in the dc link.

2.3 A high-voltage line is series compensated, the degree of compensation being 30%. What is the expected troublesome subharmonic?

2.4 Provide one example of an operation which could lead to ferroresonance (see Volume 1, Appendix C).

References

1. IEC Standard 61000-2-1, Part 2. Part 2, Environment, Section 1. Description of the environment. Electromagnetic environment for low frequency conducted disturbances and signaling in public power supply systems, 1990.
2. IEEE Task Force on Harmonic Modeling and Simulation. Interharmonics: Theory and measurement, *IEEE Trans. Power Delivery*, 22(4), 2007.
3. M.B. Rifai, T.H. Ortmeyer, W.J. McQuillan, Evaluation of current interharmonics from AC drives, *IEEE Trans. Power Delivery*, 12(3), 1094–1098, July 2000.

4. R. Yacamini, Power system harmonics—Part 4, interharmonics, *IEEE Power Eng J*, 182–193, 1996.
5. J.D. Anisworth. Non-characteristic frequencies in AC/DC converters, *International Conference on Harmonics in Power Systems, UMIST*, pp. 76–84, Manchester 1981.
6. L. Hu, R. Yacamini, Harmonic transfer through converters and HVDC links, *IEEE Trans*, PE-7 (3), 214–222, 1992.
7. L. Hu, L. Ran, Direct method for calculation of AC side harmonics and interharmonics in an HVDC system, *IEEE Proc Generat Trans Distrib*, 147(6), 329–332, 2000.
8. L. Hu, R. Yacamani, Calculation of harmonics and interharmonics in HVDC systems with low DC side impedance, *IEE Proc C*, 140(6), 469–476, 1993.
9. B.R. Pelly, *Thyristor Phase-Controlled Converters and Cycloconverters, Operation Control and Performance*. John Wiley, New York, 1971.
10. E.W. Gunther, Interharmonics recommended updates to IEEE 519, *IEEE Power Engineering Society Summer Meeting*, Chicago, IL, pp. 920–924, July 2002.
11. C.O. Gercek, M. Ermis, A. Ertas, K.N. Kose, O. Unsar, Design implementation and operation of a new C-type 2nd order harmonic filter for electric arc and ladle furnaces, *IEEE Trans Ind Appl*, 47(4), 1242–1227, 2011.
12. IEEE P219.1. Draft Guide for Applying Harmonic Limits on Power Systems, 2004.
13. I. Yilmaz, Ö. Salor, M. Ermiş, I. Çadirci, Field-data-based modeling of medium frequency induction melting furnaces for power quality studies, *IEEE Trans Ind Appl*, 48(4), 1212–224, 2012.
14. IEC Standard 61000-4-12. Part 4–12. Testing and measurement techniques—Flicker meter. Functional and design specifications, 2010.
15. IEEE Standard 519. IEEE Recommended Practices and Requirements for Harmonic Control in Power Systems, 1992, also see Revision of 2014.
16. IEEE Standard 1453. Recommended Practice for Analysis of Fluctuating installations, 2015.
17. IEC Standard 6100-3-3. Electromagnetic Compatibility (EMC) Part 3–3: Limits-Section 3. Limitations of voltage changes voltage fluctuations and flicker in public low-voltage supply systems, for equipment with rated current ≤16 A per phase and not subjected to conditional connection, 2008.
18. IEC Standard 6100-3-8. Electromagnetic Compatibility (EMC) Part 3–8: Limits-Section 8. Signaling on low-voltage electrical installations—Emission levels, frequency bands and electromagnetic disturbance levels, 1997.
19. IEC Standard 61000-3-11. Electromagnetic Compatibility (EMC) Part 3–11: Limits-Limitations of voltage fluctuations and flicker in low-voltage power supply systems for equipment with rated current ≤72 A, 2000.
20. S.M. Halpin, V. Singhvi, Limits of interharmonics in the 1–100 Hz range based upon lamp flicker considerations, *IEEE Trans Power Delivery*, 22(1), 270–276, 2007.
21. M. Göl, O. Salor, B. Alboyaci, B. Mutluer, I. Cadirci, M. Ermis, A new field data-based EAF model for power quality studies, *IEEE Trans Ind Appl*, 46(3), 1230–1241, 2010.
22. S.R. Mendis, D.A. González, Harmonic and transient overvoltages analyses in arc furnace power systems, *IEEE Trans Ind Appl*, 28(2), 336–342, 1992.
23. R.W. Fei, J.D. Lloyd, A.D. Crapo, S. Dixon. Light flicker test in the United States, *IEEE Trans Ind Appl*, 36(2), 438–443, 2000.
24. C.D. Schauder, L. Gyugyi, STATCOM for electric arc furnace compensation, *EPRI Workshop*, Palo Alto, 1992.
25. M. Routimo, A Makinen, M. Salo, R. Seesvuori, J. Kiviranta, H. Tuusa, Flicker mitigation with hybrid compensator, *IEEE Trans Ind Appl*, 44(4), 1227–1238, 2008.
26. A. Nabae and M. Yamaguchi, Suppression of flicker in an arc furnace supply system by an active capacitance—A novel voltage stabilizer in power systems, *IEEE Trans Ind Appl*, 31(1), 107–111, 1992.
27. F.Z. Peng, H. Akagi, A. Nabae, A new approach to harmonic compensation in power systems—A combined system shunt and series active filters, *IEEE Trans Ind Appl*, 26(6), 983–990, 1990.

28. D. Zhang, W. Xu, A. Nassif, Flicker source identification by interharmonic power direction. *Proceedings Canadian Conference Electrical Computer Engineering,* Saskatoon, Canada, pp. 249–222, May 1–4, 2002.

29. P.G.V. Axelberg, M.H.J. Bollen, I.Y.H. Gu, Trace of flicker sources by using the quantity of flicker power, *IEEE Trans Power Delivery,* 23(1), 462–471, 2008.

30. S. Perera, D. Robinson, S. Elphick, D. Geddy, N. Browne, V. Smith, V. Gosbell, Synchronized flicker measurements for flicker transfer evaluation in power systems, *IEEE Trans Power Delivery,* 21(3), 1477–1482, 2006.

31. E. Altintas, O. Sailor, I. Cadirci, M. Ermis, A new flicker tracing method based on individual reactive current components of multiple EAFs at PCC, *IEEE Trans Ind Appl,* 46(2), 1746–1724, 2011.

32. T. Heydt, Identification of harmonic sources by a state estimation technique, *IEEE Trans Power Delivery,* 4(1), 269–276, 1989.

33. E.L. Owen, Torsional coordination of high speed synchronous motors: Part 1, *IEEE Trans Ind Appl,* 17, 267–271, 1981.

34. C.B. Meyers, Torsional vibration problems and analysis of cement industry drives, *IEEE Trans Ind Appl,* 17(1), 81–89, 1981.

35. J.C. Das, *Transients in Electrical Systems Analysis Recognition and Mitigation.* McGraw-Hill, New York, 2010.

36. N.G. Hingorani, A New Scheme for Subsynchronous Resonance Damping of Torsional Oscillations and Transient Torques—Part I IEEE PES Summer Meeting, Paper no. 80 SM687–4, Minneapolis, 1980.

37. N.G. Hingorani, K.P. Stump, A New Scheme for Subsynchronous Resonance Damping of Torsional Oscillations and Transient Torques—Part II IEEE PES Summer Meeting, Paper no. 80 SM688-2, Minneapolis, 1980.

38. J.C. Das, Subsynchronous resonance-series compensated HV lines and converter cascades, *Int J Eng Appl,* 2(1), 1–10, 2014.

3

Estimation of Harmonics

For the purpose of harmonic load flow and harmonic filter designs, the harmonic emission from the nonlinear loads should be accurately estimated. *This is the first step* and it is not an easy task due to varied topologies and harmonic limitation at source. There are three possibilities:

- For small systems and six-pulse converters, enough publications and data exist to estimate the harmonics analytically. The harmonic emissions from 6-pulse or 12-pulse current source or voltage source converters have been widely discussed in the literature.
- For the electrical systems where the harmonic loads are in operation, the harmonics can be estimated by online measurements. These should be done under various operating conditions as the emissions will vary depending upon the variations in the processes. The harmonic resonance, if any, the time stamp of harmonics, and the current and voltage distortions at the various buses and PCC (point of common coupling) can be captured.
- The harmonic spectrums and angle of each harmonic including noncharacteristic harmonics can be obtained from the vendors. *As the harmonic emission is also a function of the operating load and system impedance, it is necessary to obtain these data under various operating conditions and system impedance variations at the point of application of the nonlinear loads.*

Online measurement of harmonics will provide reliable results when this analysis is conducted properly over a period of time; however, the measurement techniques cannot be applied for the power systems in the design stage. More often than not, the harmonic generation needs to be estimated at the design stage of the power systems, so that appropriate measures could be taken to meet the requirements of harmonic limitations as per standards. When shunt power capacitors are applied along with harmonic-producing loads, further complications of harmonic resonance can occur at one of the load-generated frequencies. These important issues must be studied at the design stage.

Enough data have been gathered for the harmonic emissions from various sources in typical installations, yet the system impact must be considered—the system source impedance at the point of connection of harmonic-producing load impacts the harmonic emission, see Chapter 8.

Six-pulse current source converters are analyzed. Figure 3.1 shows the line current waveforms. The theoretical or textbook waveform is rectangular and considers instantaneous commutation, see Figure 3.1a. The effect of commutation delay and firing angle still retains the flat-top assumption, see Figure 3.1b, but the waveform has lost the even symmetry with respect to the center of the idealized rectangular pulse. The DC current is not flat-topped and the actual waveform has a ripple, see Figure 3.1c. For lower values of the DC reactor and large phase-control angles, the current is discontinuous, see Figure 3.1d.

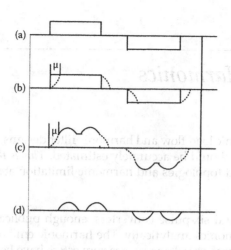

FIGURE 3.1
(a) Rectangular current waveform, (b) waveform with commutation angle, (c) waveform with ripple content, and (d) discontinuous waveform due to large delay angle control.

3.1 Waveform without Ripple Content

The harmonic emissions from six-pulse CS converters have been investigated in References [1–4]. The first approximation to the wave with overlap angle is to regard it as a trapezoidal wave shape in which the leading and trailing edges are assumed to be linear. The waveform becomes symmetrical and easy to analyze; Figure 3.2a shows the actual wave shape, while Figure 3.2b depicts its approximation.

The classical work was done by Read [2]. He provided a number of curves to illustrate the harmonic emissions based on the firing angle α for rectifiers, and β–μ for the inverters for harmonics of the order of 5th, 7th, 11th, 13th, 17th, 19th, 23rd, and 25th, respectively.

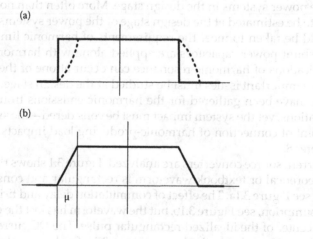

FIGURE 3.2
(a) Unsymmetrical waveform due to large delay angle and (b) approximation to the symmetrical trapezoidal waveform.

The X-axis scales the parameter $X_T + X_S$, where X_T is the percentage reactance of the rectifier transformer and X_S is the percentage impedance of the supply system. These are referred to rms current in the transformer primary lines equal to that corresponding to DC current I_d, assuming rectangular-shaped currents and no magnetizing current. Also

$$X_T + X_S = \frac{2\varepsilon}{V_{do}} \times 100 \tag{3.1}$$

where ε is the change of DC voltage from no-load to load current I_d, due to reactance drop in volts.

Example 3.1

Consider a 13.8 kV three-phase 60 Hz power source with a 2 MVA step-down transformer of 13.8–0.48 kV, serving a six-pulse CS converter. The transformer impedance is 5.75% on a 2 MVA base. If the converter base load is considered equal to the transformer KVA rating, then X_T=5.75%. The supply source impedance X_S=0.004 per unit (=0.4%). The transformer impedance predominates.

If the converter is being operated at a firing angle α=45°, then the various harmonics can be directly read from figures in References [2]. For this example, Figure 3.3 from this reference is reproduced, and the 5th harmonic can be read as approximately=19% of the fundamental.

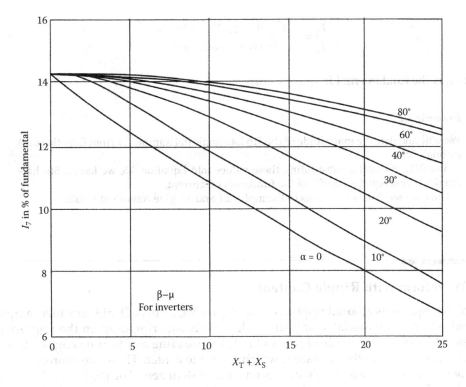

FIGURE 3.3
Fifth harmonic of primary line current $p=\leq 6$.

3.1.1 Harmonic Estimation Using IEEE 519 Equations

An estimation of the harmonics ignoring waveform ripple is provided by Equations 1.63 through 1.67 in Chapter 1 from IEEE 519. These equations are reproduced below for ease of reference:

$$I_h = I_d \sqrt{\frac{6}{\pi}} \frac{\sqrt{A^2 + B^2 - 2AB\cos(2\alpha+\mu)}}{h[\cos\alpha - \cos(\alpha+\mu)]} \tag{3.2}$$

where

$$A = \frac{\sin\left[(h-1)\dfrac{\mu}{2}\right]}{h-1} \tag{3.3}$$

and

$$B = \frac{\sin\left[(h+1)\dfrac{\mu}{2}\right]}{h+1} \tag{3.4}$$

Equation 3.3 can also be written as follows:

$$\frac{I_h}{I_1} = \frac{\sqrt{A^2 + B^2 - 2AB\cos(2\alpha+\mu)}}{h[\cos\alpha - \cos(\alpha+\mu)]} \tag{3.5}$$

where I_1 is the fundamental frequency current.

Example 3.2

We will calculate the magnitude of the 5th harmonic in Example 3.1 from Equations 3.3 through 3.5

$A=0.0417$, $B=0.0414$. Substituting these values into Equation 3.5, we have a 5th harmonic current equal to 18.95% of the fundamental current.

Thus, we see that calculations in Examples 3.1 and 3.2 give consistent results.

3.2 Waveform with Ripple Content

The above equations ignored ripple content. Figure 3.4, from IEEE 519, considers a ripple content which is sinusoidal, and a sine half-wave is superimposed on the trapezoidal waveform. There is linear rate of rise and fall at the leading and trailing edges. The two current lobes are equally displaced, which form the center. These are represented by tops of sine waves appropriately displaced from system zero. For the interval between the lobes, the current is assumed constant equal to commutation current. This current is defined as I_c.

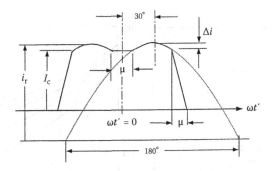

FIGURE 3.4
Trapezoidal current waveform with superimposed sinusoidal AC ripple.

The change of the current between the end of commutation and peak of each sine lobe is represented by Δi; the peak ripple current.

This ripple is expressed as a fraction of the commutation current I_c, and is defined as ripple coefficient r_c:

$$r_c = \frac{\Delta i}{I_c} \tag{3.6}$$

Also the ripple ratio r is defined:

$$r = \frac{\Delta i}{I_d} \tag{3.7}$$

where I_d is the average DC current.

3.2.1 Graphical Procedure for Estimating Harmonics with Ripple Content

Figure 3.5a through h provides the relative values of the harmonics in rms as I_h/I_1. To apply these figures, the following steps of calculation are required:

- Short-circuit current, based on supply side and transformer inductance is calculated.
- Decide the level of commutation current I_c.
- The DC voltage at the converter terminals is decided and expressed as a function of the maximum average DC voltage.
- The firing angle and the overlap angles can be read from Figure 3.6.
- Figure 3.7 is used to read off voltage ripple area, A_r.
- The ripple coefficient r_c is calculated:

$$r_c = \frac{A_r(V_{do}/X_r)}{I_c} \tag{3.8}$$

where

$$X_r = \omega L_d + 2X_c \tag{3.9}$$

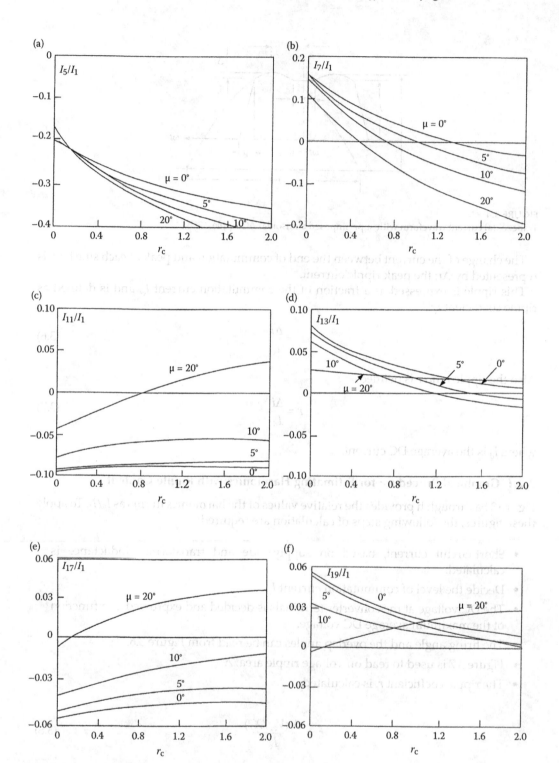

FIGURE 3.5

(a)–(h) Harmonic currents in per unit of fundamental frequency line current as a function of r_c.

(Continued)

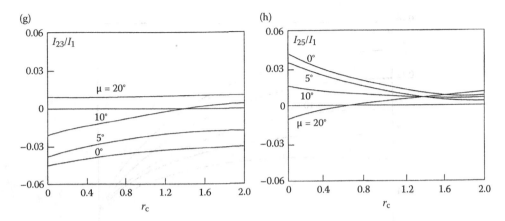

FIGURE 3.5   (CONTINUED)

(a)–(h) Harmonic currents in per unit of fundamental frequency line current as a function of r_c.

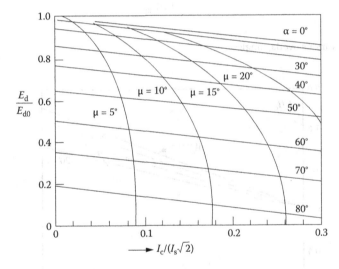

FIGURE 3.6

Converter load curves for a six-pulse current source converter.

- The harmonics can be read off from Figure 3.5a through h
- The harmonics can be converted into ampères by multiplying with a factor I_1/I_c from Figure 3.8 and then by I_c

Example 3.3

We will continue with the earlier examples to illustrate the method of calculation. Consider that $V_{do} = 2.34 \, (480/\sqrt{3}) = 648 \, V$. The short-circuit current on the 480 V bus is designated as $I_s = 39$ kA. Let $I_c = 2$ kA, and $V_d/V_{do} = 0.8$, where V_d is the DC operating voltage and V_{do} is the no load voltage; then, $I_c/I_s \sqrt{2} = 0.036$. Using these values and from Figure 3.6 for $a = 45°$, the overlap angle μ can be read as approximately 4.5°. Let us say it is 4.8°, as calculated before.

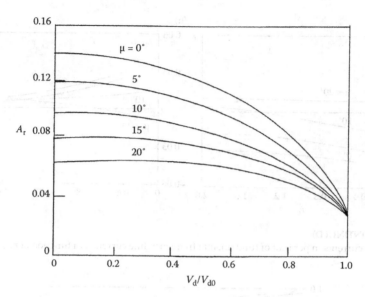

FIGURE 3.7
V_d/V_{do} versus A_r, for various overlap angles.

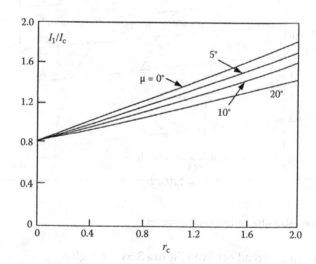

FIGURE 3.8
Fundamental line current and commutation current as a function of r_c.

Entering the values of μ and $V_d/V_{do}=0.8$ in Figure 3.7, we read $A_r=0.1$, where A_r is the voltage-ripple integral or the ripple area.

Calculate the ripple current $\Delta i=A_r V_{do}/X_r$:

$$X_r = \omega L_d + 2X_C$$

where X_r is the ripple reactance, L_d is the inductance in the DC circuit in H, and X_c is the commutating reactance.

First consider that there is no DC inductance, then $X_r=2X_c=2\times0.0071$ Ω (the reactance of the transformer and the source referred to the 480 V side), and calculated $\Delta i=4563$, then the ripple coefficient $r_c=\Delta i/I_c=2.28$.

Now, the harmonics as a percentage of the fundamental current can be calculated graphically from Figure 3.5a through h from Reference [2]. Negative values indicate a phase shift of π, and I_1 is the fundamental frequency current.

Continuing with the example, and entering Figure 3.5a for $\mu=4.8$ and $r_c=2.28$, we find that it is outside the range on the X-axis. An approximate value of 40% 5th harmonic can be read.

Note that the harmonics on the line side are impacted by the ripple content.

3.2.2 Analytical Calculations

The following equations are applicable from IEEE Standard 519 [5], also see Reference [1]:

$$I_h = I_c\frac{2\sqrt{2}}{\pi}\left[\frac{\sin\left(\frac{h\pi}{3}\right)\sin\frac{h\mu}{2}}{h^2\frac{\mu}{2}}+\frac{r_c g_h\cos\left(h\frac{\pi}{6}\right)}{1-\sin\left(\frac{\pi}{3}+\frac{\mu}{2}\right)}\right] \tag{3.10}$$

where

$$g_h=\frac{\sin\left[(h+1)\left(\frac{\pi}{6}-\frac{\mu}{2}\right)\right]}{h+1}+\frac{\sin\left[(h-1)\left(\frac{\pi}{6}-\frac{\mu}{2}\right)\right]}{h-1}-\frac{2\left[\left(\sin\left(h\left(\frac{\pi}{6}-\frac{\mu}{2}\right)\right)\sin\left(\frac{\pi}{3}+\frac{\mu}{2}\right)\right)\right]}{h} \tag{3.11}$$

where I_c is the value of the DC current at the end of the commutation and r_c is the ripple coefficient (= $\Delta i/I_c$).

In Figure 3.4, the time zero reference is at $\omega t'=0$, at the center of the current block. This is even symmetry and, therefore, only cosine terms are present. The instantaneous current is then

$$i_h = I_h\sqrt{2}\ \cos n\omega t' \tag{3.12}$$

More accurately, Equations 3.10 and 3.11 can be used.

Example 3.4

From Equation 3.11, and substituting the numerical values

$$g_h=\frac{\sin 165.6°}{6}+\frac{\sin 110.4°}{4}-\frac{2(\sin 138°\sin 62.4°)}{5}=0.03875$$

and from Equation 3.10

$$I_h = I_c\times\frac{2\sqrt{2}}{\pi}\left[\frac{\sin 300°\sin 12°}{1.0472}+\frac{2.28(0.03875)\cos 150°}{0.115}\right]=-0.726 I_c$$

The ratio of currents I_1/I_c is given in Figure 3.8. For $r_c=2.28$, it is 1.8 and therefore the 5th harmonic in terms of the fundamental current is 42.3%. This is close to 40% estimated from the graphs.

3.2.3 Effect of DC Reactor

The ripple content is a function of the source reactance (commutating reactance) and also the L_d, in the DC link circuit. For zero DC output voltage, the DC current is

$$I_{dm} = 0.218 \frac{V_{ln}}{X_d + 2X_N} \tag{3.13}$$

where X_d is the reactance in the DC circuit and X_N is the total source reactance including that of the rectifier transformer.

Example 3.5

Consider that a DC inductor of 1 mH is added in Example 3.4. Recalculate the magnitude of the 5th harmonic.

Following the same procedure as in example, $X_r=0.3912$ Ω, $\Delta i=165.6$, the ripple coefficient $r_c=\Delta i/I_c=0.08$, and the 5th harmonic reduces to approximately 20% of the fundamental current.

3.3 Phase Angle of Harmonics

For simplicity, all the harmonics may be considered cophasial. This does not always give the most conservative results, unless the system has one predominant harmonic; in which case only harmonic magnitudes can be represented. The phase angles of the current sources are functions of the supply voltage phase angle and are expressed as follows:

$$\theta_h = \theta_{h,\text{spectrum}} + h(\theta_1 - \theta_{1,\text{spectrum}}) \tag{3.14}$$

where θ_1 is the phase angle obtained from fundamental frequency load flow solution, and $\theta_{h,\text{spectrum}}$ is the typical phase angle of harmonic current source spectrum. The phase angles of a three-phase harmonic source are rarely 120° apart, as even a slight unbalance in the fundamental frequency can be reflected in a considerable unbalance in the harmonic phase angle.

When a predominant harmonic source acts in isolation, it may not be necessary to model the phase angles of the harmonics. For multiple harmonic sources, phase angles should be modeled. Figure 3.9a shows the time–current waveform of a six-pulse current source converter, when the phase angles are represented and it is recognizable as the line current of a six-pulse converter, with overlap and no ripple content; however, the waveform of Figure 3.9b has exactly the same spectra, but all the harmonics are cophasial. It can be shown that the harmonic current flow and the calculated distortions for a single-source harmonic current will be almost identical for these two waveforms. Table 3.1 shows the spectra with phase angles.

For multiple source assessment, the worst-case combination of the phase angles can be obtained by performing harmonic studies with one harmonic-producing element modeled at a time. The worst-case harmonic level, voltage, or current is the arithmetical summation of the harmonic magnitudes in each harmonic model. This will be a rather lengthy study. Alternatively, all the harmonic sources can be simultaneously modeled, with proper phase angles. The fundamental frequency angles are known by prior load flow.

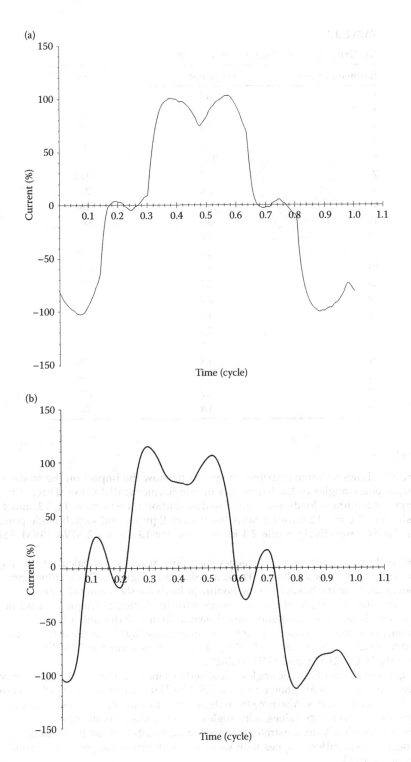

FIGURE 3.9
(a) Six-pulse current source converter current waveform, harmonics modeled at proper angles and (b) waveform with all harmonics cophasial.

TABLE 3.1

DC Drive Current Harmonic Spectrum

Harmonic Order	Magnitude	Phase Angle (°)
1	100.0	−43
2	0.3	68
3	0.4	−126
4	0.2	30
5	25.3	−30
7	5.5	−122
9	0.6	37
11	8.2	−102
13	3.9	170
17	5.0	−179
19	2.9	193
23	3.4	109
25	2.2	14
29	2.7	39
31	1.9	−59
35	2.3	−35
37	1.7	−135
41	1.8	−110
43	1.5	150
47	1.7	177
49	1.4	70

Example 3.6

Figure 3.10 shows a simple distribution system to show the impact on the modeling of proper phase angles of the harmonics on the harmonic distortion. Three different types of nonlinear loads are applied to distribution transformers T1, T2, and T3. Transformers T1 and T2 have 1.2 MVA fluorescent lighting and switch mode power supply loads, respectively, while 2.4 kV transformer T3 carries 2 MVA PWM ASD load.

First, perform a fundamental frequency load flow analysis and calculate the magnitude and angles of bus voltages. These are shown in Figure 3.10. The fundamental frequency angles at the buses carrying nonlinear loads are shown in this figure.

Now calculate the angle of the harmonics with fundamental frequency load flow angles as the base. These calculations are shown in Tables 3.2 through 3.4.

The harmonic analysis is carried out first without modeling the angles of the harmonics and the results of the current and voltage distortions are shown in Table 3.5. The total THD$_I$ at the PCC is 21.82% and THD$_v$ is 7.62%.

Next, model all the harmonic angles calculated in Tables 3.3 through 3.5. The current and voltage distortions are shown in Table 3.5. The THD$_I$ is reduced to 13.11% and the THD$_v$ is reduced to 4.98%. Also note from these tables that the distortions at individual harmonics vary over large values, with angles modeled and without angles.

This example clearly demonstrates the misleading results that can be obtained when nonlinear loads of different types with varying waveforms are applied and harmonic angles are ignored.

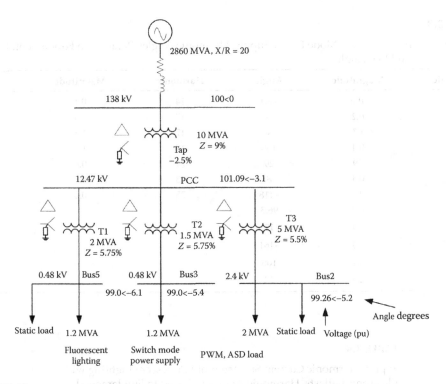

FIGURE 3.10
A distribution system for calculations of current and voltage harmonics, with and without proper phase angles, multiple nonlinear loads, Example 3.6.

TABLE 3.2

Harmonic Spectrum of a Voltage Source PWM ASD, Harmonic Angles Referred to Fundamental Frequency Load Flow Angle

Harmonic Order	Magnitude	Phase Angle
1	100	−5.2
3	0.35	−172
5	61.0	−195
7	33.8	165
9	0.50	160
11	3.84	122
13	7.78	131
15	0.41	75
17	1.28	−36
19	1.60	103
21	0.35	26
23	1.10	−54
25	0.18	−51

TABLE 3.3

Spectrum of Typical Switch Mode Power Supply, Harmonic Angles Referred to Fundamental Frequency Load Flow Angle

Harmonic	Magnitude	Angle	Harmonic	Magnitude	Angle
1	100	−6.1	14	0.1	−32
2	0.2	5	15	1.9	172
3	67	4	17	1.8	8
4	0.4	160	19	1.1	176
5	39	29	21	0.6	−14
6	0.4	8	23	0.8	173
7	13	−118	25	0.4	−41
8	0.3	96.3			
9	4.4	17			
11	5.3	−161			
12	0.1	106			
13	2.5	23			

TABLE 3.4

Typical Harmonic Current Spectrum of Fluorescent Lighting with Electronic Ballasts, Harmonic Angles Referred to Fundamental Frequency Load flow Angle

Harmonic Order	Magnitude	Angle (°)
FUND	100	−5.4
2	0.2	−104
3	19.9	−144
5	7.4	−178
7	3.2	−159
9	2.4	−171
11	1.8	−129
13	0.8	−103
15	0.4	−93
17	0.1	−44
19	0.2	141
21	0.1	160
23	0.1	−154
27	0.1	161
32	0.1	−84

When all the nonlinear loads are replaced only with one type, say fluorescent lighting, the calculations with or without modeling the harmonic angles give the same results.

Thus, even with multiple sources, the same results hold, provided their harmonic spectrums are identical in amplitude and phase angle.

Further impact is shown in Figures 3.11 through 3.13.

TABLE 3.5

Harmonic Spectrums Current and Voltage at PCC—Example

Harmonic Order	Current Harmonics		Voltage Harmonics	
	Angles Not Modeled	**Angles Modeled**	**Angles Not Modeled**	**Angles Modeled**
2	0.06	0.04	0.01	0.00
4	0.05	0.05	0.01	0.01
5	19.88	10.36	6.34	3.37
7	8.78	7.89	3.93	3.59
8	0.03	0.03	0.02	0.02
11	1.27	0.99	0.89	0.71
13	1.43	1.01	1.20	0.97
14	0.01	0.01	0.01	0.01
17	0.27	0.27	0.30	0.30
19	0.25	0.23	0.31	0.29
23	0.14	0.08	0.22	0.12
25	0.03	0.03	0.05	0.05
32	0.00	0.00	0.01	0.01

THD_v without angle model: 7.62%.
THD_1 without angle model = 21.82%.
THD_v with angle model: 4.98%.
$hTHD_1$ with angle model = 13.11%.

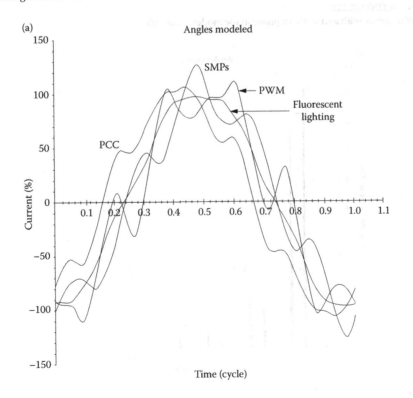

FIGURE 3.11

(a) and (b) Waveforms with and without phase angle models, Example 3.6.

(*Continued*)

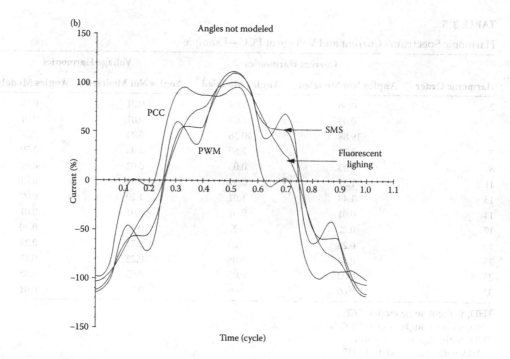

FIGURE 3.11   (CONTINUED)
(a) and (b) Waveforms with and without phase angle models, Example 3.6.

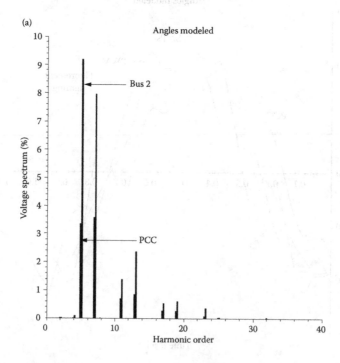

FIGURE 3.12
(a) and (b) Voltage spectrum at PCC and bus 2 with and without phase angle models, Example 3.6.

(Continued)

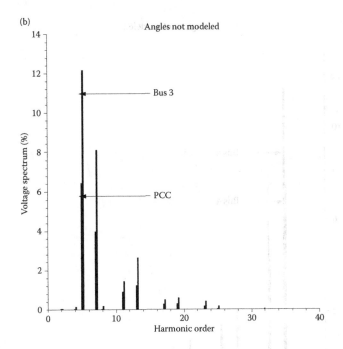

FIGURE 3.12   (CONTINUED)
(a) and (b) Voltage spectrum at PCC and bus 2 with and without phase angle models, Example 3.6.

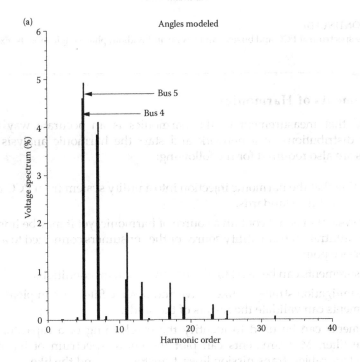

FIGURE 3.13
(a) and (b) Voltage spectrum at PCC and buses 4 and 5 with and without phase angle models, Example 3.6.

(Continued)

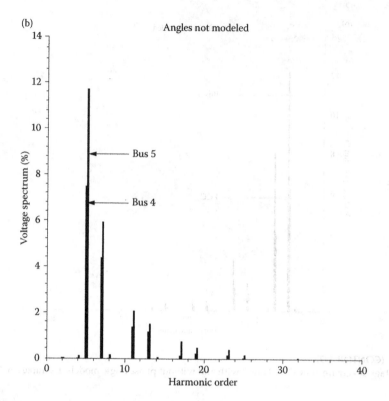

FIGURE 3.13 (CONTINUED)
(a) and (b) Voltage spectrum at PCC and buses 4 and 5 with and without phase angle models, Example 3.6.

3.4 Measurements of Harmonics

It is reiterated that measurement of the harmonics is an accurate way to estimate the harmonic distributions in a network and start the harmonic analysis study. The measurements are also required for the following:

- Ascertaining that the harmonic injection into a utility system from PCC meets the requirements of the standards.

- A power system may not contain a source of harmonic, yet it may be impacted by harmonic infiltration from utility source or the consumers connected to a common utility service point.

- The measurements can be used to identify the resonant conditions.

- After the mitigation strategies like active and passive filters are implemented, the measurements can validate the results of the study.

- Measurements can be used to identify the overloading of a capacitor bank or capacitive filter. Measurements can plot the entire spectrum of harmonics in passive filters, cables, transmission lines, transformers, and the like.

- Measurements of harmonic currents and voltages with their respective phase angles can help deriving the transfer impedance at a given location.

The PCC can be either on the high side or low side of the utility interconnecting transformer. It is also the metering point where the CTs and PTs for measurement will be available. The accuracy of these devices is discussed in a section to follow. The harmonic current measurements on the secondary can be transferred to the primary in the turns ratio of the transformer, but it is not true for harmonic voltage measurements. The 3rd harmonics will be blocked by the delta winding of the transformer. Also transformer connection affects the phase angle of the harmonics. A transformer with primary and secondary windings both connected in wye-grounded connection will allow zero sequence currents to flow in the primary system. Unbalanced triplen harmonics are not zero sequence components. The harmonics currents are expressed as a percentage of average maximum demand to calculate TDD. The phase angle of the harmonics should be included in the measurements—it provides a better picture of harmonic cancellation from various harmonic source types. All phase angles should be related to the same reference, commonly selected as the zero crossing of the fundamental frequency line-to-neutral voltage on phase *a*, Example 3.7.

3.4.1 Monitoring Duration

Monitoring duration should be long enough to capture the time stamp of the harmonics with varying processes. The time trend of the harmonics is required to be captured, from which a probability histogram can be developed. For facilities like steel mills and arc furnaces that can have harmonic emissions which vary from day to day, monitoring over longer durations is required. It is also important to evaluate the effect of different operating system conditions at harmonic levels:

- Effect of power factor capacitors and passive filters.
- Effect of outage of a filter.
- Effect of variations in the utility source; e.g., the load may be supplied over redundant feeders each having different characteristics. Utility source impedance affects harmonic generation.
- Effect of nearby consumers with significant harmonic generation.
- Effect of variations of loads in the facility.

3.4.2 IEC Standard 6100 4-7

For the purpose of measurements, IEC divides the harmonics into three categories:

1. Slowly varying (quasi-stationery)
2. Fluctuating
3. Rapidly varying; short bursts of harmonics

- Measurements up to 50th harmonic are recommended. The time intervals recommended are as follows:
- T_{vs}: Very short interval, 3 s
- T_{sh}: Short interval, 10 min
- T_L: Long interval, 1 h
- T_D: 1 day
- T_W: 1 week

For thermal effects, the rms value of each harmonic and cumulative probabilities from 1% to 99% are calculated.

For instantaneous effects, the maximum value of each harmonic and cumulative probabilities of 95% and 99% are considered.

The measurement instrument should be able to handle the category of harmonic being measured.

Category 1: Quasi-stationary harmonics can be measured with a rectangular window of 0.1–0.5 s, Appendix A, with some gap between the windows.

Category 2: Hanning window can be used, Appendix A.

Category 3: A rectangular window of 0.08 s can be used without a gap between successive windows.

3.4.3 Measurement of Interharmonics

If the measured waveform contains an interharmonic, then the measurement time interval kT, where T is the time period at fundamental frequency, should be

$$kT = mT_i \tag{3.15}$$

where T_i is the time period of interharmonic.

That is, it should be the least common multiple of the periods of fundamental component and interharmonic component. If Equation 3.15 is not satisfied, then the rms value of interharmonics as well as powers associated with it is incorrectly measured. For $m = 20$, the interharmonic will be measured with a maximum error of ±0.2%.

If at least one of the interharmonics of order h is an irrational number, then the observed waveform is not periodic. A waveform of two or more nonharmonically related frequencies may not be periodic. In such cases, kT should be infinitely large.

The monitoring equipment based on FFT will have errors. The application of proper window function and zero padding, Appendix A, can improve the performance.

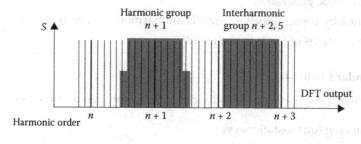

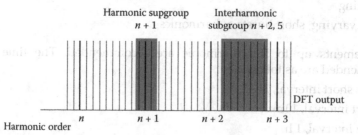

FIGURE 3.14

A histogram of voltage THD, presentation of harmonic measurement results.

IEC recommends sampling interval of the waveform to 10 and 12 cycles for 50 and 60 Hz systems, respectively, resulting in a set of spectra with 5 Hz resolution for harmonics and interharmonics. This may be simplified by summing the components between harmonics into one single interharmonic group. Figure 3.14 shows these groupings for 50 Hz [6].

$$X_{IH}^2 = \sum_{i=2}^{8} X_{10n+1}^2 \text{ for a 50 Hz system} \tag{3.16}$$

$$X_{IH}^2 = \sum_{i=2}^{10} X_{12n+1}^2 \text{ for a 60 Hz system} \tag{3.17}$$

3.5 Measuring Equipment

The measuring equipments are as follows:

- Storage oscilloscope can be used and will visually indicate the resonance and distortion. The stored data can be later on transferred to a computer [5].
- Spectrum analyzers display the power distribution of a signal as a function of frequency on a CRT or chart recorder.
- Harmonic analyzers or wave analyzers can measure the amplitude and phase angle and provide the line spectrum of the signal. The output can be recorded or monitored with analog or digital meters.
- Distortion analyzers will indicate THD directly. Digital measurements use two basic techniques: By means of a digital filter which are set up with proper bandwidth to capture small magnitude of harmonics, when the fundamental current is large. By the FFT techniques, which are powerful algorithms and with selection of proper window and bandwidth can eliminate picket fence and aliasing, Appendix A.
- Microprocessor based *online* meters are available which will continuously monitor and store the harmonic spectrums and distortion data with varying processes. This continuous measurement is helpful in ascertaining the time stamp of harmonics.

3.5.1 Specifications of Measuring Instruments

3.5.1.1 Accuracy

The minimum accuracy requirements of measuring instruments are specified in IEEE Standard 519 [5]. The instrument must perform measurement of a steady-state harmonic component with an uncertainty no more than 5% of the permissible limit. For example, if the 11th harmonic of 0.70% is to be measured in a 480 V system, it means a harmonic voltage of 1.94 V. The instrument should have an uncertainty of less than ±(0.05)(1.94)= ±0.097 V. Thus, the accurate measurements of harmonics become a problem when their order increases and the amplitudes are correspondingly reduced. The ultimate accuracy results should consider not only the measuring instrument but also the transducers used for harmonic measurement, i.e., overall measuring system.

TABLE 3.6

Minimum Required Attenuation (dB)

Injected Frequency (Hz)	Frequency-Domain Instrument	Time-Domain Instrument
60	0	0
30	50	60
120–720	30	50
720–1200	20	40
1200–2400	15	35

Source: IEEE Standard 519, IEEE recommended practice and requirements for harmonic control in electrical power systems © 1992 IEEE.

3.5.1.2 Attenuation

The second parameter is attenuation which dedicates instrument capability to separate harmonic components of different frequencies. Minimum attenuation limits for a frequency and time-domain instrument are specified in Reference [5], and are reproduced in Table 3.6. Almost all measurements can meet 60 dB (0.1% of fundamental). More expensive instruments can have 90 dB (0.00316%).

The load harmonics can be rapidly fluctuating and averaging over a period of time can give a false picture of harmonic emission and distortion. If an average over a period of 3 s is required, then the response of the output meter should be identical to first-order low-pass filter with a time constant of 1.5 ± 0.15 s.

3.5.1.3 Bandwidth

The bandwidth of the instrument will have a profound effect, especially when the harmonics are fluctuating. Instruments with a constant bandwidth for the entire range of frequencies must be used. The bandwidth should be 3 ± 0.5 Hz between 3 dB points with a minimum attenuation of 40 dB at a frequency of $f_h + 15$ Hz. When interharmonics are to be measured, a larger bandwidth will cause large positive errors.

3.5.2 Presentation of Measurement Results

The harmonic measured data are presented in the tabular form, as shown in the tables and waveforms in the time domain and spectrums in the frequency domain in previous chapters. The results can also be presented as a variation of harmonics with time, a time-trend plot.

The time-varying harmonics can be presented as a probability histogram, see Figure 3.15. This allows direct evaluation of the harmonic levels which are not exceeded 95% of the time and 99% of the time.

Figure 3.16 shows the inverse distribution curve. The data acquisition period T_D can be divided into m intervals, $mT = T_D$. Then, the mean value of current for each interval is

$$\sum_1^k \frac{I_{kh}}{k} \tag{3.18}$$

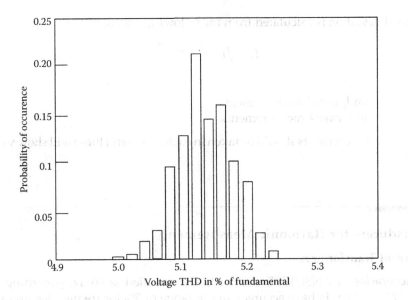

FIGURE 3.15
Harmonic histogram of voltage THD.

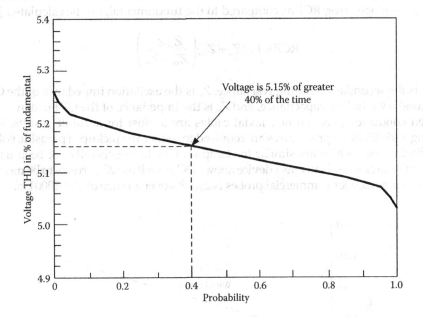

FIGURE 3.16
Probability histogram of voltage THD.

where over the subinterval T, k measurements were taken.

The mean square value is

$$\sum_{1}^{k} \frac{I_{kh}^2}{k} \qquad (3.19)$$

The standard deviation is calculated from the following equation:

$$I_h = \sqrt{I_{h,\max}^2 - I_{h,\min}^2} \qquad (3.20)$$

where

$I_{h,\max}$ = maximum I_h over k measurements
$I_{h,\min}$ = Minimum I_h over k measurements.

The harmonic measurements should be taken in each phase and these will show variations, Chapter 1.

3.6 Transducers for Harmonic Measurements

3.6.1 Current Transformers

The CT accuracies in ANSI/IEEE standards are specified at 60 Hz. According to IEEE Standard 519 [5], the CTs have accuracy in the range of 3% for frequencies up to 10 kHz. The accuracy is a function of the impedance of the CT and the external burden connected to it. Figure 3.17 shows the CT ratio correction factor (RCF) with burden and without burden. The percentage error, RCF as compared to the fundamental, can be calculated [7]:

$$RCF = 1 + (Z_s + Z_b)\left(\frac{Z_{cs}Z_e}{Z_{cs} + Z_e}\right) \qquad (3.21)$$

where Z_s is the secondary winding resistance, Z_e is the excitation impedance of the CT, Z_{cs} is the secondary winding capacitance, and Z_b is the impedance of the CT burden.

Shielded conductors, coaxial or triaxial cables are a must for accurate results. Proper grounding and shielding procedures are required to reduce the pickup of parasitic voltages.

Hall effect probes, which are similar to clamp-on CTs, have generally not been used for harmonic measurements. Hall effect device allows DC as well as AC currents to be measured; the specified accuracy for commercial probes is 2%–5% over a range of 500–1000 Hz.

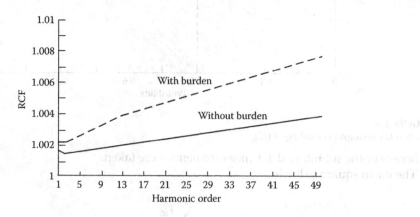

FIGURE 3.17
Ratio correction factor (RCF) of a CT with burden and without burden versus harmonics.

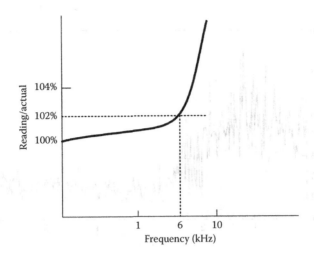

FIGURE 3.18
Potential transformer accuracy.

3.6.2 Rogowski Coils

Rogowski coils or Maxwell worms are devices where coils are wound on flexible plastic mandrels so that these can be used as clamp-on devices. Magnetic saturation is avoided when large currents up to 100 kA or DC currents are to be measured.

3.6.3 Voltage Measurements

The magnetic voltage transformers are designed to operate at 60 Hz. Harmonic frequency resonance between winding inductances and capacitances can cause large ratio and phase errors, see Figure 3.18. For harmonics of frequency less than 5 kHz, the accuracy of most potential transformers is within 3%.

Capacitive voltage transformers cannot be used for harmonic measurements, because typical resonant frequency peak appears at 200 Hz. Capacitive voltage dividers can be easily built. High-voltage bushings are provided with a capacitive tap for voltage measurements.

3.7 Characterizing Measured Data

The accuracy of the measured data is important in several applications like design of harmonic filters, imposition of stress on electrical equipment and it is not as simple as may be made out from the above description. IEEE Task Force publication in two parts [8,9] provides valuable insight into the problem of measurement of nonstationary voltages and current waveforms, characterization of recorded data, harmonic summation and cancellation in systems with multiple nonlinear loads and probabilistic harmonic power flow. It starts with two typical recorded data and then analyzing them. These recorded data from Reference [9] at one of the sites (site A) are shown in Figure 3.19a and b.

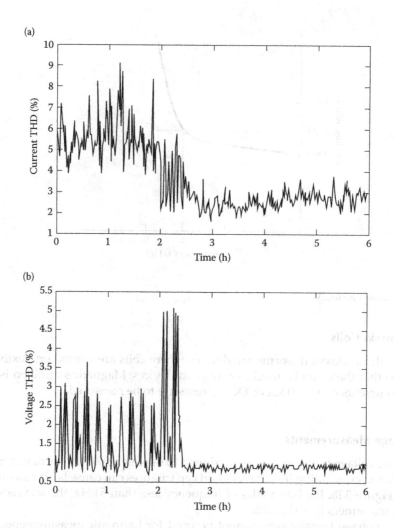

FIGURE 3.19
(a) Signature of current signal and (b) voltage signal; actual measurements. (IEEE Task Force on Probabilistic Aspects of Harmonics (Y. Baghzouz-Chair), Time varying harmonics part II—Harmonic simulation and propagation, *IEEE Trans Plasma Sci* © 2002 IEEE.)

This data-representing site A shows a customer's 13.8 kV bus having rolling mill equipped with 12-pulse DC drives and tuned harmonic filters. After 2.5 h, the current and voltage distortions go down because the rolling mill is not operating and the background distortion is captured. Site B measurements, not reproduced here, show that while the current distortion is high, the corresponding voltage distortion is low because of *stiff system*. *This is a very important concept.*

The common techniques in harmonic estimating are based on DFT (Appendix A), which gives accurate results provided:

- The signal is stationary, but practically it is not.
- The sampling frequency is greater than twice the frequency being measured, Nyquist theorem, see Appendix A.

- The number of periods sampled is an integer.
- The waveform does not contain frequencies that are noninteger multiples of fundamental frequency, i.e., interharmonics.

When interharmonics are present, multiple periods need to be sampled, discussions in Section 3.4.3. Time variations of the individual harmonics are generated by windowed Fourier transformations or short-time Fourier transform. Each harmonic spectrum corresponds to the each window section of the continuous signal. Thus, different window sizes give different spectra. Appendix A discusses aliasing, leakage, and picket fence effects. Both leakage and picket fence effect can be mitigated by spectral windows.

The irregularities in harmonic recorded data may fail to conform to a coherent pattern, yet some patterns will show a deterministic component. In such cases, the signal can be expressed as a *deterministic component and random component. As an example, the signal shown in Figure 3.19a can be decomposed into two parts as shown in Figure 3.20a and b. For further reading, see References [8,9].*

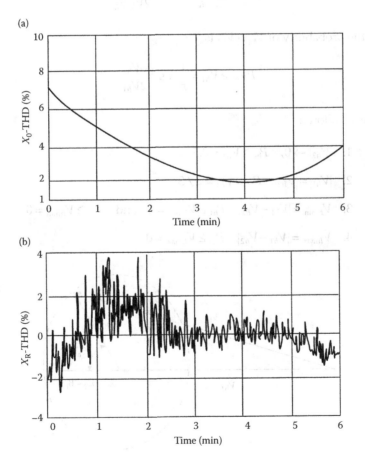

FIGURE 3.20
(a) and (b) Breakdown of current signal in Figure 3.19a in to definitive X_D and random X_R components.

3.8 Summation of Harmonic Vectors with Random Angles

The current harmonic amplitudes and phase angles vary. While the amplitudes of harmonics are commonly measured, the phase angles are not. Probabilistic concepts are provided in Reference [10].

If we have two vectors V_{h1} and V_{h2} and different phase angles, θ_{h1} and θ_{h2}, then the angle θ_h between them can be written as follows:

$$\theta_h = \theta_{h2} - \theta_{h1} = \cos^{-1} \frac{V_{hs}^2 - V_{h1}^2 - V_{h2}^2}{2V_{h1}V_{h2}} \tag{3.22}$$

The locus of V_{hs} is shown in Figure 3.21.

If the phase-angle difference is a uniform distribution around the circle, then the probability of PV_{hp} over PV_{hs} is defined as follows:

$$PV_{hp} \geq PV_{hs} = \frac{1}{\pi} \cos^{-1} \frac{V_{hs}^2 - V_{h1}^2 - V_{h2}^2}{2V_{h1}V_{h2}} \tag{3.23}$$

Accordingly, the probability of $V_{hs} > V_{h1}$ is

$$PV_{hs} \geq V_{h1} = \frac{1}{\pi} \cos^{-1} \frac{-V_{h2}}{2V_{h1}} \tag{3.24}$$

Four cases are as follows:

1. $|V_{h2}| \to 0, \quad P_{hs} \geq V_{h1} = 0.5$

2. $|V_{h1}| = |V_{h2}| \quad P_{hs} \geq V_{h1} = 2/3$

3. $V_{h\min} = |V_{h1} - V_{h2}| \quad P_{hs} \geq V_{h\min} = 1 \quad \text{and} \quad P_{hs} \geq V_{h\min} = 0$

4. $V_{h\max} = |V_{h1} + V_{h2}| \quad P_{hs} \geq V_{h\max} = 0$

$$(3.25)$$

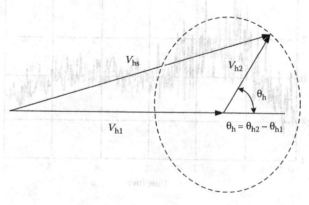

FIGURE 3.21
Summation of two vectors with random phase angles.

The uniform distribution of phase-angle difference is

$$f(\theta) = \frac{1}{\pi}, \quad \theta \in [0, \pi]$$ (3.26)

Their expected mean, $E(\theta)$, is $\pi/2$, and the standard deviation $\sigma(\theta)$ is $\pi/(2\sqrt{3})$. The statistically phase-angle difference at negative standard deviation is

$$\theta_{-\sigma} = E(\theta) - \sigma(\theta) = \frac{\pi}{2}\left(1 - \frac{1}{\sqrt{3}}\right) = 38.04°$$ (3.27)

Then, the summation with known negative deviation becomes

$$V_{hs} = \sqrt{V_{h1}^2 + V_{h2}^2 + 2V_{h1}V_{h2}\cos 38.04°}$$

$$= \sqrt{V_{h1}^2 + V_{h2}^2 + 1.575\,V_{h1}V_{h2}}$$ (3.28)

Then, the probability is

$$PV_{hp} \geq V_{hs} = \frac{38.04°}{180°} = 0.2113$$ (3.29)

IEC 61000-3-6 [11] gives the following summation equation:

$$V_{hs} = \sqrt[\lambda]{\sum_i V_{hi}^\lambda}$$ (3.30)

where $\lambda=1$ for $h < 5$, $\lambda=1.4$ for $5 \leq h \leq 10$, and $\lambda=2$ for $h > 10$. These are exponents for summation of two vectors. For multisources, Equation 3.30 is still valid but λ is different. Having found the pdfs of the variables in the above equations, the pdf of the sum of the probabilistic harmonic vectors can be found.

Problems

3.1 Graphically calculate the 7th harmonic in Example 3.3.

3.2 Analytically calculate the 7th harmonic in Example 3.4.

References

1. A.D. Graham, E.T. Schonholzer. Line harmonics of converters with DC-motor loads. *IEEE Trans Ind Appl*, 19(1), 84–93, 1983.

2. J.C. Read. The calculation of rectifier and inverter performance characteristics. *JIEE*, 92(29), 495–509, 1945.
3. M. Grötzbach, R. Redmann. Line current harmonics of VSI-Fed adjustable-speed drives. *IEEE Trans Ind Appl*, 36, 683–690, 2000.
4. M. Grötzbach, R. Redmann. Analytical predetermination of complex line current harmonics in controlled AC/DC converters. *IEEE Trans Ind Appl*, 33(3), 601–611, 1997.
5. IEEE Standard 519. IEEE recommended practice and requirements for harmonic control in electrical power systems, 1992.
6. E.W. Gunther. Interharmonics recommended updates to IEEE 519. *IEEE Power Engineering Society Summer Meeting*, pp. 950–954, July 2002.
7. P.E. Sutherland. Harmonic measurements in industrial power systems. *IEEE Trans Ind Appl*, 31(1), 175–183, 1995.
8. IEEE Task Force on Probabilistic Aspects of Harmonics (Y. Baghzouz-Chair). Time varying harmonics part I—Characterizing harmonic data. *IEEE Trans Power Deliv*, 13(3), 938–944, 1998.
9. IEEE Task Force on Probabilistic Aspects of Harmonics (Y. Baghzouz-Chair). Time varying harmonics part II—Harmonic simulation and propagation. *IEEE Trans Plasma Sci*, 17(1), 279–285, 2002.
10. Y. Xiao, X. Yang. Harmonic summation and assessment based on probability distribution. *IEEE Trans Power Deliv*, 27(2), 1030–1032, 2012.
11. IEC Standard 61000-3-6. Electromagnetic compatibility (EMC) Part 3: Limits-Section 6: Assessment of emission limits for distorting loads in MV and HV power systems, 2nd ed, 2008.

4

Harmonic Resonance, Secondary Resonance, and Composite Resonance

Resonance is an important factor affecting the system harmonic levels. A load-generated harmonic can be magnified many times due to harmonic resonance. A harmonic resonance without application of shunt capacitor banks can practically be ignored in industrial systems, but not so in the distribution and transmission systems. Distribution system frequency response is affected by the shunt capacitances and the system inductances of the transmission lines and cables. A number of small capacitor banks can be applied in a distribution system and these will give rise to multiple resonant frequencies. The transmission systems have complex frequency response and cable, line, and transformer capacitances must be modeled. It is common to apply large capacitor banks and Static Var Compensators (SVCs) on transmission systems, and when these are switched, the frequency response characteristics change.

The harmonic resonance in a power system cannot be tolerated and must be avoided. The magnified harmonics will have serious effects such as equipment heating, harmonic torque generation, nuisance operation of protective devices, derating of electrical equipment, damage to the shunt capacitors due to overloading, and can precipitate shutdowns.

4.1 Resonance in Series and Parallel Circuits

Amplification of the voltages and currents can occur in RLC series and parallel resonant circuits.

4.1.1 Series RLC Circuit

The impedance of a series circuit (Figure 4.1a) is

$$z = R + j\left(\omega L - \frac{1}{\omega C}\right) \tag{4.1}$$

At a certain frequency, say f_0, z is minimum when

$$\omega L - \frac{1}{\omega C} = 0 \quad \text{or} \quad \omega = \omega_0 = \frac{1}{\sqrt{LC}} \tag{4.2}$$

The value of the current at resonance is

$$I_r = \frac{V_1}{R} = \frac{V_2}{R} \tag{4.3}$$

The current is limited only by the resistance, and if it is small, current can be high.

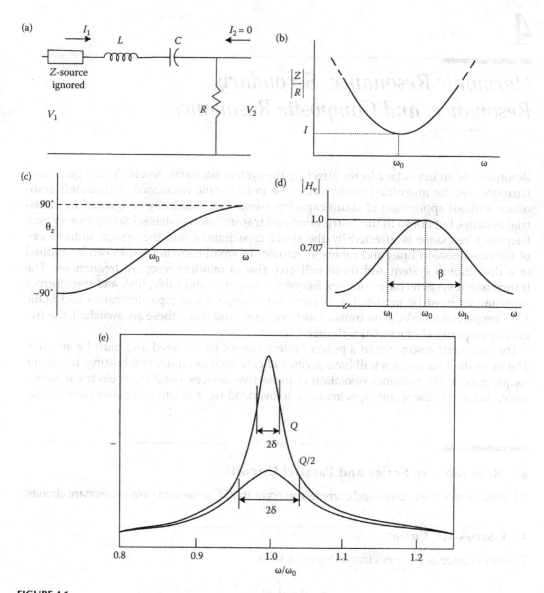

FIGURE 4.1
(a) The series circuit; (b) the frequency response, modulus only; (c) the angle; (d) half power points and frequencies; (d) performance of RLC circuit as a function of Q; and (e) resonance in terms of Q.

At resonance the voltage across the inductor is

$$V_L = \frac{j\omega_0 LV}{R} = jQV \tag{4.4}$$

Likewise the voltage across the capacitor is

$$V_c = -jQV \tag{4.5}$$

These voltages are equal in magnitude and opposite in sign. The voltages developed in L and C are larger than the system voltage by factor Q, and Q can be large for sharpness of tuning.

Figure 4.1b shows the frequency response; the capacitive reactance inversely proportional to frequency is higher at low frequencies, whereas the inductive reactance directly proportional to frequency is higher at higher frequencies. Thus, the reactance is capacitive and angle Z is negative below f_0, and above f_0, the circuit is inductive and angle Z is positive (Figure 4.1c). The voltage transfer function $H_v = V_2/V_1 = R/Z$ is shown in Figure 4.1d. This curve is reciprocal of Figure 4.1b. The half power frequencies are expressed as

$$\omega_h = \frac{R}{2L} + \sqrt{\left(\frac{R}{2L}\right)^2 + \frac{1}{LC}} = \omega_0 \left(\sqrt{1 + \frac{1}{4Q_0^2}} + \frac{1}{2Q_0} \right)$$

$$\omega_l = -\frac{R}{2L} + \sqrt{\left(\frac{R}{2L}\right)^2 + \frac{1}{LC}} = \omega_0 \left(\sqrt{1 + \frac{1}{4Q_0^2}} - \frac{1}{2Q_0} \right)$$

(4.6)

The bandwidth β shown in Figure 4.1d is given by

$$\beta = \frac{R}{L} = \frac{\omega_0}{Q_0} \left(= \omega_h - \omega_l \right)$$

(4.7)

or

$$Q_0 = \frac{\omega_0 L}{R}$$

The quality factor Q is defined as

$$Q_0 = 2\pi \left(\frac{\text{Maximum energy stored}}{\text{Energy dissipated per cycle}} \right)$$

(4.8)

Q factor is called the *figure of merit* given by ratio of maximum energy stored per cycle and energy dissipated per cycle multiplied by (2π). The energy stored in a reactor is $2\pi L I_m^2$ and dissipated in its series resistance R is $R I_m^2/2$. Here I_m is the peak current. Then Q is

$$Q = \frac{2\pi L I_m^2}{R I_m^2 / f} = \frac{\omega L}{R}, \text{ as before}$$

(4.9)

A capacitor has some losses and is represented by an ideal capacitor in parallel with a high resistance. Then Q for a capacitor becomes

$$Q = \frac{2\pi C V_m^2}{E_m^2 / rf} = \omega C R$$

(4.10)

where V_m is the maximum voltage across the parallel combination of resistor and capacitor. The Q is, therefore, a frequency-dependent function as R does not vary with the frequency.

As R of a capacitor is very large in resonant circuits, Q of the inductor is the controlling factor. Q of a resonant circuit implies the measured value of the circuit.

$$\omega_h \omega_i = \frac{1}{LC} = \omega_0^2 \tag{4.11}$$

Resonant circuits are frequency selective; it is desired that the circuit respond to a narrow band of frequencies and should have no response to the other frequencies. Close to resonance, we can write the circuit impedance of a series circuit as

$$Z = R + j\sqrt{\frac{L}{C}}\left(\omega\sqrt{LC} - \frac{1}{\omega\sqrt{LC}}\right) \tag{4.12}$$

$$Z = R + j\sqrt{\frac{L}{C}}\left(\frac{\omega}{\omega_0} - \frac{\omega_0}{\omega}\right)$$

$$= R\left[1 + jQ\left(\frac{\omega}{\omega_0} - \frac{\omega_0}{\omega}\right)\right] \tag{4.13}$$

Introduce a new variable δ, defined as

$$\delta = \left(\frac{\omega}{\omega_0} - \frac{\omega_0}{\omega}\right) \text{ or } \frac{\omega}{\omega_0} = 1 + \delta \tag{4.14}$$

where δ is the fractional deviation of the actual frequency from the resonant frequency. Then

$$Z = R\left[1 + jQ\left(1 + \delta - \frac{1}{1+\delta}\right)\right] \tag{4.15}$$

$$(1+\delta)^{-1} \approx 1 - \delta + \delta^2$$

Thus,

$$Z \approx R\left[1 + jQ\delta(2 - \delta)\right]$$

This gives the impedance of the series resonant circuit for small deviations from the resonant frequency.

How well a series resonant circuit will perform as a frequency selector is illustrated in Figure 4.1e (a circuit of high Q gives more selective curve). The band width can be defined in cycles, at the frequency at which the power in the circuit is one-half the maximum power.

At half power,

$$\frac{P}{2} = \frac{I_r^2 R}{2} \tag{4.16}$$

where I_r is the current at resonance, which is in peak value.

Then,

$$\frac{I_r}{\sqrt{2}} = \frac{1}{\sqrt{2}}\frac{V}{R} = \frac{V}{\sqrt{R^2 + X^2}} \tag{4.17}$$

That is,

$$R = X \tag{4.18}$$

The resistance and reactance are equal at half power frequencies and may be found from Equation 4.17:

$$1 = Q\delta_{1/2}(2 - \delta_{1/2}) \tag{4.19}$$

For most selective circuits, value of $\delta_{1/2}$ at half power frequency is small with respect to factor 2:

$$2Q\delta_{1/2} = 1 \tag{4.20}$$

The frequency deviation of each half power point from resonance will be $\delta_{(1/2)}$, and therefore, deviation between two half power frequencies will be $2(\delta_{1/2})$. Then bandwidth in cycles is

$$\beta = 2\delta_{1/2}f_0 = \frac{f_0}{Q} \tag{4.21}$$

Here, Q is for the complete series circuit, inductor capacitor, and source. Note that in Figure 4.1a, source impedance is not shown. The total impedance of the series circuit is the *total* resistance, and this means that series circuit should be connected to voltage source of low resistance.

In practical application of the series resonant circuit, single-tuned (ST) harmonic filters are essentially a series resonant circuit (Chapter 8).

4.1.2 Parallel RLC Circuit

In a parallel RLC circuit (Figure 4.2a), the impedance is high at resonance frequency:

$$y = \frac{1}{R} + \frac{1}{j\omega L} + j\omega C \tag{4.22}$$

Thus, the resonant condition is

$$-\frac{1}{\omega L} + \omega C = 0 \quad \text{or} \quad \omega = \omega_\alpha = \frac{1}{\sqrt{LC}} \tag{4.23}$$

Compare Equation 4.23 for parallel resonance with Equation 4.12 for the series circuit.

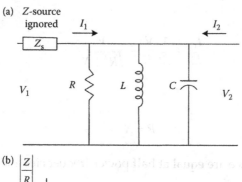

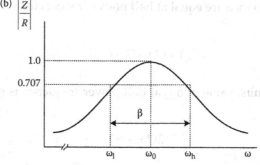

FIGURE 4.2
(a) Parallel RLC circuit and (b) the frequency response, modulus only.

The magnitude Z/R is plotted in Figure 4.2b. Half power frequencies are indicated in the plot. The bandwidth is given by

$$\beta = \frac{\omega_a}{Q_a} \tag{4.24}$$

where the quality factor for the parallel circuit is given by

$$Q_a = \frac{R}{\omega_a L} = \omega_a RC = R\sqrt{\frac{C}{L}} \tag{4.25}$$

A practical inductor will have some series resistance and a capacitor has some power losses.

A physical explanation of the Q factor can be stated as the efficiency with which the energy is stored. Capacitors and inductors are energy storage elements.

4.2 Practical LC Tank Circuit

The resistance of the inductor was ignored in parallel resonant circuit (Figure 4.2a). While the capacitor may be treated as an ideal capacitor, the inductor must be modeled with some

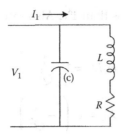

FIGURE 4.3
Tank resonant circuit.

practical value of resistance. The circuit shown in Figure 4.3 is called *tank circuit*. Consider that it is excited by a source of zero resistance. Then, its admittance can be written as

$$Y = \frac{R}{R^2 + \omega^2 L^2} - j\left(\frac{\omega L}{R^2 + \omega^2 L^2} - \omega C\right) \tag{4.26}$$

For resonance, the circuit must have a unity power factor, that is,

$$\left(\frac{\omega_0 L}{R^2 + \omega_0^2 L^2} - \omega_0 C\right) = 0 \tag{4.27}$$

This shows that the resonance occurs at

$$\omega_0 = \sqrt{\frac{1}{LC} - \frac{R^2}{L^2}} \tag{4.28}$$

This means resonance cannot occur if

$$\frac{R^2}{L^2} > \frac{1}{LC} \tag{4.29}$$

In contrast, a series circuit can be resonant at all values of resistance; only Q factor and sharpness of tuning will change.

Equation 4.27 can be written as

$$\omega_0 = \sqrt{\frac{1}{LC}}\sqrt{1 - \frac{R^2 C}{L}} = \sqrt{\frac{1}{LC}}\sqrt{1 - \frac{1}{Q^2}} \tag{4.30}$$

Thus, it differs from the series circuit by the factor

$$\sqrt{1 - \frac{1}{Q^2}} \tag{4.31}$$

If Q is large, say >10, the error is less than 1%. Also resonance is not possible with $Q < 1$. From Equation 4.30,

$$\omega_0^2 LC = 1 - \frac{1}{Q^2}$$

$$X_L = X_C\left(1 - \frac{1}{Q^2}\right)$$

(4.32)

Thus, unlike series resonant circuit, inductive and capacitive branches are not exactly equal.

The impedance variation with frequency can be examined similar to series circuit. We can write the admittance in Equation 4.32 as

$$Y = \frac{R\left(1 - \frac{j\omega L}{R} + \frac{\omega^2 CL^2}{R} + \omega CR\right)}{R^2 + \omega^2 L^2}$$

(4.33)

Writing

$$\frac{\omega L}{R} = \frac{\omega_0 L}{R}\frac{\omega}{\omega_0} = Q(1+\delta)$$

$$\omega^2 LC = \omega_0^2 LC(1+\delta)^2 = \left(1 - \frac{1}{Q^2}\right)(1+\delta)^2$$

(4.34)

and substituting in Equation 4.33, we get

$$Y = \frac{1 - jQ(1+\delta)\left\{1 - \left(1-\frac{1}{Q^2}\right)(1+\delta)^2\left(1 - \frac{1}{Q^2(1+\delta)^2}\right)\right\}}{R\left[1 + Q^2(1+\delta)^2\right]}$$

(4.35)

This can be simplified assuming $Q > 10$:

$$Y = \frac{1 - jQ(1+\delta)\left\{1 - (1+\delta)^2\right\}}{RQ^2(1+\delta)^2}$$

(4.36)

or

$$Z = \frac{RQ^2(1+\delta)^2}{1 + jQ\delta(1+\delta)(2+\delta)}$$

(4.37)

At resonance, $\delta = 0$, and therefore,

$$Z_r \approx R(1+Q^2) \approx RQ^2$$

(4.38)

Then

$$\frac{Z}{Z_r} = \frac{(1+\delta)^2}{1+jQ\delta(1+\delta)(2+\delta)} \tag{4.39}$$

Considering that close to resonance $\delta \ll 1$

$$\frac{Z}{Z_r} = \frac{1}{1+j2Q\delta} = A < \theta^{\circ} \tag{4.40}$$

The maximum possible impedance of the circuit does not occur at unity power factor. The square of the admittance is

$$|Y|^2 = \frac{1-2\omega^2 LC + \omega^2 C^2 (R^2 + \omega^2 L^2)}{R^2 + \omega^2 L^2} \tag{4.41}$$

This can be minimized with respect to ω by

$$\frac{d|Y|^2}{d\omega} = 0 \tag{4.42}$$

This gives

$$\omega = \left[\frac{1}{LC} \sqrt{1 + \frac{2CR^2}{L} - \frac{R^2}{L^2}} \right] \tag{4.43}$$

Therefore, the maximum impedance as the frequency is varied does not occur at unity power factor. If Q is large, then it will reduce to that condition. See References [1–3] for further reading.

4.3 Harmonic Resonance

Harmonic resonance is described in the previous section, and it occurs at a certain harmonic or harmonics. We call a resonance a "harmonic resonance" when it takes place at one of the nonlinear load-generated harmonics. When shunt capacitors are connected for power factor improvement, voltage support, or reactive power compensation, these act in parallel with the system impedance.

This means that the system impedance considered inductive (with some resistance in series) is in parallel with a capacitor. As discussed in previous sections, resonance occurs at a frequency where the inductive and capacitive reactance is equal; the impedance of the

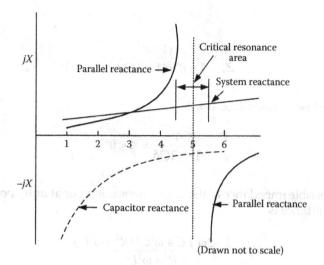

FIGURE 4.4
Illustration of harmonic resonance.

combination is infinite for a lossless system. The impedance angle changes abruptly as the resonant frequency is crossed. The inductive impedance of the power source and distribution (utility, transformers, generators, and motors) as seen from the point of application of the capacitors equals the capacitive reactance of the power capacitors at the resonant frequency:

$$j2\pi f_n L = \frac{1}{j2\pi f_n C} \tag{4.44}$$

where f_n is the resonant frequency. Figure 4.4 shows that the resonance occurs at the fifth harmonic, which is commonly generated by the electronic loads. The resonance point depends upon relative values of the inductance and capacitance. If the resonant frequency is 300 Hz, $5j\omega L = 1 / (5j\omega C)$, where ω pertains to the fundamental frequency (Example 4.2). When excited at the resonant frequency, a harmonic current magnification occurs in the parallel tuned circuit, though the exciting input current is small, as illustrated in Example 4.2. This magnification can be many times the exciting current and can even exceed the fundamental frequency current. It overloads the capacitors, may result in nuisance fuse operation, severe amplification of the harmonic currents resulting in waveform distortions, which has consequent deleterious effects on the power system components.

Resonance with one of the load-generated harmonics is the major effect of harmonics in the presence of power capacitors. This condition has to be avoided in any application of power capacitors. More conveniently, the resonant condition in the power system can be expressed as

$$h = \frac{f_n}{f} = \sqrt{\frac{KVA_{sc}}{kvar_c}} \tag{4.45}$$

where h is the order of harmonics, f_n is the resonant frequency, f is the fundamental frequency. KVA_{sc} is the short-circuit duty at the point of application of the shunt capacitors, and $kvar_c$ is the shunt power capacitor rating in kvar.

Consider that the short-circuit level at the point of application of power capacitors in 500 MVA. Resonance at 5th, 7th, 11th, 13th,..., harmonics will occur for a shunt capacitor size of 20, 10.2, 4.13, 2.95, ..., Mvar, respectively. The smaller the size of the capacitor, the higher the resonant frequency.

The short-circuit level in a system is not a fixed entity. It varies with the operating conditions. In an industrial plant, these variations may be more pronounced than in the utility systems, as a plant generator or part of the plant rotating loads may be out of service, depending upon the variations in processes and operation. The resonant frequency in the system will *float around*.

The lower order harmonics are more troublesome from a resonance point of view. As the order of harmonics increases, their magnitude reduces. Sometimes, a harmonic analysis study is limited to 25th or 29th harmonic only; however, possible resonances at higher frequencies may be missed.

It can be concluded that

- The resonant frequency will swing around depending upon the changes in the system impedance, for example, switching a tie-circuit on or off and operation at reduced load. Some sections of the capacitors may be switched out altering the resonant frequency.

- An expansion or reorganization of the distribution system may bring out a resonant condition where none existed before.

- Even if the capacitors in a system are sized to escape current resonant condition, immunity from future resonant conditions cannot be guaranteed owing to system changes, for example, increase in the short-circuit level of the utility system. Also see Reference [4].

An elementary tool to accurately ascertain the resonant frequency of the system in the presence of capacitors and nonlinear loads is to run a frequency scan on a digital computer (Chapter 7). Fundamentally, the frequency is applied in incremental steps, say at 2 Hz increments, for the range of harmonics to be studied to calculate resonance frequency accurately (a larger step reduces computer time but gives a 50% tolerance band with respect to the frequency step used).

System component models used for power frequency applications, for example, transformers, generators, reactors, and motors, are modified for higher frequencies (Chapters 7 and 8).

4.3.1 Harmonic Resonance in Industrial Power Systems

A number of examples in the following chapters will illustrate the harmonic resonance in industrial systems. Examples 7.1 and 7.2 may be seen.

4.3.2 Resonance at Even Harmonics

Resonance can occur at even harmonics too, though it may seem unusual; this situation is illustrated in Reference [5]. Note that arc furnaces and induction melting generate predominant second harmonic.

4.3.3 Transmission Systems

A distributed parameter long line frequency scan will show a number of resonant frequencies. See Chapter 7.

4.3.4 Elusiveness of Resonance in Power Systems

The resonance problem in power systems is a potential problem and a serious one. There are many documented cases of resonance, component failures, and plant shutdowns. It may not surface immediately and may appear at certain operating condition of the power system. It may sometimes "come" and "go" without attracting immediate attention, as it may be a "partial" resonance—thereby implying that the resonant impedance did not exactly produce $Q\delta = 0$ situation. Thus, most of the time, a resonant condition is "elusive." It may require long-term on-line measurements to establish the disturbing element in the system. References [5,6] provide some documented cases.

Harmonic distortions may come and go with time of the day. The patterns can be recognized and correlated with particular type of disturbing load such as mass transit systems, rolling mills, and arc furnaces. Arc lighting harmonics can be recognized from dusk to dawn patterns.

Figure 4.5 is an illustration of such a harmonic distortion caused by the load pattern of a consumer. Note the correlation of the harmonic distortion with the load pattern [4].

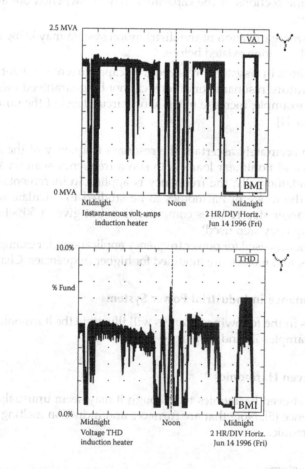

FIGURE 4.5
Voltage distortion versus time on a distribution system, and the load profile of a particular customer that is causing harmonic distortion. (IEEE P519.1/D9a, IEEE guide for applying harmonic limits on power systems, Unapproved draft © 2004 IEEE.)

4.3.5 Shifted Resonance Frequencies of ST Filters

Chapter 8 discusses ST filters, which are commonly applied for harmonic control and these are very efficient in shunting the desired harmonic from the power system. However, it is shown in Chapter 8 that the harmonic resonance is not eliminated, but the harmonic resonance frequency shifts to a lower value with respect to the tuned frequency of the ST filter.

4.3.6 Switched Capacitors

For the power factor improvement, switched capacitors are normally applied at low or medium voltage. As the load increases and power factor drops, a capacitor is switched in. In the presence of nonlinear loads, this will give rise to multiple resonant frequencies. Consider a four-step switching. With reference to Figure 4.6, when the first capacitor bank is switched, it gives rise to resonant frequency f_1. As the load increases and the power factor drops, the second capacitor bank is switched, and the resonant frequency shifts to f_2. Similarly, when the third capacitor bank is switched, frequency shifts to f_3, and finally to f_4 when the last capacitor is switched in. These frequencies will vary with the size of capacitor bank being switched in each step. Here the objective is to keep power factor within certain operating limits as the load varies.

This gives rise to two problems:

- Transients due to back-to-back switching and duties on the switching devices, Volume 1
- With shifting resonant frequencies, multiple resonances can occur as the frequency shifts from f_1 to f_4. These will be difficult to control.

If each switched capacitor is turned into a ST filter, tuned to the same frequency, then there will be only one resonant frequency to deal with. With the proper design of the ST filters, the resonance at a load-generated harmonic can be escaped. There is one more consideration, that is, that ST filters tuned to the same frequency may drive each other due to component tolerances. Sometimes a slight purposeful detuning is adopted in ST parallel filter designs.

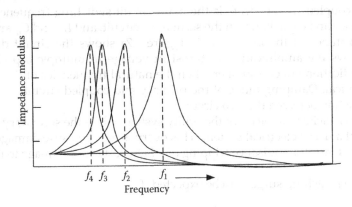

FIGURE 4.6
Multiple resonance frequencies, capacitor banks switched in sequence to maintain load power factor.

4.3.7 Nearby Harmonic Loads

A utility source may serve more than one customer from the same source. A consumer may design a filter to take care of his harmonic producing loads, and yet there can be problems because of adjacent harmonic producing loads. *The effects of harmonics can be present at considerable distances from their origins.*

The disturbing loads of the other consumers, especially on the same close by service, can make the filter designs ineffective and even overload these filters. The presence of capacitors along with filters can create more distortion and degrade the performance of filter, sometimes creating additional resonance frequencies. Monitoring of each customer loads and the distortion these are producing becomes important. See also Reference [6].

4.3.8 Subsynchronous Resonance Series Compensated Lines

See Chapter 2.

4.3.9 Ferroresonance

See Volume 1, Appendix C.

4.4 Secondary Resonance

When there are secondary circuits which have resonant frequencies close to the switched capacitor bank, the initial surge can trigger oscillations in the secondary circuits that are much larger than the switched circuit. The ratio of these frequencies is given by

$$\frac{f_c}{f_m} = \sqrt{\frac{L_m C_m}{L_s C_s}} \tag{4.46}$$

where f_c is the coupled frequency, f_m is the main circuit switching frequency, L_s and C_s are the inductance and capacitance in the secondary circuit, and L_m and C_m are the inductance and capacitance in the main circuit. Figure 4.7a shows the circuit diagram and Figure 4.7b shows the amplification of transient voltage in multiple capacitor circuits [7–9]. The amplification effect is greater when the natural frequencies of the two circuits are almost identical. Damping ratios of the primary and coupled circuits will affect the degree of interaction between the two circuits.

The switched capacitor can be on in the utility system, while the secondary capacitor in service can be at a user's electrical system. This overvoltage can cause damage to the secondary capacitors, result in nuisance tripping of ASDs, and cause damage to the sensitive equipment.

The maximum switching surges can be expected if

$$L_m C_m = C_s(L_S - L_m) \tag{4.47}$$

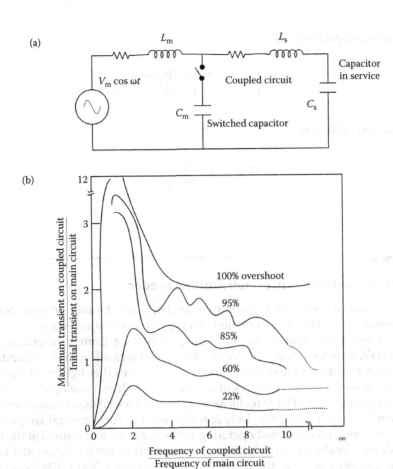

FIGURE 4.7
(a) Circuit configuration to illustrate secondary resonance and (b) overvoltages on switching depending upon coupled circuit parameters.

The angular frequencies of the circuit are as follows:

$$\beta_1 = \left(\frac{\alpha}{2} - \sqrt{\frac{\alpha^2}{4} - \beta} \right)$$

(4.48)

$$\beta_2 = \left(\frac{\alpha}{2} + \sqrt{\frac{\alpha^2}{4} - \beta} \right)$$

where

$$\alpha = \frac{1}{L_m C_m} + \frac{1}{L_s C_s} + \frac{1}{L_s C_m}$$

(4.49)

$$\beta = \frac{1}{L_m C_m L_s C_s}$$

The voltage across capacitor C_s is

$$V_s = V_m \left[1 - \frac{\beta_2^2 \cos\beta_1 t - \beta_1^2 \cos\beta_2 t}{\beta_2^2 - \beta_1^2} \right] \tag{4.50}$$

and the maximum voltage is

$$V_{S,\,max} = V_m \frac{\beta_2^2 + \beta_1^2}{\beta_2^2 - \beta_1^2} \tag{4.51}$$

4.5 Multiple Resonances in a Distribution Feeder

Multiple resonances can occur in a distribution feeder, due to location of capacitor banks. This is illustrated with following simulation of a typical distribution feeder.

Figure 4.8 shows a distribution feeder circuit, emanating from a substation, with a 66-12.47 kV, 6 MVA transformer. A 12.47 kV overhead line takes off from the secondary side of the substation transformer and serves loads as shown in this figure. To support the voltages in the system, shunt capacitors of the ratings as shown are applied at the load points. With adjustment of 6 MVA transformer to provide 2.5% voltage boost to the 12.47 kV secondary windings, and the applied capacitor banks, the fundamental frequency load flow shows that approximately rated operating voltages can be maintained at the loads.

The 12.47 kV line conductors of AAC 4/0 are spaced 3 ft in flat formation at a height of 25 ft from the ground level, and the distribution line carries a 3/#10 AWG steel ground wire. The soil resistivity is 100 Ω/m. With these line parameters, the line constants are calculated using a computer routine and are shown in Figure 4.9.

The impedance modulus and phase angle plots are depicted in Figures 4.10 and 4.11, respectively. These show multiple harmonic resonances, and these do not occur at integer multiple of the fundamental frequency.

4.6 Part-Winding Resonance in Transformer Windings

Part-winding resonance in transformers occurs mostly due to switching surges, current chopping in circuit breakers (low inductive currents) and VFT (very fast transients) in gas insulated substations. Failure of 4 (four) autotransformers of 500 and 765 kV systems of American Electric Power in 1968, 1971 led to the investigations of winding resonance phenomena [10–12]. When an exciting oscillating overvoltage arise due to line or cable switching or faults and coincides with one of the fundamental frequencies of part of a winding in the transformer, high overvoltages and dielectric stresses can occur. Low-amplitude and high-oscillatory switching transients cannot be suppressed by surge arrestors, but can couple to the transformer windings. This has been discussed in Volume 1 and further discussions are not repeated.

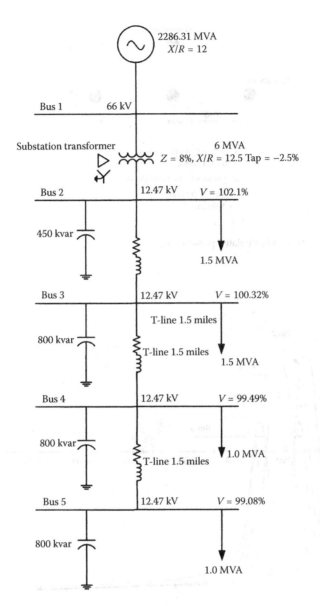

FIGURE 4.8
A distribution feeder with multiple capacitor locations and nonlinear loads for the study of harmonic resonance.

With respect to harmonics exciting a part-winding resonance, generally, the first natural resonant frequency of the transformer is above 5 kHz. Leaving aside VFTs, the resonant frequencies in core type transformer lie between 5 kHz and a few hundred kHz. The resonant frequencies may not vary much between transformers of different manufacturers and cannot be altered much, though efforts have been directed in this direction too. Generally, efforts are made to avoid network conditions which can produce oscillating voltages. (In-phase switching and resistance switching are discussed in Volume 1.) In the design stage, a lumped equivalent circuit of winding inductances and capacitance can be analyzed and for the transformers in operation the frequency response can be measured.

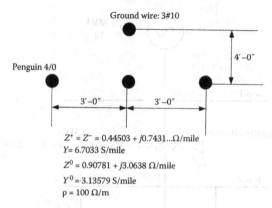

$Z^+ = Z^- = 0.44503 + j0.7431...\Omega/\text{mile}$

$Y = 6.7033\ \text{S/mile}$

$Z^0 = 0.90781 + j3.0638\ \Omega/\text{mile}$

$Y^0 = 3.13579\ \text{S/mile}$

$\rho = 100\ \Omega/m$

FIGURE 4.9

A 12.47 kV line configuration and calculated parameters.

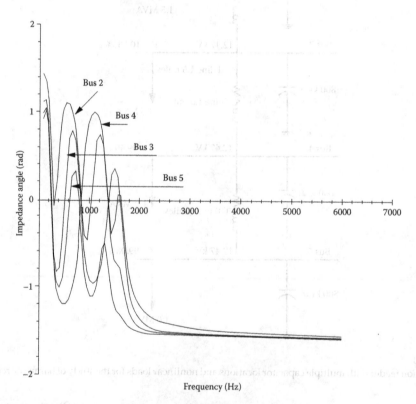

FIGURE 4.10

Impedance angle versus frequency for distribution feeder in Figure 4.8.

Figure 4.12 shows input impedance of a transformer as seen from the high-voltage terminal, showing a number of natural resonant frequencies.

Under steady-state conditions, it is unlikely that harmonics (generally under 3 kHz) will produce a part-winding resonant condition. However, on switching a capacitor bank, high inrush currents and frequencies occur; IEEE limits on back-to-back switching of a shunt capacitor bank with "definite purpose circuit breakers" are discussed in Volume 1.

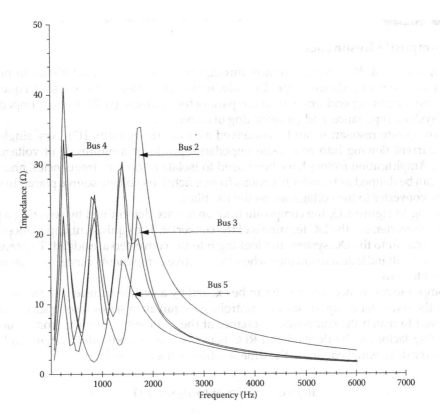

FIGURE 4.11
Impedance modulus versus frequency for distribution feeder in Figure 4.8.

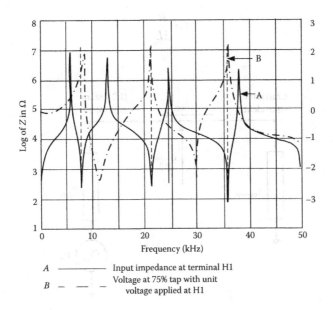

A ————— Input impedance at terminal H1

B — — — — Voltage at 75% tap with unit voltage applied at H1

FIGURE 4.12
A transformer terminal impedance as seen from the high-voltage winding side over frequency.

4.7 Composite Resonance

The impedance of AC system interacts through the converter characteristics to present entirely different impedance to the DC side, which gives rise to resonance frequencies. These frequencies depend upon all three parameters, namely (1) AC system impedance, (2) DC system impedance, and (3) switching of converter.

The composite resonance can be conceived as a matrix quantity [13]. Any single harmonic current flowing into composite impedance produces a multitude of voltage harmonics. Amplification factors have been used to isolate resonance frequencies, and these factors can be defined as transfer functions from a fictitious voltage source placed in series with the converter to the voltage across the DC filter.

Referring to Figure 4.13, the composite frequencies are determined by selecting a point near the converter, say the DC terminals of the converter. The equivalent impedances looking each way (into the DC system and looking into the converter) are added. The resulting impedance will indicate a resonance when the reactive components cancel each other and are equal to zero.

A composite resonance frequency can be excited by a system transient or by an unbalance in the converter components and controls. An equivalent series resonance circuit can be derived to match the composite resonance at the resonant frequency, from which the amplifying factor can be derived. An RLC network is a second-order system, and from elementary differential equation, the time solution of the current is

$$i(t) = e^{-\sigma t} \left(A_1 \cos \omega_r t + A_2 \sin \omega_r t \right) \tag{4.52}$$

where

$$\sigma = \frac{R}{2L} \tag{4.53}$$

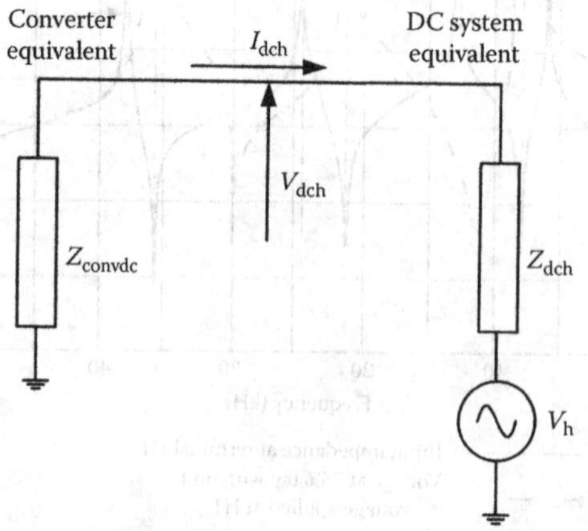

FIGURE 4.13
Equivalent of converter and DC system.

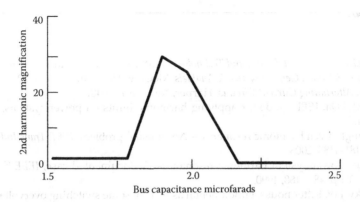

FIGURE 4.14
Amplification of the second harmonic with respect to bus capacitance.

is the damping factor, and A_1 and A_2 are calculated from initial conditions. Figure 4.14 shows the magnification of the second harmonic based on bus capacitance. At some value, the inductive and capacitive components are canceling each other or have approximately the same values.

A test case is CIGRE benchmark model of HVDC link [14]. Three cases of composite circuit series impedances are discussed. The model is designed to represent parallel resonance on the AC side and a series resonance at the fundamental frequency on the DC side. The rectifier DC terminals are chosen as the point where the system impedances are added. Two examples consider constant control gains at the rectifier, with slightly negative and positive damping factors. A third case is where a high-pass filter is placed in parallel with the Π control path, to increase the damping factor at composite resonance, without affecting the transient response at lower frequencies.

Problems

4.1 A 1.5 Mvar capacitor bank is applied to a power system. If the short-circuit level at the point of application of the capacitor banks is 250, 500, 750, and 1000 MVA, what are the approximate resonant frequencies?

4.2 An inductance of 0.1 H, a capacitor of 2.814 μF, and a resistance of 0.01 Ω are connected in series. What is the resonant frequency and what is the impedance at the resonant frequency? Now the same inductance and capacitance are connected in parallel with a resistance of 0.01 Ω in series. What is the resonant frequency and the impedance of the circuit?

4.3 In a series resonant circuit, $X_c = 300\,\Omega$, and resonance occurs at the 10th harmonic. Find the value of the reactor and plot impedance of the series circuit as frequency varies from 60 to 600 Hz. Consider a Q of 50.

4.4 Repeat Problem 4.3 for a parallel resonant circuit.

4.5 Calculate half power frequencies in Problem 4.3.

References

1. J.O. Bird. *Electrical Circuit Theory and Technology.* Butterworth Heinmann, Oxford, UK, 1997.
2. W.R. LePage, S. Sealy. *General Network Analysis,* McGraw Hill, New York, 1952.
3. M.B. Reed. *Alternating Current Circuits,* Harper, New York, 1948.
4. IEEE P519.1/D9a. IEEE guide for applying harmonic limits on power systems, Unapproved draft, 2004.
5. P.C. Buddingh. Even harmonic resonance—An unusual problem, *IEEE Trans Ind Appl,* vol. 39, no.4, pp. 1181–1186, 2003.
6. G. Lemieux. Power system harmonic resonance—A documented case, *IEEE Trans Ind Appl,* vol. 26, no. 3, pp. 483–488, 1990.
7. J. Zaborszky, J.W. Rittenhouse. Fundamental aspects of some switching overvoltages in power systems, *IEEE Trans Power App Syst,* vol. PAS-82, no. 64, pp. 815–822.
8. A. Kalyuzhny, S. Zissu, D. Shein. Analytical study of voltage magnification transients due to capacitor switching, *IEEE Trans Power Deliv,* vol. 24, no. 2, pp. 797–805, 2009.
9. J.C. Das. Analysis and control of large–shunt-capacitor-bank switching transients, *IEEE Trans Ind Appl,* vol. 41, no. 6, pp. 1444–1451, 2005.
10. H.B. Margolis, J.D. Phelps, A.A. Carlomagno, A.J. McElroy. Experience with part-winding resonance in EHV autotransformers; diagnosis and corrective measures, *IEEE Trans Power Appar Syst,* vol. PAS-94, no. 4, pp. 1294–1300, 1975.
11. A.J. McElroy. On significance of recent transformer failures involving winding resonance, *IEEE Trans Power Appar Syst,* vol. PAS-94, no. 4, pp. 1301–1316, 1975.
12. P.A. Abetti. Transformer models for determination of transient voltages, *AIEEE Trans,* vol. 72, no. 2, pp. 468–480, 1953.
13. S.R. Naidu, R.H. Lasseter. A study of composite resonance in AC/DC converters, *IEEE Trans Power Deliv,* vol. 18, no. 3, pp. 1060–1065, 2003.
14. M. Szechtman, T. Weiss, C.V. Thio. First benchmark model for HVDC control studies, *Electra* vol. 135, pp. 55–75, 1991.

5

Harmonic Distortion Limits

5.1 Standards for Harmonic Limits

Many countries have enacted their standards for harmonic limitations. IEC standard series 61000 provide internationally accepted information for the control of harmonics and interharmonics. This consists of a number of standards. EN standards represent European Norms approved by CENELEC [1]. The European standardization bodies are as follows:

CEN, Comitè Europeèn de Normalization

CENELEC, Comitè Europeèn de Normalization Electro-technique

ETSI, European Telecommunication Standards Institute

5.1.2 IEEE Standard 519

In North America, the harmonic limits described in IEEE 519 [2] prevail. Also these limits are accepted in other countries too. In the various studies and examples contained in this book, the limits specified in IEEE 519 are followed without any further lowering of these limits, irrespective of the power system under consideration and the study. Note that IEEE 519 indirectly allows higher current distortion limits from large consumers, as the short-circuit levels in their systems will, *generally*, be higher, though this may not always be true. IEC concept talks about two levels: planning level and compatibility level. *IEEE 519 was revised in 2014, and only some harmonic emission limits as compared to the 1992 version have been changed as discussed in this chapter.*

5.2 IEEE 519 Current and Voltage Limits

The standard provides recommended harmonic indices:

- Depth of notches, total notch area, and distortion of bus voltage distorted by commutation notches at low-voltage systems
- Individual and total current distortion
- Individual and total voltage distortion

The standard realizes that the harmonic effects differ substantially depending upon the characteristics of equipment being affected and the severity of harmonic effects cannot be perfectly correlated with a few simple indices. Harmonic characteristics of utility circuit seen from point of common coupling (PCC) are not accurately known. Good engineering judgments are required on a case-by-case basis, and the recommendations in standards *in no way override such judgments.*

Also it is acknowledged that strict adherence to harmonic limits will not always prevent problems, particularly when the limits are approached.

The harmonic indices, current, and voltage are defined at the PCC [2]. The concept of PCC is explained further in Section 5.3. The following points are of importance when interpreting the harmonic limits specified in IEEE 519 [2].

- HVDC systems and SVCs owned and operated by the utility are excluded from the definition of the PCC. Harmonic measurements are recommended at PCC. It is assumed that the system is characterized by the short-circuit impedance and that the effect of capacitors is neglected. The recommended current distortion limits are concerned with total demand distortion (TDD), which is defined as the total root-sum square harmonic current distortion as a percentage of maximum demand load current (15 or 30 min demand).

- The limits for the current distortion that a consumer must adhere to are shown in Tables 5.1 and 5.2. The ratio I_{sc}/i_1 is the ratio of the short-circuit current available at the PCC to the maximum fundamental frequency current and is calculated on the basis of the average maximum demand for the preceding 12 months. As the size of user load decreases with respect to the size of the system (size of the system is rather vague, here meant by the stiffness of the power system given by the three-phase symmetrical short-circuit current at the PCC), the percentage of the harmonic current that the user is allowed to inject into the utility system increases.

- To calculate the ratio I_{sc}/i_1, it is implied that a three-phase short-circuit current calculation should be carried out followed by the load flow calculation. Note that here the intention is not to calculate the short-circuit duties on the switching devices, like circuit breakers and fuses. Normally, the short-circuit calculation procedures detailed in ANSI/IEEE standards are followed without post multiplying factors for the circuit breaker duties. It is not only establishing the current demand at PCC, but also fundamental frequency load flow should establish that

TABLE 5.1

Current Distortion Limits for Systems (120 V–69 kV)

	Maximum Harmonic Current Distortion of Fundamental Current (%)					
	Harmonic Order (Odd Harmonics)[a]					
I_{sc}/I_L[b]	<11	$11 \leq h < 17$	$17 \leq h < 23$	$23 \leq h < 35$	$35 \leq h \leq 50$	TDD
<20[c]	4.0	2.0	1.5	0.6	0.3	5.0
20–50	7.0	3.5	2.5	1.0	0.5	8.0
50–100	10.0	4.5	4.0	1.5	0.7	12.0
100–1000	12.0	5.5	5.0	2.0	1.0	15.0
>1000	15.0	7.0	6.0	2.5	1.4	20.0

Note: Current distortions that occur in a DC offset, e.g., half-wave converters are not allowed. For general sub-transmission systems (69,001 V through 161,000 V), the limits are 50% of the limits shown in Table 5.1.

Source: IEEE Standard 519, *IEEE recommended practice and requirements for harmonic control in electrical systems* © 1992 IEEE, Revision: 2014.

[a] Even harmonics are limited to 25% of the odd harmonic limits above.

[b] I_{sc} = Maximum short-circuit current at PCC; I_L = maximum load current (fundamental frequency) at PCC.

[c] All power generation equipment is limited to these values of current distortion regardless of I_{sc}/I_L.

TABLE 5.2

Current Distortion Limits for Systems (>161 kV)

$I_{sc}/I_L{}^b$	Maximum Harmonic Current Distortion in % of I_L					
	Harmonic Order (Odd Harmonics)[a]					
	<11	$11 \leq h < 17$	$17 \leq h < 23$	$23 \leq h < 35$	$35 \leq h \leq 50$	TDD
<25[c]	1.0	0.5	0.38	0.15	0.1	1.5
<50	2.0	1.0	0.75	0.3	0.15	2.5
>50	3.0	1.5	1.15	0.45	0.22	3.75

Note: Current distortions that occur in a DC offset, e.g., half-wave converters are not allowed.

Source: IEEE Standard 519, *IEEE recommended practice and requirements for harmonic control in electrical systems* © 1992 IEEE, Revision: 2014.

[a] Even harmonics are limited to 25% of the odd harmonic limits above.

[b] I_{sc}=Maximum short-circuit current at PCC; I_L=maximum load current (fundamental frequency) at PCC.

[c] All power generation equipment is limited to these values of current distortion regardless of I_{sc}/I_L (added in 2014).

the operating voltages at all power system buses are within acceptable limits. The short-circuit calculations and fundamental load flow calculations are discussed in Volumes 1 and 2, respectively.

- Tables 5.1 and 5.2 are applicable to six-pulse rectifiers. For higher pulse numbers, the limits of the characteristic harmonics are increased by a factor of $\sqrt{P/6}$ *provided* that the amplitudes of noncharacteristic harmonics are less than 25% of the limits specified in the tables.

- An overriding article is provided that the transformer connecting the user to the utility system should not be subjected to harmonic currents in excess of 5% of the transformer rated current. Where this requirement is not met, a higher rated transformer should be provided. The derating of transformers serving nonlinear loads has been discussed in Chapter 6 with solved examples of calculations.

- The injected harmonic currents may create resonance with the utility's system and a consumer must insure that harmful series and parallel resonances are not occurring. The utility's source impedance may have a number of resonant frequencies, see Chapter 7, and it is necessary to model the utility's system in greater detail for harmonic analysis calculations.

- The limit of harmonic current injection does not limit a user's choice of converters or selection of harmonic-producing equipment technology. Neither does it lay down limits on the harmonic emission from the nonlinear or electronic equipment. It is left to the user how he would adhere to the limits of harmonic current injection, whether by choice of an alternative technology, use of passive or active filters, or any other harmonic mitigating device. This is in contrast to the IEC which has laid out the maximum emission limits from the equipment, for example IEC 61000-3-2 [3].

- The current distortion limits assume that there will be some diversity between harmonic currents injected by different consumers. This diversity can be with respect to time, phase angle of harmonics, and harmonic spectrums.

- The objectives of current limits are to limit the maximum individual frequency voltage harmonic to 3%.

- An important qualification is that TDD is for maximum demand load current, 15 or 30 min demand. The limits specified are to be used for the worst case condition for normal operation, conditions lasting for more than 1 h. For shorter periods during startups or unusual conditions, the limits may be exceeded by 50%.

- Ideally, the harmonic distortion caused by a single consumer should be limited to an acceptable level *at any point in the power system*. And the entire power system should be operated without substantial distortion anywhere in the system. The objectives of the current limits are to limit the maximum individual frequency voltage harmonic to 3% of the fundamental and the voltage total harmonic voltage distortion (THD) to 5% for systems without a major parallel resonance at one of the injected harmonic frequencies.

- The ideal of ensuring that the harmonic limits are met throughout a system may not be achieved in practice. The intent of the IEEE limits applies to the PCC only. Downstream, the user equipment may be able to tolerate higher harmonic distortion limits and operate satisfactorily. IEEE Draft Guide [4] for applying harmonic limits on power systems is based on IEEE 519.

5.3 PCC

PCC is the point of metering of power supply from the utility to a consumer, or any point as long as both the consumer and utility can either access the point for direct measurement of the harmonic indices meaningful to both or can estimate the harmonic indices at the point of interference through mutually agreeable methods. *The PCC is also the point where another consumer can be served from the same system.* The PCC could be located at the primary or secondary of the supply transformer, depending whether or not multiple customers are supplied from the transformer. A utility may supply more than one consumer from the secondary of a transformer.

As stated in Chapter 1, the delta transformer windings are a sink to the zero sequence currents and third harmonics. Thus, if the primary of the transformer is declared as the PCC, these harmonics will not enter into the evaluations.

The utilities may accept higher harmonic limits from the customers than those specified in the standards. Consider that in a consumer's installation, all the TDD limits, TDD-total, and distortion on individual harmonics are met, except that 11th harmonic is 5% higher than IEEE 519 limits. This situation may be acceptable to the utility.

5.4 Applying IEEE 519 Limits

Harmonics from small consumers with a limited amount of disturbing load may not require detailed analysis. A technique developed in IEC Standard 61000-3-6 [5] uses weighting factors.

The weighted disturbing power is calculated as follows:

$$S_{Dw} = \sum_i (S_{Di} \times W_i) \qquad (5.1)$$

For all the disturbing loads in the facility, S_{Di} is the power rating of an individual disturbing load in kVA, and W_i is the weighting factor.

If

$$S_{Dw}/S_{SC} < 0.1\% \tag{5.2}$$

where S_{SC} is the short-circuit capacity at the PCC in kVA, if Equation 5.2 is satisfied, then automatic acceptance occurs, without detailed analysis. This criterion is proposed in IEC Standard 61000-3-6. Weighting factors are given in Table 5.3.

An even simpler evaluation is advocated in IEC 61000-3-6 [5]. If the majority of the load is one of the types shown in the first three rows of Table 5.3, a detailed analysis is only necessary if the nonlinear load is more than 5% of the facility load.

For the other types of loads with current distortions less than 50%, the nonlinear load can be as high as 10%. The waveforms of the disturbing loads are shown in Figure 5.1.

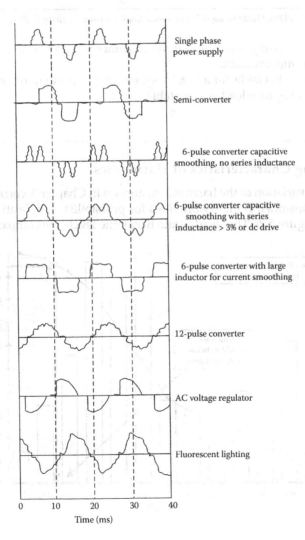

Single phase power supply

Semi-converter

6-pulse converter capacitive smoothing, no series inductance

6-pulse converter capacitive smoothing with series inductance > 3% or dc drive

6-pulse converter with large inductor for current smoothing

12-pulse converter

AC voltage regulator

Fluorescent lighting

Time (ms)

FIGURE 5.1
Waveforms of various disturbing loads.

TABLE 5.3

Weighting Factors

Type of Load	Current Distortion	Weighting Factor
Single-phase power supply	80%, high 3rd	2.5
Semiconverters, see Chapter 1	High 2nd, 3rd, 4th at partial load	2.5
6-pulse converter, capacitive smoothing, no series reactor	80%	2.0
6-pulse converter, capacitive smoothing, series inductance >3%, or DC drive	40%	1.0
6-pulse converter, with large reactor for current smoothing	28%	0.8
12-pulse converter	15%	0.5
AC voltage regulator	Varies with firing angle	0.7
Fluorescent lighting	20%	0.5

Source: IEEE Draft P519.1/D9a, *Guide for applying harmonic limits on power systems* © 2004 IEEE.

A harmonic analysis study is required if the customer is planning to provide capacitors for the power factor improvement.

As an application of this table, for a 13.8 kV system, with short-circuit level of 1000 MVA at the PCC, a 2 MVA 12-pulse load is acceptable.

5.5 Time Varying Characteristics of Harmonics

We alluded to time variation of the harmonic distortion in Chapter 3, corresponding to the load pattern of a consumer. IEEE 519 provides for probabilistic applications of harmonic distortion limits (Figure 5.2). The steady-state harmonic levels are compared to measured

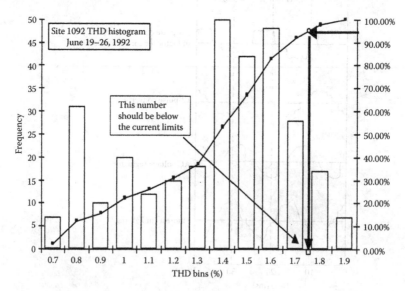

FIGURE 5.2

Probability plot illustrating the variable nature of harmonic levels. (IEEE Draft P519.1/D9a, *Guide for applying harmonic limits on power systems* © 2004 IEEE.)

level that is not exceeded 95% of the time (95% probability point). This is consistent with the compatibility level of IEC 61000-2-2 [6]. IEEE 519 also states that higher harmonic limits up to 150% are acceptable for short periods of time, 1 h per day, that is approximately 4% of the time. This limitation is consistent with design limits of 95% probability level not being exceeded. The higher limit can be compared with the measured harmonic level that is not exceeded 99% of the time (99% probability level).

It is rather rare that the harmonic limits will behave like a rectangular pulse of long duration; practically, this may consist of a number of pulses of shorter duration. These two profiles will have different impact on motor or transformer heating (Figure 5.3).

This requires that the harmonics are measured over a considerable period of time (Chapter 3). The three factors are as follows:

- The total duration of harmonic bursts is the summation of all the time intervals in which the measured levels exceed a particular level.

- The maximum duration of a single burst is the longest time interval in which the measured levels exceed a particular level.

- The measurements should account for the presence of capacitors, harmonic filters, outage of a filter bank, effect of alternate sources from the utility with different short-circuit levels, and the like.

Both the parameters in bulleted items 1 and 2 above are considered. Figure 5.4 shows a time duration plot versus TDD. The 6% TDD is exceeded for 8 min. The probability of a specified harmonic being exceeded can be evaluated with histograms and cumulative probability density and distribution plots (Chapter 3).

As the harmonic levels are continuously changing, a single sample cannot be used to evaluate compliance with harmonic limits. Harmonic-producing loads may be swathed in or out or these may operate for a specific duration during the day depending upon the processes.

Table 5.4 provides magnitude/duration limits that can be used to evaluate time varying nature of harmonics. It is understood that the actual variations may not exactly follow the patterns shown in this table.

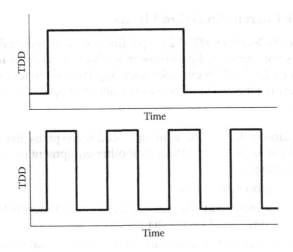

FIGURE 5.3
Two different harmonic emission trends, which will have a different heating impact on equipment.

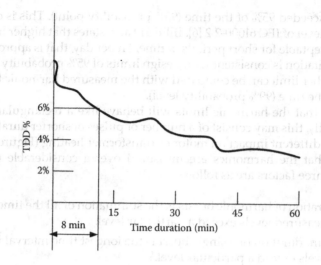

FIGURE 5.4
Plot of TDD versus duration.

TABLE 5.4

Short-Duration Harmonic Limits Based on a 24 h Measurement Period

Acceptable Harmonic Distortion Level (Individual or TDD)	Maximum Duration of Single Harmonic Burst $T_{maximum}$	Total Duration of All Harmonic Bursts T_{Total}
3.0 × (design limits)	$1s < T_{maximum} < 5s$	$15s < T_{Total} < 60s$
2.0 × (design limits)	$5s < T_{maximum} < 10\,min$	$60s < T_{Total} < 40\,min$
1.5.0 × (design limits)	$10\,min < T_{maximum} < 30\,min$	$40\,min < T_{Total} < 120\,min$
1.0 × (design limits)	$30\,min < T_{maximum}$	$120\,min < T_{Total}$

Source: IEEE Draft P519.1/D9a, *Guide for applying harmonic limits on power systems* © 2004 IEEE.

5.6 IEC Harmonic Current Emission Limits

The IEC standards 61000-3-2 and 61000-2-2 [3,6] define limits on harmonic current injected into a public distribution network by nonlinear appliances with an input current less than or equal to 16 A (at 220 V). This classifies such appliances into four classes (A–D). For each class, harmonic current emission limits are established up to the 39th harmonic. The classes are as follows:

Class A: General-purpose loads. These are balanced three-phase loads, the line currents differing by no more than 20% and all other equipment except as described in the following categories.

Class B: Portable appliances and tools.

Class C: Lighting equipment, with the exception of light dimmers (Class A) and self-ballasted lamps having class D wave shapes.

Class D: Appliances with input current with an assigned special wave shape and an active power input $P < 600$ W measured according to the method illustrated in the standard. This special wave shape is shown in Figure 5.5.

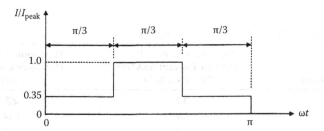

FIGURE 5.5
IEC Class D equipment waveshape. Each half-cycle of input current is within the envelope for at least 95% of the time.

The standard lays down methods for testing individual appliance harmonic emissions in the regulated frequency range. The harmonic limits were developed to limit the impact of these loads on the overall system. A system impedance of $R=0.4\ \Omega$ and $X=0.25\ \Omega$ (50 Hz) is used to evaluate the impact of these loads on local voltage distortion levels.

Tables 5.5 through 5.7 show the limits for Class A, C, and D equipment. Limits for Class B equipment are 1.5 times of that given for Class A equipment. The electronic equipment falls in Class D.

TABLE 5.5

IEC 61000-3-2 Limits for Class A Equipment (General-Purpose Loads)

Harmonic Order	Odd Harmonics Maximum Permissible Current in A	Harmonic Order	Even Harmonics Maximum Permissible Current in A
3	2.3	2	1.08
5	1.14	4	0.43
7	0.77	6	0.3
9	0.4	8–40	$0.23(8/n)$
11	0.33		
13	0.21		
15–39	$0.15(15/n)$		

TABLE 5.6

IEC 61000-3-2 Limits for Class C Equipment (Lighting)

Harmonic Order	Maximum Value Expressed as a % of the Fundamental Input Current of Luminaries
2	2.0%
3	$30\% \times PF$
5	10%
7	7%
9	5%
11–39	3%

146

Harmonic Generation Effects Propagation and Control

TABLE 5.7

Harmonic Emission Limits for Class D Appliances

Harmonic Order	Maximum Admissible Harmonic Current (mA/W) 75W < P < 600 W	Maximum Admissible Harmonic Current (A) P > 600 W
3	3.4	2.3
5	1.9	1.14
7	1.0	0.77
9	0.5	0.4
11	0.35	0.33
$13 \le n \le 39$	$3.85/n$	$0.15\,(15/n)$

Source: IEC 61000-3-2. *Limits for Harmonic Current Emissions (Equipment Input Current Less Than or Equal to 16A per Phase)*, 2009.

5.7 Voltage Quality

5.7.1 IEEE 519

While a user can inject only a certain amount of harmonic current into the utility system as discussed in previous sections, the utilities and power producers must meet requirements of a certain voltage quality to the consumers. The recommended voltage distortion limits are given in Table 5.8 [2]. The index used is THD as a percentage of nominal fundamental frequency voltage. The limits are system design values for *worst case* for normal operation, conditions lasting for more than 1 h. For shorter periods, during startups or unusual conditions, the limits can be exceeded by 50%. If the limits are exceeded, harmonic mitigation through use of filters or stiffening of the system through parallel feeders is recommended.

The limits of distortion on most medium voltage systems are 5%. The consumer may accept a higher voltage distortion depending on the sensitivity of his loads to voltage distortions. Note that the IEEE 519 does not lay down any limits of voltage distortion at the PCC *which a consumer must adhere to.* The utility company has to maintain a certain voltage quality at the PCC. An example of a filter design in Chapter 8 shows that for weak power systems, the voltage distortions at PCC are a bigger problem. The consumer nonlinear load can seriously distort the voltage at PCC.

TABLE 5.8

Harmonic Voltage Limits (Line-to-Neutral) for Power Producers (Public Utilities or Cogenerators)

	Harmonic Distortion in % at PCC			
	$V \le 1.0$ kV	1 kV < V ≤ 69 kV	69 kV < V ≤ 161 kV	161 kV < V
Maximum for individual harmonic	5.0	3.0	1.5	1.0
Total harmonic distortion (THD)%	8.0	5.0	2.5	1.5[a]

Source: IEEE Standard 519, *IEEE recommended practice and requirements for harmonic control in electrical systems* © 1992. Revision: 2014.

[a] High-voltage systems can have up to 2.0% THD where the cause is an HVDC terminal whose effects would have attenuated at points in the network where future users may be connected.

5.7.2 IEC Voltage Distortion Limits

IEC 61000-2-2 [6] provides compatibility levels for various types of power quality characteristics. The compatibility level for the distortion on low-voltage systems is 8%.

Table 5.9 gives the voltage harmonic distortion limits in public low-voltage networks
(IEC-61000-2-2). These limits are the same as in IEC 61000-2-4 for Class 2. For Class 3, as per
this standard the limits are shown in Table 5.10.

The THD (voltage) is ≤8% for Class 2 and ≤10% for Class 3.

Class 2: PCC and IPC (in-plant point of coupling) in industrial environment in general.

Class 3: Applies to IPCs in industrial environment.

TABLE 5.9

IEC 61200-2-2 Harmonic Voltage Limits in Public Low-Voltage
Network. Also IEC 61000-2-4. Harmonic Voltage Limits in
Industrial Plants, Class 2

Odd Harmonics		Even Harmonics		Triplen Harmonics	
h	% Vh	h	% Vh	h	% Vh
5	6	2	2	3	5
7	5	4	1	9	1.5
11	3.5	6	0.5	15	0.3
13	3	8	0.5	≥21	0.2
17	2	10	0.5		
19	1.5	≥12	0.2		
23	1.5				
25	1.5				
≥29	x				

$x = 0.2 + 12.5/h$, for $h = 29.31, 35$, and 37; $V_h = 0.63, 0.60, 0.56$, and 0.54%.

TABLE 5.10

IEC 61200-2-4 Harmonic Voltage Limits Class 3

Odd Harmonics		Even Harmonics		Triplen Harmonics	
h	% Vh	h	% Vh	h	% Vh
5	8	2	3	3	6
7	7	4	1.5	9	2.5
11	5	≥6	1	15	2
13	4.5			21	1.75
17	4			≥27	1
19	4				
23	3.5				
25	3.5				
≥29	y				

$y = \sqrt[5]{11/h}$

For $h = 29, 31, 35$ and 37; $V_h = 3.1, 3.0, 2.8$, and 2.7%.

5.7.3 Limits on Interharmonics

IEC 61000-4-15 [7] has established a method of measurement of harmonics and interharmonics utilizing a 10- or 12-cycle window for 50 and 60 Hz systems. This results in a spectrum with 5 Hz resolution. The 5 Hz bins are combined to produce various groupings and components for which limits and guidelines can be referenced. The IEC limits the interharmonic voltage distortion to 0.2% for the frequency range from DC to 2 kHz, see also Chapter 3.

IEEE 519-2014, revision of 1992 now describes voltage interharmonic limits for PCC voltages less than 1 kV, based on flicker [2]; see Chapter 2 for a figure and table. The effect of interharmonics on other equipment and systems such as generator mechanical systems, motors transformers, signaling and communication systems, and filters are not described. A reader may also see Reference [8].

It may be necessary to impose severe restrictions on interharmonics if there are concerns of torsional interactions at nearby generating plants. In other cases, interharmonics need not be treated differently from the integer harmonics.

For recommendations of current distortion limits from EAF, see Chapter 2.

Example 5.1

The harmonic current and voltage spectrum of a 12-pulse LCI inverter, operating at $\alpha = 15°$, are presented in the first two columns of Table 5.11. The demand current is 1200 A, which is also the inverter maximum operating current. The available short-circuit current at the PCC is 12 kA. Calculate the distortion limits. Are these exceeded?

Table 5.11 is extended to show permissible limits of the TDD at each of the harmonics, and individual and total permissible current distortions are calculated. From Table 5.1 for $I_{sc}/I_1 < 20$ (actual $I_{sc}/I_L = 10$). The noncharacteristic harmonics are reduced to 25% and the characteristic harmonics are multiplied by a factor of the square root of p over 6, i.e., $\sqrt{2}$.

TABLE 5.11

Calculations of TDD, 12-Pulse Converter, Example 5.1

h	I_h(A)	Harmonic Distortion %	IEEE Limits %	V_h (V)
5	24.32	2.027	1.0	20.86
7	13.43	1.119	1.0	16.13
11	65.42	5.452	2.82	123.46
13	39.69	3.331	2.82	88.52
17	1.87	0.156	0.375	5.45
19	0.97	0.081	0.375	3.16
23	8.45	0.704	0.846	33.34
25	8.54	0.711	0.846	36.62
29	0.98	0.081	0.15	4.87
31	0.76	0.063	0.15	4.04
35	5.02	0.418	0.423	30.14
37	3.27	0.273	0.423	20.75
41	0.23	0.019	0.075	1.62
43	0.23	0.019	0.075	1.70
47	3.45	0.287	0.423	27.81
49	2.58	0.215	0.423	21.69
TDD		6.887%	5%	4.08%

Table 5.11 shows that current distortion limits on a number of harmonics are exceeded. Also, the TDD is greater than permissible limits. The total THD (voltage) is below the 5% limit, but the distortion at the 11th harmonic is 3.32%, which exceeds the maximum permissible limit of 3% on an individual voltage harmonic.

In this example, the distortion limits exceed the permissible limits, though a 12-pulse converter is used, and the base load equals the converter load, i.e., the entire load is nonlinear. Generally, the harmonic-producing loads will be some percentage of the total load demand and this will reduce TDD as it is calculated on the basis of total fundamental frequency demand current.

5.8 IEEE 519 (2014)

5.8.1 Measurement Window Width

Any instrument for accessing harmonic levels for comparison with recommended limits in IEEE 519 must comply with specifications of IEC 61000-4-7 and 61000-4-30 [9,10]. The width of the measurement window used by digital instruments employing DFT techniques should be 12 cycles (approximately 200 ms) for 60 Hz power systems and 10 cycles for 50 Hz power systems. With this window width spectral components will be available every 5 Hz. The harmonic component magnitude is considered to be a value at the center frequency (60, 120, 180, etc.) combined with two adjacent 5 Hz bin values. The three values are combined into a single rms value that defines the harmonic magnitude for the particular center frequency component.

5.8.2 Very Short Time Harmonic Measurements

Very short duration harmonic values are assessed for 3 s based on an aggregation of 15 consecutive 12 cycle windows for 60 Hz.

$$F_{n,vs} = \sqrt[2]{\frac{1}{15}\sum_{i=1}^{15} F_{n,i}^2} \tag{5.3}$$

where F represents the current or voltage, n is the harmonic order, i is a simple counter, and subscript vs is used to denote very short.

5.8.3 Short-Time Harmonic Measurements

Short-duration harmonic values are assessed for 10 min based on an aggregation of 200 consecutive very short time values for a specific frequency component.

$$F_{n,sh} = \sqrt[2]{\frac{1}{200}\sum_{i=1}^{200} F_{(n,vs),i}^2} \tag{5.4}$$

where F represents the current or voltage, n is the harmonic order, i is a simple counter, and subscript sh is used to denote short.

5.8.4 Statistical Evaluation

Then the specified limits are qualified as follows:

Very short and short-time values must be accumulated for 1 day and 1 week, respectively. For very short time measurements, the 99th percentile value, that is, the value that is exceeded for 1% of the measurement period, should be calculated for each 24-h period for comparison with the recommended limits in Tables 5.1, 5.2, and 5.8. For short harmonic measurements, the 95th and 99th percentile values should be calculated for a 7-day period for comparison with the limits specified in the aforesaid tables. These statistics are used for both current and voltage harmonics except that the 99th percentile short-time value is not recommended for voltage harmonics.

5.9 Commutation Notches

Commutation notches are shown in Figure 5.6a. As we examined in Chapter 1, commutation produces two primary notches per cycle plus four secondary notches of lesser magnitude due to notch reflection from other leg of the bridge. The full-wave diode bridge with capacitor load operates in discontinues mode and does not produce notches. For low-voltage systems, the notch depth, the total notch area of the line-to-line voltage at the PCC, and THD should be limited as shown in Table 5.12. The notch area is given by

$$A_N = V_N t_N \tag{5.5}$$

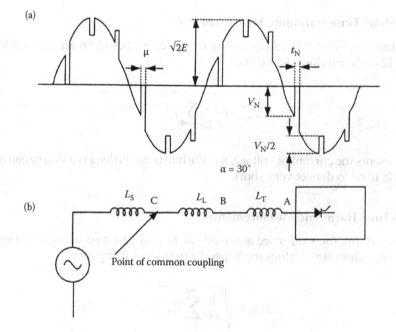

FIGURE 5.6
(a) Notching in current source converters with DC link reactors and (b) the notch area varies at various points shown in this figure.

TABLE 5.12

Low-Voltage System Classification and Distortion Limits

	Special Applications	General Systems	Dedicated Systems
Notch depth	10%	20%	50%
THD (voltage)	3%	5%	10%
Notch area (A_n)	16,400	22,800	36,500

Source: IEEE Standard 519. *IEEE Recommended Practice and Requirements for Harmonic Control in Electrical Systems,* 1992. Revision: 2014.

The notch area for voltages other than 480 V is given by multiplication by a factor $V/480$. Notch area is given in V µs at rated voltage and current. Special applications include hospitals and airports. A dedicated system is exclusively dedicated to the converter load.

where A_N is the notch area in volt-microseconds, V_N is the depth of the notch in volts, line-to-line (L–L) of the deeper notch in the group, and t_N is the width of the notch in microseconds.

Consider the equivalent circuit of Figure 5.6b and define the following inductances:

- L_t is the inductance of the drive transformer.
- L_L is the reactance of the feeder line.
- L_s is the inductance of the source.

The notch depth depends where we look into the system. The primary voltage goes to zero at the converter terminals, point A in Figure 5.6b, and the depth of the notch is the maximum. At B the depth of the notch in per unit of the notch depth at A is given by

$$\frac{L_s + L_L}{L_s + L_t + L_L} \tag{5.6}$$

If point C in Figure 5.6b is defined as the PCC, then the depth of the notch at the PCC in per unit with respect to notch depth at the converter is

$$\frac{L_s}{L_L + L_s + L_t} \tag{5.7}$$

For actual values multiply Equations 5.6 or 5.7 with e, where e is the instantaneous voltage (L–L) prior to the notch. That is, $V_N = e$ multiplied by Equations 5.6 or 5.7.

The impedances in the converter circuit are acting as a sort of potential divider. The width of the notch is given by the commutation angle μ and by the expression

$$t_N = \frac{2(L_L + L_t + L_s)I_d}{e} \tag{5.8}$$

The area of the notch at the converter terminals is given by

$$A_N = 2I_d(L_L + L_t + L_s)$$

A relationship between line notching and distortion factor is given by

$$V_h = \sqrt{\frac{2V_N^2 t_N + 4(V_N/2)^2 t_N}{1/f}} = \sqrt{3V_N^2 t_N f} \tag{5.9}$$

where f is the power system fundamental frequency. If a factor ρ equal to the ratio of total inductance to common system inductance is written as

$$\rho = \frac{L_L + L_t + L_s}{L_L} \tag{5.10}$$

then

$$V_{NMAX} = \frac{\sqrt{2}E_L}{\rho} \tag{5.11}$$

and

$$THD_{MAX} = 100\sqrt{\frac{3\sqrt{2}.10^{-6}A_N f}{\rho E_L}} = 0.074\sqrt{\frac{A_N}{\rho}}\% \tag{5.12}$$

In Equation 5.9, the two deeper and four less deep notches per cycle (Figure 5.6a) are considered. Also see Reference [2]. Equation 5.12 is good for 480 V systems only.

Example 5.2

Consider a system configuration as shown in Figure 5.7. It is required to calculate the notch depth and notch area at buses A, B, and C.

The inductances throughout the system are calculated and are shown in Figure 5.7. The source inductance (reflected at 480 V), based on the given short-circuit data, is 0.64 μH. It is a stiff system at 13.8 kV, and the source reactance is small.

A 1 MVA transformer reactance referred to the 480 V side, based on the X/R ratio, $X_t = 5.638\% = 0.1299$ Ω. This gives an inductance of 34.4 μH at 480 V. Similarly, for a 1.5 MVA transformer, the inductance is 23.2 μH. The feeder inductance is given or it can be calculated, based on given cable/bus duct data. The notch depth at bus C as a percentage of depth at the converter $= (34.4 + 0.64)/(34.4 + 0.64 + 30) = 54\%$. Referring to Table 5.12, this exceeds the limits even for a dedicated system.

Consider that the converter is supplying a motor load of 500 hp at 460 V and the DC current is continuous and equal to 735 A. The notch area at the converter AC terminals is then given by

$$2I_d(L_s + L_{T1} + L_{F1}) = 2 \times 735(0.64 + 34.3 + 30)$$
$$= 95,609 \text{ V}\mu s$$

Therefore, the notch area at bus $C = 0.54 \times 95,609 = 51,628$ V μs. From Table 5.12, the limits for dedicated systems and general systems are 36,500 and 22,800, respectively. The calculated value is much above these limits.

The notch area at bus A due to the converter at bus C is

$$\frac{0.64}{0.64 + 34.4 + 30}(95,609) = 940.8 \text{ V}\mu s$$

In a likewise calculation, the notch depth at bus $B = 44.28\%$, area $= 35,061$ V μs. The notch area at bus A due to the converter load at bus $B = 940.8$ V μs.

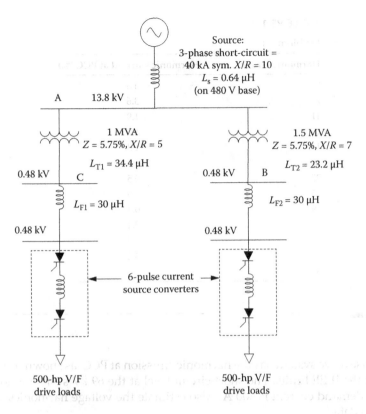

FIGURE 5.7
A system configuration for the calculation of notch area.

The total notch area at bus A can be considered as a sum of the notch areas due to converter loads at buses B and C. This assumes that the notch widths due to commutation are additive, which is a very conservative assumption.

The example shows that the calculated values are much above the permissible values. The standard requirements can be met if the cable inductance is increased or drive isolation transformer is provided. This means that, in this example, the source plus transformer inductance should be much lower than the cable plus drive transformer inductance. In the above example, no drive isolation transformer is considered. However, note that the distortion at PCC increases with increasing reactance and this may have to be reduced for controlling the distortion.

This example suggests that problems of notching can be mitigated by

- Decreasing the source impedance behind the PCC bus.
- Increasing the impedance ahead of (on the load side) of the PCC bus. Isolation transformers and line reactors are commonly used and serve the same purpose.
- Providing second-order high-pass filters is another option, which can also provide reactive power compensation and improve the voltage profile on low-power factor loads. The high-pass filter will provide a low impedance source of commutating current to reduce notching.

A case of distribution system oscillations due to notching is illustrated in Reference [4]. A 25 kV bus, which serves drive system loads and is connected to 144 kV system, experiences oscillations close to the 60th harmonic and failure of surge capacitors.

TABLE P5.1

Problem 5.1

Harmonic Order	Harmonic Current at PCC (%)
5	3.8
7	3.6
11	1.9
13	1.8
17	1.3
19	1.2
23	0.5
25	0.5
29	0.4
31	0.4
33	0.3
35	0.2
37	0.1

Problems

5.1 A six-pulse drive system has the harmonic emission at PCC, as shown in Table P5.1 Tabulate the IEEE limits. The short-circuit level at the 69 kV system is 10 kA and the load demand current is 600 A. Also estimate the voltage harmonics based on the given data.

5.2 Calculate the distortion at point C in Example 5.2.

References

1. EN 50160. *Voltage Characteristics of Electricity Supplied by Public Distribution Systems*, Brussels, 1994.
2. IEEE Standard 519. IEEE recommended practice and requirements for harmonic control in electrical systems, 1992. Revision: 2014.
3. IEC 61000-3-2. Limits for harmonic current emissions (equipment input current less than or equal to 16A per phase), 2009.
4. IEEE Draft P519.1/D9a. Guide for applying harmonic limits on power systems, 2004.
5. IEC 61000-3-6. Part 3–6, Limits: Assessment of emission limits for the connection of distorting installations to MV, HV and EHV power systems, 2008.
6. IEC 61000-2-2. Part 2–2, Environment, Section 1. Description of the environment. Compatibility levels for low frequency conducted disturbances and signaling in public low-voltage power supply systems, 2002.
7. IEC 61000-4-15. Part 4–15, Testing and measurement techniques; Flicker meter—Functional and design specifications, 2010.
8. E.W. Gunther. Interharmonics recommended updates to IEEE 519. *IEEE Power Engineering Society Summer Meeting*, pp. 950–954, July 2002.
9. IEC 61000-4-7. General guide on harmonic and interharmonic measurement and instrumentation, for power supply systems and equipment connected thereto, 2009.
10. IEC 61000-4-30. Power quality measurement methods, 2015.

6

Effects of Harmonics

Harmonics have deleterious effects on electrical equipment. These can be itemized as follows (see Reference [1]):

1. Capacitor bank failure because of reactive power overload, resonance, and harmonic amplification. Nuisance fuses operation.

2. Excessive losses, heating, harmonic torques, and oscillations in induction and synchronous machines, which may give rise to torsional stresses.

3. Increase in negative sequence current loading of synchronous generators, endangering the rotor circuit and windings.

4. Generation of harmonic fluxes and increase in flux density in transformers, eddy current heating, and consequent derating.

5. Overvoltages and excessive currents in the power system resulting from resonance.

6. Derating of cables due to additional eddy current heating and skin effect losses.

7. Inductive interference with telecommunication circuits.

8. Signal interference and relay malfunctions, particularly in solid-state and microprocessor-controlled systems.

9. Interference with ripple control and power line carrier systems, causing misoperation of the systems, which accomplish remote switching, load control, and metering.

10. Unstable operation of firing circuits based upon zero voltage crossing detection and latching.

11. Interference with large motor controllers and power plant excitation systems.

12. Flicker.

In Volume 1, we discussed the high-frequency inrush currents on switching of shunt capacitor banks. Nonlinear loads in the presence of capacitors can bring about a resonant condition with one of the load-generated harmonics as stated above and discussed in Chapter 4. This also has the following additional effects:

- Increase the transient inrush current of the transformers and prolong its decay rate [2].

- Increase the duty on the switching devices.

- Possibility of part-winding resonance exists if the predominant frequency of the transient coincides with natural frequency of the transformer.

6.1 Rotating Machines

6.1.1 Pulsating Fields and Torsional Vibrations

Harmonics will produce elastic deformation, e.g., shaft deflection, parasitic torques, vibrations noise, additional heating, and lower the efficiency of rotating machines.

The movement of the harmonics is with or against the direction of the fundamental. Criterion of forward or reverse rotation is established from $h = 6m \pm 1$, where h is the order of harmonic and m is any integer. If $h = 6m + 1$, the rotation is in the forward direction but at $1/h$ speed. Thus, 7th, 13th, 19th, ..., harmonics rotate in the same direction as the fundamental rotates. Harmonics of the order 1, 4, 7, 10, 13, ..., are termed positive sequence harmonics (Chapter 1).

If $h = 6m - 1$, the harmonic rotates in a reverse direction to the fundamental. Thus, 5th, 11th, 17th, ..., are the reverse rotating harmonics. Harmonics of the order 2, 5, 8, 11, 14, ..., are the negative sequence harmonics.

The magnitude of harmonic current in a three-phase induction motor may be calculated from the expression

$$I_h = \frac{V_h}{h\omega_0 L_{1h}} \tag{6.1}$$

where I_h is the hth harmonic current, V_h is the hth harmonic voltage, and L_{1h} is the stator and rotor leakage inductance at hth harmonic referred to the stator. The effective inductance tends to decrease as h increases. Approximately, L_{1h} is equal to L_1 (stator leakage reactance), which is the minimum value when internal bar inductance is negligible. A harmonic impedance of an induction motor is derived in Chapter 7, see also Volume 2 for the induction motor models.

With certain assumptions, the harmonic losses can be defined as [1]

$$\frac{P_h}{P_{RL}} = k \sum_{h=5}^{h=\infty} \frac{V_h^2}{h^{3/2} V_1^2} \tag{6.2}$$

where

$$k = \frac{(T_s/T_R)E}{(1-S_R)(1-E)} \tag{6.3}$$

P_h is the harmonic loss, P_{RL} is the loss at the rated point with sinusoidal supply, T_s is the starting torque, T_R is the rated torque, S_R is the slip, and E is the efficiency. In NEMA class C motors [3], k can range up to 25 or higher.

A motor distortion index (MDI) is defined as

$$MDI = \frac{1}{V_1} \left(\sum_{h=5}^{h=\infty} \frac{V_h^2}{h^{3/2}} \right)^{1/2} \tag{6.4}$$

The use of Equation 6.4 permits a convenient comparison of different motor designs, but it will not evaluate localized heating. A similar ratio can be derived for rotor heating only. The motors with large deep bars or double cage will have highest harmonic heating.

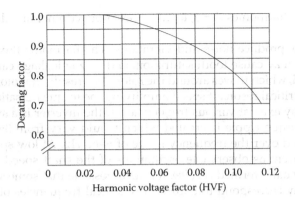

FIGURE 6.1
Proposed derating curve due to harmonic voltage factor (HVF—same as THDv), all machines, NEMA Part 30 [3].

In a detailed analysis, the effect of the harmonics on motor losses should consider the subdivision of losses into windage and friction, stator copper loss, core loss, rotor copper loss, and stray loss in the core and conductors and the effect of harmonics on each of these components. The effective rotor and stator leakage inductance decreases and the resistance increases with frequency. The equivalent circuits of the induction motors for positive and negative sequence are derived in Volume 2, and the effect of negative sequence currents will be more pronounced at higher frequencies. The rotor resistance may increase—four to six times the DC value, while leakage reactance may reduce to a fraction of the fundamental frequency value. The stator copper loss increases in proportion to the square of the total harmonic current plus an additional increase due to skin effect on resistance at higher frequencies. Harmonics contribute to magnetic saturation and the effect of distorted voltage on core losses may not be ignored. Major loss components influenced by harmonics are stator and rotor copper losses and stray losses. A harmonic factor of 11% gives approximately 25% derating of general purpose motors.

Figure 6.1 from NEMA [3] shows derating factor with respect to harmonic voltage factor, which is another name for the voltage distortion factor.

6.1.1.1 Synchronous Machines

In a synchronous machine, the frequency induced in the rotor is the net rotational difference between the fundamental frequency and the harmonic frequency. Fifth harmonic rotates in reverse with respect to stator, and with respect to the rotor, the induced frequency is that of sixth harmonic. Similarly, the forward rotating seventh harmonic with respect to stator produces sixth harmonic in the rotor. The interaction of these fields produces a pulsating torque at 360 Hz and results in the oscillations of the shaft. Similarly, the harmonic pair 11 and 13 produces a rotor harmonic of 12th. If the frequency of the mechanical resonance exists close to these harmonics during starting, large mechanical forces can occur.

The same phenomena occur in induction motors. Considering slip of the induction motors, the positive sequence harmonics, $h = 1, 4, 7, 10, 13, \ldots$, produce a torque of $(h-1+s)\,\omega$ in the direction of rotation, and the negative sequence harmonics, $h = 2, 5, 8, 11, 14, \ldots$, produce a torque of $-(h+1-s)\,\omega$ opposite to that of rotation. Here s is the slip of the induction motor.

It is possible that harmonic torques are magnified due to certain combinations of stator and rotor slots and cage rotors are more prone to circulation of harmonic currents as compared with the wound rotors.

The zero sequence harmonics ($h = 3, 6, ...$) do not produce a net flux density. These produce ohmic losses.

All parasitic fields produce noise and vibrations. The harmonic fluxes superimposed upon the main flux may cause tooth saturation, and zigzag leakage can generate unbalanced magnetic pull, which moves around the rotor. As a result, the rotor shaft can deflect and run through a critical resonant speed amplifying the torque pulsations.

Torque ripples may exist at various frequencies. If the inverter is a six-step type, then a sixth harmonic torque ripple is created which would vary from 36 to 360 Hz when the motor is operated over the frequency range of 6–60 Hz. At low speeds such torque ripple may be apparent as observable oscillations of the shaft speed or as torque and speed pulsations, usually termed *cogging*. It is also possible that some speeds within the operating range may correspond to natural mechanical frequencies of the load or support structure. At such frequencies, amplification can occur, giving rise to large dynamic stresses. Operation other than momentary, i.e., during starting, should be avoided at these speeds.

The oscillating torques in synchronous generators can simulate the turbine generator into complex coupled mode of vibration that result in torsional oscillations of rotor elements and flexing of turbine buckets. If the frequency of a harmonic coincides with the turbine-generator torsional frequency, it can be amplified by the rotor oscillation.

A documented case of failure of a large generator is described in Reference [4]. A control loop within an SVC unit in a nearby steel mill resulted in a modulation of 60 Hz waveform. This created upper and lower sidebands, producing 55 and 65 Hz current components. The reverse phase rotation manifested itself as a 115 Hz stimulating frequency on the rotor, which drifted between 114 and 118 Hz. This excited sixth mode natural frequency of the rotor shaft, creating large torsional stresses.

The NGH-SSR (after Narain Hingorani Subsynchronous Resonance Suppressor) [5,6] scheme can minimize subsynchronous electrical torque and hence mechanical torque and shaft twisting, limit build up of oscillations due to subsynchronous resonance, and protect series capacitors from overvoltages.

6.1.2 Subharmonic Frequencies and Subsynchronous Resonance

We discussed subharmonic frequencies in Chapter 2 in conjunction with a series compensation of transmission lines. Generally, the transient currents excited by subharmonic resonant frequencies damp out quickly due to positive damping. This is a stable subharmonic mode. Under certain conditions, this can become unstable. We know that in a synchronous machine, the positive sequence subharmonic frequencies will set up a flux that rotate in the same direction as the rotor, and its slip frequency $= f_e - f$, where f_e is the frequency of the subharmonic. As f_e is $<f$, it is a negative slip and contributes to the negative damping. The synchronous machine can convert mechanical energy into electrical energy associated with subharmonic mode. If the negative damping is large, it can swap the positive resistance damping in the system and a small disturbance can result in large levels of currents and voltages. The subharmonic torque brought about by the difference frequency $f_e - f$ rotates in a backward direction with respect to the main field and if this frequency coincides with one of natural torsional frequencies of the machine rotating system, damaging torsional oscillations can be excited. This phenomenon is called subsynchronous resonance.

6.1.3 Increase of Losses

The effect of the harmonics on motor losses should consider the subdivision of losses into windage and friction, stator copper loss, core loss, rotor copper loss and stray loss in the core and conductors and the effect of harmonics on each of these components. The effective rotor and stator leakage inductance decreases and the resistance increases with frequency. The effects are similar as discussed for negative sequence in Volume 2, but these will be more pronounced at higher frequencies. The rotor resistance may increase four to six times the DC value while leakage reactance may reduce to a fraction of the fundamental frequency value. The stator copper loss increases in proportion to the square of the total harmonic current plus an additional increase due to higher resistance at skin effect at higher frequencies. Harmonics contribute to magnetic saturation and the effect of distorted voltage on core losses may not be significant. Major loss components influenced by harmonics are stator and rotor copper losses and stray losses. A harmonic factor of 11% gives approximately 25% derating of general purpose motors [6].

6.1.4 Effect of Negative Sequence Currents

Synchronous generators have both continuous and short-time unbalanced current capabilities, which are shown in Tables 6.1 and 6.2 [7,8]. These capabilities are based upon 120 Hz negative sequence currents induced in the rotor due to continuous unbalance or unbalance under fault conditions. In the absence of harmonics, unbalance loads and impedance asymmetries (i.e., nontransposition of transmission lines) require that the generators should be able to supply some unbalance currents. When these capabilities are exploited

TABLE 6.1

Requirements of Unbalanced Faults on Synchronous Machines

Type of Synchronous Machine	Permissible $I_2^2 t$
Salient pole generator	40
Synchronous condenser	30
Cylindrical rotor generators	
Indirectly cooled	30
Directly cooled (0–800 MVA)	10
Directly cooled (801–1600 MVA)	10–(0.00625) (MVA–800)

TABLE 6.2

Continuous Unbalance Current Capability of Generators

Type of Generator and Rating	Permissible I_2 (%)
Salient pole, with connected amortisseur windings	10
Salient pole, with nonconnected amortisseur windings	5
Cylindrical rotor, indirectly cooled	10
Cylindrical rotor, directly cooled to 960 MVA	8
961–1200 MVA	6
1201–1500 MVA	5

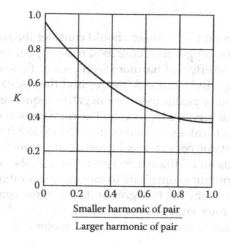

FIGURE 6.2
Ratio K for average loss to maximum loss based on harmonic pair. From M.D. Ross, J.W. Batchelor, 1943. Operation of non-salient-pole type generators supplying a rectifier load. *AIEE Trans, 62*, 667–670.

for harmonic loading, the variations in loss intensity at different harmonics versus 120 Hz should be considered. The following expression can be used for equivalent heating effects of harmonics translated into negative sequence currents:

$$I_{2\,eqiv}=\left[\left(\frac{6f}{120}\right)^{1/2}(K_{5,7})(I_5+I_7)^2+\left(\frac{12f}{120}\right)^{1/2}(K_{11,13})(I_{11}+I_{13})^2+\cdots\right]^{1/2} \quad (6.5)$$

where $K_{5,7}$, $K_{11,13}$, … are correction factors to convert from maximum rotor surface loss intensity to average loss intensity [9] (from Figure 6.2), f is the fundamental frequency, and I_5 and I_7 are the harmonic currents in pu values.

Example 6.1

Consider a synchronous generator, with continuous unbalance capability of 0.10 pu (Table 6.2). It is subjected to fifth and seventh harmonic loading of 0.07 and 0.04 pu, respectively. Is the unbalance capability exceeded?
From Figure 6.2 and harmonic ratio 0.04/0.07 = 0.57, $K_{5,7}$ = 0.43. From Equation 6.5

$$I_{2eqv}=\left[\sqrt{3}(0.43)(0.07+0.04)\right]^{1/2}=0.095$$

The continuous negative sequence capability is not exceeded. The example shows that the sum of harmonic currents in pu can exceed the generator continuous negative sequence capability, that is, in this case 11% which exceeds 10% capability. Thus, calculation with simple summation is inaccurate.

6.1.5 Insulation Stresses

The high-frequency operation of modern PWM converters with insulated gate bipolar transistors (IGBTs) is discussed in Chapter 1. It subjects the motors to high dv/dt. It has an adverse effect on the motor insulation and contributes to the motor bearing currents and shaft voltages. The rise time of the voltage pulse at the motor terminals influences the voltage stresses on the motor windings. As the rise time of the voltage becomes higher, the motor windings behave like a network of capacitive elements in series. The first coils of the phase windings are subject to overvoltages, as shown in Figure 6.3, which shows ringing. There has been a documented increase in the insulation failure rate caused by turn-to-turn shorts or phase-to-ground faults due to high dv/dt stress [10]. The common remedies are to provide inverter grade insulation or to add filters. The soft switching slows initial rate of rise.

NEMA [3] has established limitations on voltage rise for general-purpose NEMA design A and B induction motors and definite-purpose inverter-fed motors. Windings designed for definite-purpose inverter grade motors use magnet wires with increase build, and these polyester-based wires exhibit higher breakdown strength.

- The stator winding insulation system of general-purpose motors rated ≤ 600 V shall withstand $V_{peak} = 1$ kV and rise time ≥2 μs and for motors rated >600 V these limits are $V_{peak} \leq 2.5$ pu and rise time ≥1 μs.
- For definite-purpose inverter-fed motors with base voltage rating ≤600 V, $V_{peak} \leq$ 1600 V and the rise time is ≥0.1 μs. For motors with base voltage rating >600 V, $V_{peak} \leq 2.5$ pu and the rise time ≥0.1 μs. V_{peak} is of single amplitude and 1 pu is the peak of the line-to-ground voltage at the maximum operating speed point.

The derating due to harmonic factor, effect on motor torque, starting current, and power factor are described in NEMA [3].

The motor windings can be exposed to higher than normal voltages due to neutral shift and common mode voltages [11], and in some current source inverters, it can be as high as 3.3 times the crest of the nominal sinusoidal line-to-ground voltage.

Harmonics also impose higher dielectric stresses on the insulation of other electrical apparatuses. Harmonic overvoltages can lead to corona, void formation, and degradation.

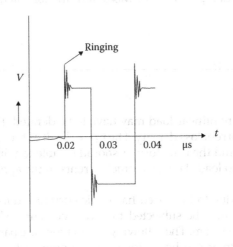

FIGURE 6.3
Pulse-width modulation, ringing in output voltage waveform due to traveling wave phenomena in interconnecting cables to adjustable speed drive motors.

6.1.6 Bearing Currents and Shaft Voltages

PWM inverters give rise to additional shaft voltages due to voltage and current spikes superimposed on the phase quantities during inverter operation. These will cause currents to flow through the bearings. If shaft voltages higher than 300 mV peak occur, the motor should be equipped with insulated bearings or shaft should be grounded.

6.1.7 Effect of Cable Type and Length

When the motor is connected through long cables, the high dv/dt pulses generated by PWM inverters cause traveling wave phenomena on the cables, resulting in reinforcement of incident and reflected waves due to impedance discontinuity at the motor terminals. And the voltages can reach twice the inverter output voltage. An analogy can be drawn with long transmission lines and traveling wave phenomena. The incident traveling wave is reflected at the motor terminals, and reinforcement of incident and reflected waves occurs. Due to dielectric losses and cable resistance, damped ringing occurs as the wave is reflected from one end of the cable to the other. The ringing frequency is a function of cable length and wave propagation velocity, and is of the order of 50 kHz–2 MHz [12]. Figure 6.3 is a representation of the ringing frequency. The type of cable between the motor and the drive system is important.

By adding output filters, the cable charging current as well dielectric stresses on the motor insulation can be reduced. The common filter types are as follows:

- Output line inductors
- Output limit filter
- Sine wave filter
- Motor termination filter

An output inductor reduces dv/dt at the inverter and the motor. The ringing and overshoot may also be reduced. Output limit filters may consist of laminated core inductor or ferrite core inductor. A sine wave filter is a conventional low-pass filter formed from an output inductor, a capacitor, and a damping resistor. Motor termination filters are first-order resistor/capacitor filters.

6.2 Transformers

A transformer supplying nonlinear load may have to be derated. Harmonics effect transformer losses and eddy current loss density. The upper limit of the current distortion factor is 5% of the load current and the transformer should be able to withstand 5% overvoltage at rated load and 10% at no load. The harmonic currents in the applied voltage should not exceed these limits.

In addition to derating due to increased harmonic current induced eddy current loss, a drive system transformer may be subjected to severe current cycling and load demand, depending upon the drive system. The following calculations are based upon Reference [13].

The losses in a transformer can be divided into (1) no-load losses and (2) load losses. The load losses consist of copper loss in windings and stray load losses. The stray load losses can be subdivided into the losses in the windings and losses in the nonwinding components of the transformer, that is, core clamps, structures, and tank. The total transformer loss P_{LL} is

$$P_{LL} = P + P_{EC} + P_{OSL} \tag{6.6}$$

where P is the I^2R loss, P_{EC} is the winding eddy current loss, and P_{OSL} is other stray loss. The other stray load loss will increase proportional to the square of the current. However, these will not increase proportional to the square of the frequency, as in the winding eddy losses. The studies show that eddy current loss in bus bars, connections, and structural parts increases by a harmonic exponential factor of 0.8 or less. The effects of these losses vary depending upon the type of transformer.

For the liquid-immersed transformers, the top oil rise θ_{TO} will increase as the total load losses increase due to harmonic loading. Unlike dry type transformers where P_{OSL} is ignored, it must be considered for oil-immersed transformers as it impacts the top oil temperature.

If the rms value of the current including harmonics is the same as the fundamental current, I^2R loss will be maintained the same. If the rms value due to harmonics increases, so does the I^2R loss.

$$I_{(pu)} = \left[\sum_{h=1}^{h=\max} (I_{h(pu)})^2 \right]^{1/2} \tag{6.7}$$

The eddy current loss P_{EC} is assumed to vary in proportion to the square of the electromagnetic field strength. Square of the harmonic current or the square of the harmonic number may be considered to be representative of it. Due to skin effect the electromagnetic flux may not penetrate conductors at high frequencies. The leakage flux has its maximum concentration between interface of the two windings:

$$P_{EC(pu)} = P_{EC-R(pu)} \sum_{h=1}^{h=\max} I_{h(pu)}^2 h^2 \tag{6.8}$$

where P_{EC-R} is the winding eddy current loss under rated conditions and I_h (pu) is the per unit rms current at harmonic h. To facilitate actual field measurements define winding eddy current loss at measured current and the power frequency by another term P_{EC-O}. Then we can write the equation

$$P_{EC} = P_{EC-O} \times \frac{\displaystyle\sum_{h=1}^{h=h_{\max}} I_h^2 h^2}{I^2} = P_{EC-O} \times \frac{\displaystyle\sum_{h=1}^{h=h_{\max}} I_h^2 h^2}{\displaystyle\sum_{h=1}^{h=h_{\max}} I_h^2} \tag{6.9}$$

where I is the rms load current. Define harmonic loss factor F_{HL} for the windings as P_{EC}/P_{EC-O} in Equation 6.9:

$$F_{HL} = \frac{\displaystyle\sum_{h=1}^{h=h_{\max}} \left[\frac{I_h}{I} \right]^2 h^2}{\displaystyle\sum_{h=1}^{h=h_{\max}} \left[\frac{I_h}{I} \right]^2} \tag{6.10}$$

In the above equation, I can be substituted with I_1, where I_1 is the rms fundamental load current. It can be shown that *whether we normalize with respect to I or I_1, the F_{HL} calculation gives the same results.*

The heating due to other stray losses is not a consideration for dry-type transformers, since the heat generated is dissipated by the cooling air. With $P_{OSL} = 0$, all the stray loss is assumed to occur in the windings. P_{LL} can be written as

$$P_{LL(pu)} = \sum_{h=1}^{h=max} I_{h(pu)}^2 \times \left(1 + F_{HL} P_{EC-R(pu)}\right) pu \tag{6.11}$$

To adjust the per unit loss density in the individual windings, the effect of F_{HL} must be known on each winding. The per unit value of the nonsinusoidal current for the dry-type transformers which will make the result of Equation 6.11 equal to the design value of the loss density in the highest loss region for rated frequency and for rated current is given by the following equation:

$$I_{max(pu)} = \left[\frac{P_{LL-R(pu)}}{1 + \left[F_{HL} \times P_{EC-R(pu)}\right]}\right]^{1/2} \tag{6.12}$$

6.2.1 Calculations from Transformer Test Data

The calculations of P_{EC-R} can be made from the transformer test data. The maximum eddy current loss density is assumed 400% of the average value for that winding. The division of eddy current loss between the windings is as follows:

- 60% in inner winding and 40% in outer winding in all transformers having a self-cooled rating of <1000 A regardless of turns ratio.
- 60% in inner winding and 40% in outer winding in all transformers having a turns ratio of 4:1 or less.
- 70% in inner winding and 30% in outer winding in all transformers having a turns ratio >4:1 and also having one or more windings with a maximum self-cooled rating of >1000 A.
- The eddy current loss distribution within each winding is assumed to be nonuniform.

The stray loss component of the load loss is calculated by the following expression:

$$P_{TSL-R} = \text{Total load loss} - \text{Copper loss}$$
$$= P_{LL} - K\left[I_{(1-R)}^2 R_1 + I_{(2-R)}^2 R_2\right] \tag{6.13}$$

In this expression for copper loss, R_1 and R_2 are the resistances measured at the winding terminals (i.e., H1 and H2 or X1 and X2) and should not be confused with winding resistances of each phase; $K = 1$ for single-phase transformers and 1.5 for three-phase transformers.

Sixty-seven percent of the stray loss is assumed to be winding eddy losses for dry-type transformers:

$$P_{EC-R} = 0.67 P_{TSL-R} \tag{6.14}$$

33% of the total stray loss is assumed to be winding eddy losses for liquid-immersed transformers:

$$P_{EC-R} = 0.33 P_{TSL-R} \tag{6.15}$$

The other stray losses are given by

$$P_{OSL-R} = P_{TSL-R} - P_{EC-R} \tag{6.16}$$

As the low-voltage winding is the inner winding, maximum P_{EC-R} is given by

$$\max P_{EC-R(pu)} = \frac{K_1 P_{EC-R}}{K \left(I_{(2-R)}\right)^2 R_2} \text{ pu} \tag{6.17}$$

where K_1 is the division of eddy current loss in the inner winding, which is equal to 0.6 or 0.7 multiplied by the maximum eddy current loss density of 4.0 per unit, i.e., 2.4 or 2.8, depending upon the transformer turns ratio and the current rating and K has already been defined depending the number of phases.

Example 6.2

A delta–wye connected dry-type isolation transformer of 23.9–2.4 kV is required for a 2.3 kV 3000 hp drive motor connected to a load commutated inverter with the following current spectrum:

$I = 709$ A, $I_5 = 123$ A, $I_7 = 79$ A, $I_{11} = 31$ A, $I_{13} = 20$ A, $I_{17} = 11$ A, $I_{19} = 7$ A, $I_{23} = 6$ A, $I_{25} = 5$ A

Calculate the harmonic loading. The following data are supplied by the manufacturer: $R_1 = 1.052$ Ω, $R_2 = 0.0159$ Ω, and the total load loss = 23,200 W at 75°C. Calculate the derating when supplying the same harmonic current spectrum.

The primary 23.9 kV winding current is 71.9 A. The copper loss in the windings is

$$1.5[(1.052)(71.19)^2 + (0.0159)(709)^2] = 19986.3 \text{ W}$$

From the given loss data

$$P_{TSL-R} = 23,200 - 19,986.3 = 3213.7 \text{ W}$$

The winding eddy loss is then

$$P_{EC-R} = 3213.7 \times 0.67 = 2153.2 \text{ W}$$

The transformer has a ratio of >4:1, but secondary current is <1000 A. Then, from Equation 6.17 the maximum $P_{EC-Rmax}$ is

$$P_{EC-Rmax} = \frac{2.4 \times 2153.2}{1.5(0.0159)(709)^2} = 0.431 \text{ pu}$$

Table 6.3 is constructed, based upon the transformer fundamental current of 709 A as the base current; the steps of calculation are obvious. The first five columns of this table apply to the calculations in this example. Then $F_{HL} = 2.7964/1.04 = 2.69$

$$P_{LL(pu)} = 1.04 \times (1 + 0.43 \times 2.69) = 2.24$$

TABLE 6.3

Calculations of Derating of a Transformer due to Harmonic Loads

h	I_h/I	$(I_h/I)^2$	h^2	$(I_h/I)^2 h^2$	$h^{0.8}$	$(I_h/I)^2 h^{0.8}$
1	2	3	4	5	6	7
1	1.0	1.0	1.0	1.0	1.0	1.0
5	0.171	0.029	25	0.725	3.62	0.105
7	0.111	0.012	49	0.588	4.74	0.057
11	0.043	0.0018	121	0.2178	6.81	0.012
13	0.028	0.0006	169	0.1014	7.78	0.005
17	0.015	0.00022	289	0.0636	9.65	0.002
19	0.010	0.00010	361	0.0361	10.54	0.001
23	0.008	0.000064	529	0.0339	12.28	0.0008
25	0.007	0.000049	625	0.0306	13.13	0.0006
3		1.04		2.7964		1.183

Then from Equation 6.12, the maximum permissible value of nonsinusoidal current is

$$I_{\max(pu)} = \sqrt{\frac{1.43}{1+2.69 \times 0.43}} = 0.81 \text{ pu}$$

Thus, the transformer capability with given nonsinusoidal load current harmonic composition is approximately 81% of its sinusoidal load current capability, which is equal to 574.3 A. To properly apply a transformer for this application raises the transformer kVA by approximately 24%.

6.2.2 Liquid-Filled Transformers

The liquid-filled transformers are similar to dry-type transformers except that effects of all stray losses are considered. For self-cooled OA (ONAN) mode, the top oil temperature rise above ambient is given by

$$\theta_{TO} = \theta_{TO\text{-}R} \left[\frac{P_{LL} + P_{NL}}{P_{LL\text{-}R} + P_{NL}} \right]^{0.8} \text{°C} \tag{6.18}$$

where $\theta_{TO\text{-}R}$ is the top oil temperature rise over ambient under rated conditions. P_{NL} is the no-load loss and P_{LL} is the load loss under rated conditions Also

$$P_{LL} = P + F_{HL}P_{EC} + F_{HL\text{-}STR}P_{OSL} \tag{6.19}$$

where $F_{HL\text{-}STR}$ is the harmonic loss factor for the other stray loss and P is the I^2R portion of the load loss.

The winding hottest spot conductor rise is given by

$$\theta_g = \theta_{g\text{-}R} \left(\frac{1 + F_{HL}P_{EC\text{-}R(pu)}}{1 + P_{EC\text{-}R(pu)}} \right)^{0.8} \tag{6.20}$$

where θ_g and $\theta_{g\text{-}R}$ are the hottest spot conductor rise over top oil under harmonic loading and under rated conditions, respectively.

For liquid-immersed transformers, Equation 6.17 becomes

$$\theta_{g1} = \theta_{g1-R}\left(\frac{1+2.4F_{HL}P_{EC-R(pu)}}{1+2.4P_{EC-R(pu)}}\right)^{0.8} \,°C \qquad (6.21)$$

or

$$\theta_{g1} = \theta_{g1-R}\left(\frac{1+2.8F_{HL}P_{EC-R(pu)}}{1+2.8P_{EC-R(pu)}}\right)^{0.8} \,°C \qquad (6.22)$$

where θ_{g1} is the hottest spot HV conductor rise over top oil temperature, and θ_{g1-R} is the hottest spot HV conductor rise over top oil under rated conditions.

Example 6.3

Consider that the transformer in Example 6.2 is liquid immersed, ONAN, self-cooled. The top oil temperature rise and winding hottest spot conductor rise are required to be calculated. In addition to the data in Example 6.2, the no-load loss of the transformer is 5000 W.

From Example 6.2:

$$P_{OSL-R} = P_{TSL-R} - P_{EC-R} = 3213.7 - 2153.2 = 1060.5 \text{ W}$$

Assume the following temperature rises, which are normal for 55°C rated transformers:

HV and LV windings average = 55°C
Top oil rise = 55°C
Hottest spot conductor rise = 65°C

The harmonic loss factor is 2.69 as calculated in Example 6.2. Columns 6 and 7 are added to Table 6.3, and the harmonic loss factor for the other stray loss is the summation of Column 7 divided by the summation of column 3, i.e., 1.183/1.04 = 1.1375. Assuming that harmonic load distribution is equal to 100% of the fundamental current, the calculated losses are

No load = 5000 W
I^2R loss = 19986.3 W
Winding Eddy = 2153.2 W
Other stray = 1060.5 W

The total gives a rated loss of 28,200 W. This checks with the given loss data by the manufacturer = 5000 + 23,200 = 28,200 W.

From Table 6.3, the rms current is given by the square root of third column summation, that is, $\sqrt{1.04} = 1.02$. The total losses must be corrected to reflect rms current and also load factor. Here assume that the load factor is 100%. Therefore, correction for the rms current is

$$P_{LL} = I^2_{rms-pu}L^2_f = 1.02^2 \times 1^2 = 1.04$$

Correct the losses using harmonic multipliers calculated above:

Winding eddy = $1.04 \times 2.69 \times 2153.2 = 6023.8$ W
Other stray loss = $1.04 \times 1.1375 \times 1060.5 = 1254.6$ W

I^2R loss $= 1.04 \times 19986.4 = 20785.9$ W
Thus, the total losses are now $= 33064$ W

Then the top oil temperature from Equation 6.18 is

$$\theta_{TO} = 55 \times \left(\frac{33064}{28200}\right)^{0.8} = 62.47°C$$

The rated inner winding losses corrected for rms current are

$$1.04 \times 1.5(0.0159)(709)^2 = 12468 \text{ W}$$

The low-voltage winding currents are less than 1000 A. 60% of the winding eddy loss is assumed to occur in the LV winding, and the maximum eddy loss at the hottest region is assumed to be four times the average eddy loss. The hottest spot conductor rise over top oil temperature is calculated using Equation 6.21:

$$\theta_g = (65-55) \times \left(\frac{12468 + 6023.8 \times 2.4}{11988 + 2153 \times 2.4}\right)^{0.8} = 14.34°C$$

Then the hottest spot temperature is $62.47 + 14.34 = 76.81°C$.

This exceeds the maximum of 65°C by 11.81°C. The transformer on continuous operation will be seriously damaged. This demonstrates that liquid-filled or air-cooled transformers cannot be loaded to the rated fundamental frequency rating when supplying nonlinear loads, and the derating must be calculated in each case.

6.2.3 Underwriter's Laboratories *K*-Factor of Transformers

The Underwriter's Laboratories (UL) standards [14, 15] also specify transformer derating (*K*-factors) when carrying nonsinusoidal loads. The UL *K*-factor is given by the expression

$$K\text{-factor} = \sum_{h=1}^{h=h_{max}} I_{h\text{-pu}}^2 h^2 \tag{6.23}$$

This can be shown to be equal to

$$K\text{-factor} = \left(\frac{\displaystyle\sum_{h=1}^{h=max} I_h^2}{I_R^2}\right) F_{HL} \tag{6.24}$$

The harmonic loss factor is a function of the harmonic current distribution and independent of its relative magnitude. The UL *K*-factor is dependent upon both the magnitude and distribution of the harmonic current. For a transformer with harmonic currents specified as per unit of the rated transformer secondary currents, the *K*-factor and harmonic loss factor will have the same numerical values. That is, the *K*-factor is equal to harmonic factor only when square root of the sum of harmonic currents squared equals the rated secondary current of the transformer.

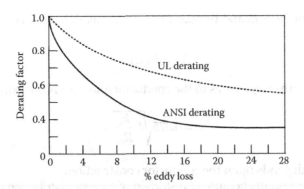

FIGURE 6.4
Relative derating of transformers: ANSI method and UL *K*-factor.

There is a marked difference between the derating in ANSI/IEEE recommendations and UL *K*-factor. If a transformer has 3% eddy current loss and a *K*-factor of 5, then the eddy current loss increases to 3×5 = 15%. The UL derating ignores eddy current loss gradient. Figure 6.4 shows higher derating with the ANSI method of calculation for a six-pulse harmonic current spectrum, assuming a theoretical magnitude of harmonics of 1/*h*, for a *K*-factor load of approximately 9.

Example 6.4

Calculate the UL *K*-factor for the transformer loading harmonic spectrum of Example 6.2.
From Table 6.3, the UL *K*-factor = 2.7964×1.04 = 2.91. If the transformer was rated in terms of rms current, then the *K*-factor = 2.7964.

6.3 Cables

A nonsinusoidal current in a conductor causes additional losses. The AC conductor resistance is changed due to skin and proximity effects. Both these effects are dependent upon frequency, conductor size, cable construction, and spacing. We observed that even at 60 Hz the AC resistance of conductors is higher than the DC resistance (Volume 1). With harmonic currents, these effects are more pronounced. The AC resistance is given by

$$R_{ac}/R_{dc} = 1 + Y_{cs} + Y_{cp} \tag{6.25}$$

where Y_{cs} is due to conductor resistance resulting from skin effect and Y_{cp} is due to proximity effect. The skin effect is an AC phenomenon, where the current density throughout the conductor cross section is not uniform and the current tends to flow more densely near the outer surface of the conductor than toward the center. This is because an AC flux results in induced EMFs which are greater at the center than at the circumference, so that potential difference tends to establish currents that oppose the main current at the center and assist it at the circumference. The result is that the current is forced to the outside, reducing the effective area of the conductor. The effect is utilized in high ampacity hollow conductors and tubular bus bars to save material costs. The skin effect is given by [16]

$$Y_{cs} = F(x_s) \tag{6.26}$$

where Y_{cs} is due to skin effect losses in the conductor and $F(x_s)$ is the skin effect function.

$$x_s = 0.875 \sqrt{f \frac{k_s}{R_{dc}}} \tag{6.27}$$

where the factor k_s depends upon the conductor construction.

The proximity effect occurs because of distortion of current distribution between two conductors in close proximity. This causes concentration of current in parts of the conductors or bus bars closest to each other (currents flowing in forward and return paths). The expressions and graphs for calculating the proximity effect are given in Reference [16]. The increased resistance of conductors due to harmonic currents can be calculated, and the derated capacity is

$$\frac{1}{1 + \sum_{h=1}^{h=\max} I_h^2 R_h} \tag{6.28}$$

where I_h is the harmonic current and R_h is the ratio of the resistance at that harmonic with respect to the conductor DC resistance.

For a typical six-pulse harmonic spectrum the derating is approximately 3%–6%, depending upon the cable size (Figure 6.5). See also Reference [17].

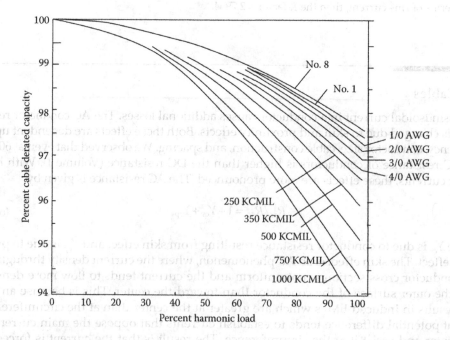

FIGURE 6.5
Derating of cables with typical six-pulse converter circuit.

6.4 Capacitors

The main effect of harmonics on capacitors is that a resonance condition can occur with one of the load-generated harmonics. Capacitors may be located in an industrial plant or close to such a plant which has significant harmonic-producing loads. This location is very possibly subject to harmonic resonance and should be avoided, or the capacitors can be used as harmonic filters. On a subtransmission system where the capacitors are located far from the harmonic-producing loads, the propagation of harmonics through the interconnecting systems needs to be studied. The linear loads served from a common feeder, which also serves nonlinear loads of some other consumers, may become susceptible to harmonic distortion. There have been two approaches: one to consider capacitor placement from a reactive power consideration and then study the harmonic effects, and second to study the fundamental frequency voltages, reactive power, and harmonic effects simultaneously. A consumer system that does not have harmonic-producing loads can be subject to harmonic pollution due to harmonic loads of other consumers in the system.

The capacitors can be severely overloaded due to harmonics, especially under resonant conditions, and can even be damaged. The capacitors are intended to be operated at or below their rated voltage and frequency. The capacitors are capable of continuous operation under contingency system and capacitor bank conditions provided that none of the following is exceeded [18,19].

110% of the rated rms voltage. If harmonics are present, this means

$$V_{rms} \le 1.1 = \left[\sum_{h=1}^{h=h_{max}} V_h^2 \right]^{1/2} \tag{6.29}$$

The crest not to exceed $1.2 \times \sqrt{2}$ times the rated rms voltage, including harmonics but excluding transients:

$$V_{crest} \le 1.2 \times \sqrt{2} \sum_{h=1}^{h=h_{max}} V_h \tag{6.30}$$

The rms current including harmonics should not exceed 135% of nominal rms current based upon rated kvar and rated voltage, including fundamental and harmonic currents:

$$I_{rms} \le 1.35 = \left[\sum_{h=1}^{h=h_{max}} I_h^2 \right]^{1/2} \tag{6.31}$$

135% of rated kvar, i.e., if harmonics are present:

$$kvar_{pu} \le 1.35 = \left[\sum_{h=1}^{h=h_{max}} (V_h I_h) \right] \tag{6.32}$$

Unbalances within a capacitor bank, due to capacitor element failures and/or individual fuse operations, result in overvoltages on some capacitor units. Typically individual capacitor voltage is allowed to increase by 10% before unbalance detection removes capacitor banks from service. In a filter bank failure of a capacitor unit will cause detuning. The limitations of the capacitor bank loadings become of importance in the design of capacitor filters.

A capacitor tested according to IEEE Std. 18 will withstand a combined total of 300 applications of power frequency terminal-to-terminal overvoltages without superimposed transients or harmonic content; the magnitude and duration are shown in Figure 6.6a, transient peak overvoltage capability of capacitor banks is shown in Figure 6.6b [19].

The capacitor unit is also expected to withstand transient currents inherent in the operation of power systems, which include infrequent high lightning currents and discharge

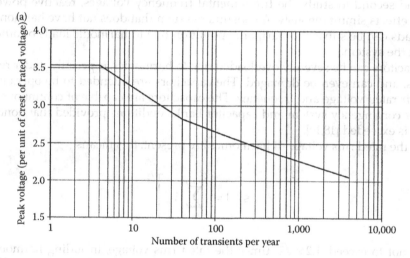

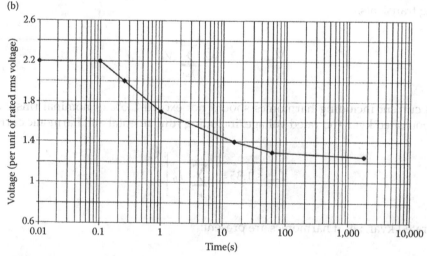

FIGURE 6.6
(a) Maximum contingency power frequency overvoltage capability of capacitor units and (b) transient peak overvoltage capability [18].

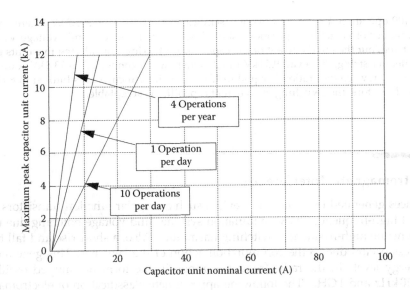

FIGURE 6.7
Transient current capabilities of capacitor units for regularly occurring transients.

currents due to nearby faults, see Chapter 4. For frequent back-to-back switching, the peak capacitor unit current should be held to a value lower than that shown in Figure 6.7 [19]. The capacitor bank current is the number of capacitor unit current multiplied by the number of capacitor units or strings in parallel. In this figure, the curves are based upon straight lines from origin to 12 kA, and the slopes are 1500 times the nominal current for four operations per year, 800 times the nominal current for one operation per day, and 400 times the nominal current for 10 operations per day.

Example 6.5

A capacitor bank is to be constructed for operation at 13.8 kV nominal voltage using 400 kvar capacitor units. Select the rated voltage of a unit capacitor to form a wye-ungrounded capacitor bank, based on the following:

- The operating voltage can be 5% higher due to load regulation.
- The phase unbalance system detection is arranged to shutdown the capacitor bank in case the voltage rises by 10%.
- For control of the power factor 400 switching operations per year are required, the switching transient overvoltage is calculated as 25 kV peak.
- The transient inrush current for back-to-back switching is limited to 6 kA.

As the capacitor units are to be connected in wye-ungrounded configuration, the rated line-to-neutral voltage is 7967.7 V. Although the power capacitors have the continuous overvoltage capability of 10%, and this capability can be utilized to keep the bank in service if one of the parallel fuses operates, it is desirable to reserve this margin for conservatism and for deterioration over a period of time. Adding 15% voltage margin, capacitor-unit voltage becomes 9162.9 V.

Although we calculated a voltage rating of 9162.9 V, the next standard capacitor unit of 9.54 kV can be selected (see Chapter 7 for standard capacitor unit ratings).

For 400 operation, $k = 2.4$, and using 9.54 kV capacitor unit, $V_{tr} = 32.37$ kV. Therefore, the number of switching operations should be reduced. The dynamic voltage V_d is not stated, but the rated current of the capacitor unit string, considering five units in parallel per string, 400 kvar, 9.54 kV rated voltage when connected to 13.8 kV source voltage in wye configuration is equal to 174 A. The inrush current is limited to 6 kA peak. Therefore, the switching overcurrent capability is acceptable.

6.5 Electromagnetic Interference

Disturbances generated by switching devices such as bipolar junction transistors (BJTs), IGBTs, and high-frequency PWM modulation systems, and voltage notching due to converters, generate high-frequency switching harmonics. Also, a short rise and fall time of 0.5 μs or less occurs due to the commutation action of the switches. This generates sufficient energy levels in the radio-frequency range in the form of damped oscillations between 10 kHz and 1 GHz. The following approximate classification of electromagnetic disturbance by frequency can be defined:

Below 60 Hz: subharmonic

60 Hz–2 kHz: harmonics

16–20 kHz: acoustic noise

20–150 kHz: range between acoustic and radio-frequency disturbance

150 kHZ–30 MHz: conducted radio frequency disturbance

30 MHz–1 GHz: radiated disturbance

The high-frequency disturbances are referred to as electromagnetic interference (EMI). The radiated form is propagated in free space as electromagnetic waves, and the conducted form is transmitted through the power lines, especially at distribution level. Conducted EMI is much higher than radiated noise. Two modes of conducted EMI are recognized, symmetrical or differential mode and asymmetrical or common mode. The symmetrical mode occurs between two conductors that form a conventional return circuit and common mode propagation takes place between a group of conductors and ground or other group of conductors.

The noise at frequencies around a few kilohertz can interfere with audiovisual equipment and electronic clocks. Shielding and proper filtering are the preventive measures. Lightning surges, arcing-type faults, and operation of circuit breakers and fuses also produce EMI.

6.6 Overloading of Neutral

The line current of switched mode power supplies flows in pulses (Chapter 1). Also, the pulse burst modulation technique (see Chapter 1) gives rise to neutral currents. At low level of current, the pulses are nonoverlapping in a three-phase system, i.e., only one phase

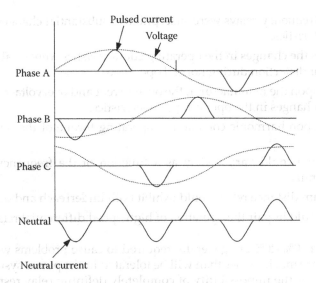

FIGURE 6.8
Summation of neutral current in a three-phase four-wire system.

of a three-phase system carries current at any one time. The only return path is through the neutral and as such the neutral can carry the summed currents of the three phases. Its rms value is therefore 173% of the line current, see Figure 6.8. As the load increases, the pulses in the neutral overlap and the neutral current as a percentage of the line current reduces. The third harmonic is the major contributor to the neutral current; other triplen harmonics have insignificant contributions. A minimum of 33% third harmonic is required to produce 100% neutral current in a balanced wye system.

The nonlinear loads may not be perfectly balanced and single-phase unbalanced loads can create higher unbalance currents in the neutral. The National Electric Code (NEC), published by NFPA-National Fire Protection Association [20], recommends that, where the major portion of loads consists of nonlinear loads, the neutral shall be considered as a current-carrying conductor. Normally, a reduced neutral cross section compared to phase conductors is specified in NEC. In some installations, the neutral current may exceed the maximum phase current. The single-phase branch circuits can be run with a separate neutral for each phase rather than using a multiwire branch circuit with a shared neutral.

6.7 Protective Relays and Meters

Harmonics may result in possible relay misoperation. Relays that depend upon crest voltage and/or current or voltage zeros are affected by harmonic distortion on the wave. The excessive third harmonic zero sequence current can cause ground relays to false trip. A Canadian study [1] documents the following effects:

1. Relays exhibited a tendency to operate slower and/or with higher pickup values rather than to operate faster and/or with lower pickup values.

2. Static underfrequency relays were susceptible to substantial changes in the operating characteristics.

3. In most cases the changes in the operating characteristics were small over a moderate range of distortion during normal operation.

4. Depending upon the manufacturer, the overcurrent and overvoltage relays exhibited various changes in the operating characteristics.

5. Depending upon harmonic content, the operating torque of the relays could be reversed.

6. Operating time could vary widely as a function of the frequency mix in the metered quantity.

7. Balanced beam distance relays could exhibit both underreach and overreach.

8. Harmonics could impair the operation of high-speed differential relays.

Harmonic levels of 10%–20% are generally required to cause problems with relay operation. These levels are much higher than will be tolerated in the power systems.

Reference [1] states the impossibility of completely defining relay responses because of the variety of relays in use and the variations in the nature of distortions that can occur, even if the discussion is limited to 6-pulse and 12-pulse converters. Not only the harmonic magnitudes and predominant harmonic orders vary, but relative phase angles can also vary. The relays will respond differently to two wave shapes which have the same characteristics magnitudes but different phase angles relative to the fundamental.

Metering and instrumentation is affected by harmonics. Close to resonance, higher harmonic voltages may cause appreciable errors. A 20% fifth harmonic content can produce a 10%–15% error in a two-element three-phase watt transducer. The error due to harmonics can be positive, negative, or smaller with third harmonics, depending upon the type of the meter. The presence of harmonics reduces the reading on power factor meters.

Modern MMPRs use filters for current and voltage waveforms. These may utilize various measuring techniques—digital sampling, digital filtering, asynchronous sampling, and rms measurements. A microprocessor relay using a digital filter is immune to the effect of harmonics, because it extracts the fundamental from the waveform. A relevant observation is that the electromechanical relays are no longer being applied in the industry. These have many limitations from the protective relaying consideration, for example, fixed time–current characteristics and high maintenance and calibration costs and downtime because of moving parts. Contrary to that, microprocessor-based relays have programmable characteristics, fault diagnostics, communication, and metering functions (Volume 4).

6.8 Circuit Breakers and Fuses

Harmonic components can affect the current interruption capability of circuit breakers. The high di/dt at current zero can make interruption process more difficult. Figure 1.18 for a six-pulse converter shows that the current zero is extended and di/dt at current zero

TABLE 6.4

Reduction in Current-Carrying Capability (%) of
Molded Case Circuit Breakers, 40°C Temperature Rise

Breaker Size	Sinusoidal Current	
	300 Hz	420 Hz
70 A	9	11
225 A	14	20

Note: Based upon published data of one manufacturer.

is very high. There are no definite standards in the industry for derating. One method is to arrive at the maximum di/dt of the breaker, based upon interrupting rating at fundamental frequency, and then translate it into maximum harmonic levels assuming that the harmonic in question is in phase with the fundamental [1].

A typical reduction in current-carrying capacity of molded case circuit breakers, when in use at high-frequency sine wave currents, is shown in Table 6.4. The effect on larger breakers can be more severe, since the phenomenon is also related to skin effect and proximity effect.

Harmonics will reduce the current-carrying capacity of fuses. Also, the time–current characteristics can be altered and the melting time will change. Harmonics also affect the interrupting rating of the fuses. Excessive transient overvoltages from current-limiting fuses, forcing current to zero before a natural zero crossing, may be generated. This can cause surge arrester operation and capacitor failures.

6.9 Telephone Interference Factor

Harmonic currents and voltages can produce electric and magnetic fields that will impair the performance of communication systems. Due to proximity there will be inductive coupling with the communication systems. Relative *weights* have been established by tests for the various harmonic frequencies that indicate disturbance to voice frequency communication. This is based on disturbance produced by injection of a signal of the harmonic frequency relative to that produced by a 1 kHz signal similarly injected. The TIF weighting factor is a combination of C-message weighting characteristics, which account for relative interference effects of various frequencies in the voice band, and a capacitor which produces weighting that is directly proportional to the frequency to account for an assumed coupling function [21]. It is dimensionless quantity that is indicative of the waveform and not the amplitude. It is given by

$$\text{TIF} = \frac{\sqrt{\sum \left(X_f W_f \right)^2}}{X_t} \tag{6.33}$$

$$\text{TIF} = \sqrt{\sum \left[\frac{\left(X_f W_f \right)}{X_t} \right]^2} \tag{6.34}$$

where X_t is the total rms voltage or current, X_f is the single-frequency rms current or voltage at frequency f, and W_f is the single-frequency weighting at frequency f.

The TIF weighting function which reflects C-message weighting and coupling normalized to 1 kHz is given by

$$W_f = 5P_f f \tag{6.35}$$

where P_f is the C-message weighting at frequency f, and f is the frequency under consideration.

The single-frequency TIF values are listed in Table 6.5. The weighting is high in the frequency range 2–3.5 kHz in which human hearing is most sensitive.

Telephone interference is often expressed as a product of current and TIF, that is, IT product where I is the rms current in ampères. Alternatively, it is expressed as a product of voltage and TIF weighting, where the voltage is in kV, i.e., kV-T product.

Table 6.6 [21] gives balanced IT guidelines for converter installations. The values are for circuits with an exposure between overhead systems, both power and telephone. Within an industrial plant or commercial building, the interference between power cables and twisted pair telephone cables is low, and the interference is not normally encountered. Telephone circuits are particularly susceptible to the influence of ground return currents. The TIF weighting values are also shown in Figure 6.9 [21].

The effect of harmonics can be felt at a distance from their point of generation. Sometimes it alludes the intuition, till a rigorous study is conducted. Reference [18] details some abnormal conditions of harmonic problems. These are natural resonance of transmission lines, over excitation of transformers, and harmonic resonance in the zero sequence circuits. See Chapter 8.

TABLE 6.5

1960 Single-Frequency TIF Values

FREQ	TIF	FREQ	TIF	FREQ	TIF	FREQ	TIF
60	0.5	1020	5100	1860	7820	3000	9670
180	30	1080	5400	1980	8330	3180	8740
300	225	1140	5630	2100	8830	3300	8090
360	400	1260	6050	2160	9080	3540	6730
420	650	1380	6370	2220	9330	3660	6130
540	1320	1440	6560	2340	9840	3900	4400
660	2260	1500	6680	2460	10340	4020	3700
720	2760	1620	6970	2580	10600	4260	2750
780	3360	1740	7320	2820	10210	4380	2190
900	4350	1800	7570	2940	9820	5000	840
1000	5000						

TABLE 6.6

Balanced IT Guidelines for Converter Installations Tie Lines

Category	Description	IT
I	Levels unlikely to cause interference	up to 10,000
II	Levels that might cause interference	10,000–50,000
III	Levels that probably will cause interference	>5,0000

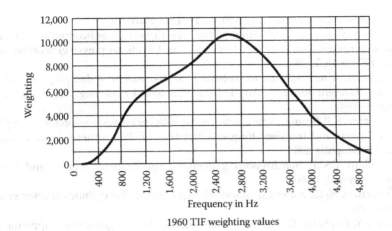

FIGURE 6.9
1960 TIF weighting values (IEEE Standard 519, IEEE recommended practice and requirements for harmonic control in electrical systems © 1992 IEEE.)

Problems

6.1 A synchronous generator has an $I_2^2 t$ of 30. Calculate the time duration for which the generator can tolerate the typical harmonic current spectrum of six-pulse converter given in Table 1.4 (Chapter 1), as a percentage of its rated full-load current, which is equal to the fundamental frequency rms current.

6.2 Calculate the derating of a 5 MVA dry type, 13.8–4.16 kV transformer, when subjected to typical harmonic spectrum of Table 1.4, as per unit of its rated current. The transformer data are $R_1 = 0.2747\ \Omega$, $R_2 = 0.02698\ \Omega$, and total load loss = 45,000 W.

6.3 In Problem 6.2, the transformer is an oil-immersed transformer. Assume the temperatures as in Example 6.3. The no-load loss is 6000 W. The transformer supplies its rated fundamental frequency current and the harmonic spectrum is as in Problem 6.2. Calculate the top oil temperature and hot-spot temperature.

6.4 In Problem 6.2, the transformer is loaded to 75% of its rating, and the spectrum currents are based upon 75% of the fundamental current. Calculate the top oil temperature and hot-spot temperature.

6.5 Calculate UL *K*-factor in Problem 6.2.

References

1. IEEE. A report prepared by Load Characteristics Task Force. The effects of power system harmonics on power system equipment and loads. *IEEE Trans Power Appar Syst*, 104, 2555–2561, 1985.
2. J.F. Witte, F.P. DeCesaro, S.R. Mendis. Damaging long term overvoltages on industrial capacitor banks due to transformer energization inrush currents. *IEEE Trans Ind Appl*, 30(4), 1107–1115, 1994.

3. NEMA. Motors and Generators, Parts 30, 31, 1993 Standard MG-1.

4. IEEE. Working Group J5 of Rotating Machinery Protection Subcommittee, Power System Relaying Committee. The impact of large steel mill loads on power generating units. *IEEE Trans Power Deliv*, 15, 24–30, 2000.

5. N.G. Hingorani. A new scheme for subsynchronous resonance damping of torsional oscillations and transient torques—Part I. *IEEE PES Summer Meeting*, Paper no. 80 SM687-4, Minneapolis, 1980.

6. N.G. Hingorani, K.P. Stump. A new scheme for subsynchronous resonance damping of torsional oscillations and transient torques—Part II. *IEEE PES Summer Meeting*, Paper no. 80 SM688-2, Minneapolis, 1980.

7. ANSI Standard C50.1. Synchronous generators, synchronous motors and synchronous machines in general, 1995.

8. ANSI Standard C50.13. American standard requirements for cylindrical rotor synchronous generators, 1965.

9. M.D. Ross, J.W. Batchelor. Operation of non-salient-pole type generators supplying a rectifier load. *AIEE Trans*, 62, 667–670, 1943.

10. A.H. Bonnett. Available insulation systems for PWM inverter fed motors. *IEEE Ind Appl Mag*, 4, 15–26, 1998.

11. J.C. Das, R.H. Osman. Grounding of AC and DC low-voltage and medium-voltage drive systems. *IEEE Trans Ind Appl*, 34, 205–216, 1998.

12. J.M. Bentley, P.J. Link. Evaluation of motor power cables for PWM AC drives. *IEEE Trans Ind Appl*, 33, 342–358, 1997.

13. IEEE Standard C57-110. IEEE recommended practice for establishing liquid-filled and dry-type power and distribution transformer capability when supplying nonsinusoidal load currents, 2008.

14. UL Standard 1561. Dry-type general purpose and power transformers, 1994.

15. UL Standard 1562. Transformers distribution, dry-type over 600V, 1994.

16. J.H. Neher, MH McGrath. The calculation of the temperature rise and load capability of cable systems. *AIEE Trans*, 76, 752–772, 1957.

17. A. Harnandani. Calculations of cable amapcities including the effects of harmonics. *IEEE Ind Appl Mag*, 4, 42–51, 1998.

18. IEEE Standard 18. IEEE standard for shunt power capacitors, 2002.

19. IEEE P1036/D13a. Draft guide for the application of shunt power capacitors, 2006.

20. NFPA. National Electric Code 2017. NFPA 70.

21. IEEE Standard 519. IEEE recommended practice and requirements for harmonic control in electrical systems, 1992.

7

Harmonic Penetrations (Propagation)

The harmonics from the point of generation will penetrate into the power system. The purpose of harmonic analysis is to ascertain the distribution of harmonic currents, voltages, and harmonic distortion indices in a power system. This analysis is then applied to the study of resonant conditions and harmonic filter designs and also to study other effects of harmonics on the power system, i.e., notching and ringing, neutral currents, saturation of transformers, and overloading of system components. On a simplistic basis, harmonic simulation is much like a load flow simulation. The impedance data from a short-circuit study can be used and modified for the higher frequency effects. In addition to models of loads, transformers, generators, motors, etc., the models for harmonic injection sources, arc furnaces, converters, SVCs, etc. are included. These are not limited to characteristic harmonics and a full spectrum of load harmonics can be modeled. These harmonic current injections will be at different locations in a power system. As a first step, a frequency scan is obtained which plots the variation of impedance modulus and phase angle at a selected bus with variation of frequency or generates R–X plots of the impedance. This enables the resonant frequencies to be ascertained. The harmonic current flows in the lines are calculated, and the network, assumed to be linear at each step of the calculations with added constraints, is solved to obtain the harmonic voltages. The calculations may include the following:

- Calculation of harmonic distortion indices
- Calculation of TIF, KVT, and IT
- Induced voltages on communication lines
- Sensitivity analysis, i.e., the effect of variation of a system component

This is rather a simplistic approach. The rigorous harmonic analysis gets involved because of interaction between harmonic-producing equipment and the power system, the practical limitations of modeling each component in a large power system, the extent to which the system should be modeled for accuracy, and the types of component and nonlinear source models. Furnace arc impedance varies erratically and is asymmetrical. Large power high-voltage DC (HVDC) converters and FACTS devices have large nonlinear loads and superimposition is not valid. Depending on the nature of the study *simplistic methods may give erroneous results.*

7.1 Harmonic Analysis Methods

There are a number of methodologies for calculation of harmonics and effects of nonlinear loads. Direct measurements can be carried out using suitable instrumentation. In a noninvasive test the existing waveforms are measured. An EPRI project describes two methods and References [1,2] provide a summary of the research project sponsored by EPRI and

BPA (Bonneville Power Administration) on HVDC system interaction from AC harmonics. One method uses harmonic sources on the AC network, and the other injects harmonic currents into the AC system using HVDC converter. In the latter case, the system harmonic impedance is a ratio of the harmonic voltage and current. In the first case, a switchable shunt impedance such as a capacitor or filter bank is required at the point where the measurement is taken. By comparing the harmonic voltages and currents before and after switching the shunt device, the network impedance can be calculated. This research project implements (1) development of data acquisition system to measure current and voltage signals, (2) development of data-processing package, (3) calculation of harmonic impedances, and (4) development of a computer model for impedance calculations. The analytical analysis can be carried out in the frequency and time domains. Another method is to model the system to use a state-space approach. The differential equations relating current and voltages to system parameters are found through basic circuit analysis. We will discuss frequency- and time-domain methods [3,4].

7.1.1 Frequency-Domain Analysis

For calculations in the frequency domain, the harmonic spectrum of the load is ascertained and the current injection is represented by a Norton's equivalent circuit. Harmonic current flow is calculated throughout the system for each of the harmonics. The system impedance data are modified to account for higher frequency, and are reduced to their Thévenin equivalent. The principal of superposition is applied. If all nonlinear loads can be represented by current injections, the following matrix equations are applicable:

$$\overline{V}_h = \overline{Z}_h \overline{I}_h \tag{7.1}$$

$$\overline{I}_h = \overline{Y}_h \overline{V}_h \tag{7.2}$$

The formation of bus impedance and admittance matrices has already been discussed. The distribution of harmonic voltages and currents are no different for networks containing one or more sources of harmonic currents. During the steady state, the harmonic currents entering the network are considered as being produced by ideal sources that operate without repercussion. The entire system can then be modeled as an assemblage of passive elements. Corrections will be applied to the impedance elements for dynamic loads, e.g., generators and motors' frequency-dependent characteristics at each incremental frequency chosen during the study can be modeled. The system harmonic voltages are calculated by direct solution of the linear matrix Equations 7.1 and 7.2.

In a power system, the harmonic injection will occur only on a few buses. These buses can be ordered last in the Y matrix and a reduced matrix can be formed. For n nodes and $n - j + 1$ injections, the reduced Y matrix is

$$\begin{vmatrix} I_j \\ \cdot \\ I_n \end{vmatrix} = \begin{vmatrix} Y_{jj} & \cdot & Y_{jn} \\ \cdot & \cdot & \cdot \\ Y_{nj} & \cdot & Y_{nn} \end{vmatrix} \begin{vmatrix} V_i \\ \cdot \\ V_n \end{vmatrix} \tag{7.3}$$

where diagonal elements are the self-admittances and the off-diagonal elements are transfer impedances as in the case of load flow calculations.

Linear transformation techniques are discussed in Volume 1. The admittance matrix is formed from a primitive admittance matrix by transformation,

$$\overline{Y}_{abc} = \overline{A}' \overline{Y}_{prim} \overline{A} \tag{7.4}$$

and the symmetrical component transformation is given by

$$\overline{Y}_{012} = \overline{T}_s^{-1} \overline{Y}_{abc} \overline{T}_s \tag{7.5}$$

The vector of nodal voltages is given by

$$\overline{V}_h = \overline{Y}_h^{-1} \overline{I}_h = \overline{Z}_h \overline{I}_h \tag{7.6}$$

For the injection of a unit current at bus k

$$\overline{Z}_{kk} = \overline{V}_k \tag{7.7}$$

where Z_{kk} is the impedance of the network seen from bus k. The current flowing in branch jk is given by

$$\overline{I}_{jk} = \overline{V}_{jk}(\overline{V}_j - \overline{V}_k) \tag{7.8}$$

where Y_{jk} is the nodal admittance matrix of the branch connected between j and k.

Variation of the bus admittance matrix, which is produced by a set of modifications in the change of impedance of a component, can be accommodated by modifications to the Y bus matrix as discussed before. For harmonic analysis the admittance matrix must be built at each frequency of interest, for component level RLC parameters for circuit models of lines, transformers, cables, and other equipment. Thus, the harmonic voltages can be calculated. A new estimate of the harmonic injection currents is then obtained from the computed harmonic voltages, and the process is iterative until the convergence on each bus is obtained. Under resonant conditions, large distortions may occur and the validity of assumption of linear system components is questionable. From Equation 7.1, we see that the harmonic impedance is important in the response of the system to harmonics. There can be interaction between harmonic sources throughout the system, and if these are ignored, the single-source model and the superposition can be used to calculate the harmonic distortion factors and filter designs. The assumption of constant system impedance is not valid, as the system impedance always changes; say due to switching conditions, operation, or future additions. These impedance changes in the system may have a more profound effect on the ideal current source modeling than the interaction between harmonic sources. A weak AC/DC interconnection defined with a short-circuit ratio (short-circuit capacity of the AC system divided by the DC power injected by the converter into converter bus) of <3 may have voltage and power instabilities, transient and dynamic overvoltages, and harmonic overvoltages [5].

7.1.2 Frequency Scan

A frequency scan is merely a repeated application in certain incremental steps of some initial value of frequency to the final value, these two values spanning the range of harmonics to be considered. The procedure is equally valid whether there are single or multiple harmonic sources in the system, so long as the principal of superimposition is held valid. Then, for unit current injection, the calculated voltages give the driving point and transfer impedance, both modulus and phase angle. The Y_{bus} contains only linear elements for each frequency. Varying the frequency gives a series of impedances which can be plotted to provide an indication of the resonant conditions. Figure 7.1 shows a frequency

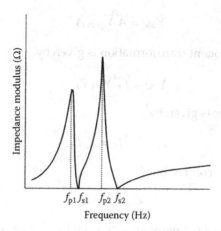

FIGURE 7.1
Frequency scan showing parallel and series resonance frequencies.

scan of impedance modulus versus frequency. The parallel resonances occur at peaks, which give the maximum impedances and the series resonances at the lowest points of the impedance plots. Figure 7.1 shows parallel resonances at two frequencies f_{p1} and f_{p2} and series resonance at f_{s1} and f_{s2}. We will see in Chapter 9 that such a frequency scan is obtained with two single-tuned shunt filters. Multiphase frequency scans can identify the harmonic resonance caused by single-phase capacitor banks.

7.1.3 Voltage Scan

A voltage scan may similarly be carried out by applying unit voltage to a node and calculating the voltages versus frequency in the rest of the system. The resulting voltages represent the voltage transfer function to all other nodes in the system. This analysis is commonly called a voltage transfer function study. The peaks in the scan identify the frequencies at which the voltages will be magnified and the lowest points indicate frequencies where these will be attenuated.

7.1.4 Phase Angle of Harmonics

For simplicity, all the harmonics may be considered cophasial. This does not always give the most conservative results, unless the system has one predominant harmonic, in which case only harmonic magnitude can be represented. See Chapter 3 for details.

7.1.5 Newton–Raphson Method

The Newton–Raphson method can be applied to harmonic current flow [6,7]. This is based on the balance of active power and reactive volt–ampères, whether at fundamental frequency or at harmonics. The active and reactive power balance is forced to zero by the bus voltage iterations.

Consider a system with $n + 1$ buses, bus 1 is a slack bus, buses 2 through $m - 1$ are conventional load buses, and buses m to n have nonsinusoidal loads. It is assumed that the active power and the reactive volt–ampère balance are known at each bus and that the nonlinearity is known. The power balance equations are constructed so that ΔP and

ΔQ at all nonslack buses are zero for all harmonics. The form of ΔP and ΔQ as a function of bus voltage and phase angle is the same as in conventional load flow, except that Y_{bus} is modified for harmonics. The specified active and reactive powers are known at buses 2 through $m - 1$, but only active power is known at buses m through n. Two additional parameters are required: the current balance and volt–ampère balance. The current balance for fundamental frequency is written and the equation is modified for buses with harmonic injections. The harmonic response of the buses 1 through $m - 1$ is modeled in admittance bus matrix.

The third equation is the apparent volt–ampère balance at each bus:

$$S_L^2 = \sum_s P_L^2 + \sum_s Q_L^2 + \sum D_L^2 \tag{7.9}$$

where the third term of the equation denotes distortion power at bus L, which is not considered as an independent variable, as it can be calculated from real and imaginary components of currents.

The final equations for the harmonic power flow become

$$
\begin{vmatrix} \Delta W \\ \Delta I^1 \\ \Delta I^5 \\ \Delta I^7 \\ \cdots \end{vmatrix}
=
\begin{vmatrix}
J^1 & J^5 & J^7 & \cdots & 0 \\
YG^{1,1} & YG^{1,5} & YG^{1,7} & \cdots & H^1 \\
YG^{5,1} & YG^{5,5} & YG^{5,7} & \cdots & H^5 \\
YG^{7,1} & YG^{7,5} & YG^{7,7} & \cdots & H^7 \\
\cdots & \cdots & \cdots & \cdots & \cdots
\end{vmatrix}
\begin{vmatrix} \Delta V^1 \\ \Delta V^5 \\ \\ \\ \Delta\alpha \end{vmatrix}
\tag{7.10}
$$

where all elements in Equation 7.10 are subvectors and submatrices partitioned from ΔM (apparent mismatches), J, and ΔU, i.e., $\Delta M = J\Delta U$.

$$\Delta V^k = \left(V_1^k \Delta\theta_1^k, \Delta V_1^k, \ldots, \Delta V_n^k\right)^t \quad k = 1,5,7,\ldots \tag{7.11}$$

where k is the order of harmonic, and θ is the phase angle.

$$\Delta\alpha = \left(\Delta\alpha_m, \Delta\beta_m, \ldots, \Delta\beta_n\right)^t \tag{7.12}$$

where α is the firing angle, and β is the commutation parameter.

ΔW = mismatch active and reactive volt–ampères

$$\Delta I^1 = \left(I_{r,m}^1 + g_{r,m}^1, I_{i,m}^1 + g_{i,m}^1, \ldots, I_{i,n}^1 + g_{i,n}^1\right)^t \tag{7.13}$$

which is equal to mismatch fundamental current, where $I_{r,m}^1, I_{i,m}^1$ are the real and imaginary parts of the current injection at bus m and $g_{r,m}^1, g_{i,m}^1$ are the real and imaginary parts of current balance equation.

ΔI^k = mismatch harmonic current at the kth harmonic

J^1 = conventional power flow Jacobian

J^k = Jacobian at harmonic k

$$= \left[\frac{0_{2(m-1),2n}}{\text{Partial derivatives of } P \text{ and } Q \text{ with respect to } V^k \text{ and } \theta^k} \right]$$

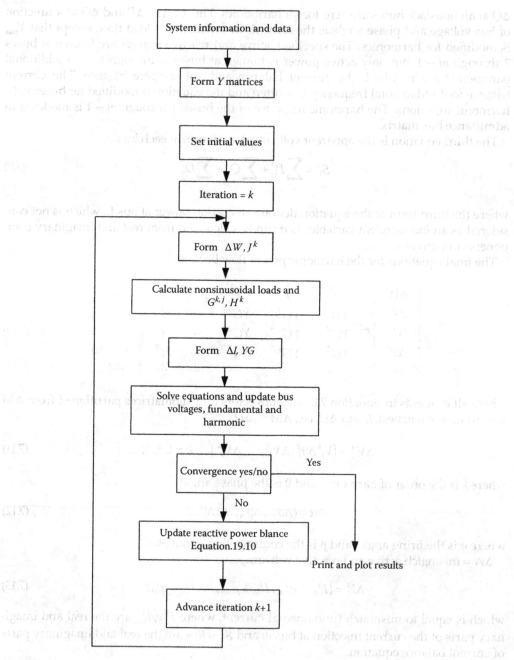

FIGURE 7.2
Harmonic power flowchart, Newton–Raphson method.

where $0_{2(m-1),2n}$ denotes a $2(m-1) \times 2n$ array of zeros.

$$(YG)^{k,j} = \begin{matrix} Y^{k,k} + G^{k,k}(k=j) \\ G^{k,j} \qquad (k \neq j) \end{matrix} \tag{7.14}$$

where $Y^{k,k}$ is an array of partial derivatives of injection currents at the kth harmonic with respect to the kth harmonic voltage, and $G^{k,j}$ are the partials of the kth harmonic load current with respect to the jth harmonic supply voltage; H^k are the partial derivatives of nonsinusoidal loads for real and imaginary currents with respect to α and β.

$G^{k,j}$ is given by

$0_{2(m-1),2(m-1)}$	$0_{2(m-1),2N}$			0			0	
	$\dfrac{\partial g^k_{r,m}}{V^j_m \partial \theta^j_m}$	$\dfrac{\partial g^k_{r,m}}{\partial V^j_m}$						
	$\dfrac{\partial g^k_{i,m}}{V^j_m \partial \theta^j_m}$	$\dfrac{\partial g^k_{i,m}}{\partial V^j_m}$						
$0_{2N,2(m-1)}$	0					0		
	0			0		$\dfrac{\partial g^k_{r,n}}{V^j_n \partial \theta^j_n}$	$\dfrac{\partial g^k_{r,n}}{\partial V^j_n}$	
						$\dfrac{\partial g^k_{i,n}}{V^j_n \partial \theta^j_n}$	$\dfrac{\partial g^k_{i,n}}{\partial V^j_n}$	

$$\tag{7.15}$$

and H^k is given by

$$H^k = \text{diag} \begin{vmatrix} \dfrac{\partial g^k_{r,t}}{\partial \alpha_t} & \dfrac{\partial g^k_{r,t}}{\partial \beta_t} \\ \dfrac{\partial g^k_{i,t}}{\partial \alpha_t} & \dfrac{\partial g^k_{i,t}}{\partial \beta_t} \end{vmatrix} \tag{7.16}$$

A flowchart is shown in Figure 7.2.

Mohmoud and Shultz [8] discuss the impedance matrix method of harmonic analysis. Equations from 7.1 through 7.7 describe the concepts of impedance method.

7.1.6 Three-Phase Harmonic Load Flow

A multiphase harmonic load flow technique is described in References [9,10], related to the *unbalance* operation of the power system. It considers network, generator, and load equations, expressing these in general form:

$$F(|x|) = 0$$

where

$$|x| = |V \; I_V \; I_{L2} \; I_M \; I_{L3} \; E_p \; y|^t$$

$$|F| = |f_1 \; f_2 \; f_3 \; f_4 \; f_5 \; f_6 \; f_7|^t \tag{7.17}$$

$$|I_u| = |I_V \; I_{L2} \; I_M \; I_{L3}|^t$$

and

$|I_V|$ is the current vector from voltage sources
$|I_{L2}|$ is the vector of single-phase PQ load currents
$|I_M|$ is the vector of machine currents
$|I_{L3}|$ is the vector of static load currents
$|E_p|$ is the vector of machine internal voltages
$|y|$ is the vector of static load parameter y

The network equation is formed as

$$|Y||V| + |I_s| + |I_u| = 0 \tag{7.18}$$

where
$|Y|$ is the node admittance matrix
$|V|$ is the vector of node voltage
$|I_s|$ is the vector of current sources leaving the node
$|I_u|$ is the vector of unknown currents leaving each node

These equations are a set of nonlinear algebraic equations solved iteratively using the Newton–Raphson method, rectangular coordinates. The load flow algorithms are discussed in Volume 2.

7.1.7 Time Domain Analysis

The simplest harmonic model is a rigid harmonic source and linear system impedance. A rigid harmonic source produces harmonics of a certain order and constant magnitude and phase, and the linear impedances do not change with frequencies. Multiple harmonic sources are assumed to act in isolation and the principal of superimposition applies. These models can be solved by iterative techniques, and the accuracy obtained will be identical to that of time-domain methods. For arc furnaces and even electronic converters under resonant conditions an ideal current injection may cause significant errors. The nonlinear and time-varying elements in the power system can significantly change the interaction of the harmonics with the power system. Consider the following:

1. Most harmonic devices that produce noncharacteristic harmonics as terminal conditions are in practice not ideal, e.g., converters operating with unbalanced voltages.

2. There is interaction between AC and DC quantities and there are interactions between harmonics of different order, given by switching functions (defined later).

3. Gate control of converters can interact with harmonics through synchronizing loops.

Time-domain analyses have been used for transient stability studies, transmission lines, and switching transients. It is possible to solve a wide range of differential equations for the power system using computer simulation and to build up a model for harmonic calculations, which could avoid many approximations inherent in the frequency domain approach. Harmonic distortions can be directly calculated, and making use of fast Fourier transform (see Appendix A), these can be converted into frequency domain. The graphical results are waveforms of zero crossing, ringing, high dv/dt, and commutation notches. The transient effects can be calculated; for example, the part-winding resonance of a transformer can be simulated. The synchronous machines can be simulated with accurate models to represent saliency, and the effects of frequency can be dynamically simulated. EMTP is one very widely used program for simulation in the time domain.

For analysis in the time domain, a part of the system of interest may be modeled in detail. This detailed model consists of three-phase models of system components, transformers, harmonic sources, and transmission lines and it may be coupled with a network model of lumped RLC branches at interconnection buses to represent the driving point and transfer impedances of the selected buses. The overall system to be studied is considerably reduced in size and time-domain simulation is simplified.

7.1.8 Switching Function

Switching function is a steady-state concept for the study of interactions between the AC and DC sides of a converter. The terminal properties of many converters can be approximated in the time domain by the converter switching function. The modulation/demodulation properties of the converter account for interaction between harmonics, generation of noncharacteristic harmonics, and propagation of DC harmonics on the AC side, and operation under unbalanced voltage or current. Consider the switching function of a converter, as shown in Figure 7.3. The switching function is 1 when DC current flows in the positive direction, −1 when it flows in the negative direction, and zero otherwise. The switching functions of three phases are symmetrical and balanced and in steady state these lag the system voltage by a converter delay angle. The AC current output of phase a is the product of the switching function and the DC current in phase a:

$$I_a = I_{dc}S_a \tag{7.19}$$

$$I_b = I_{dc}S_b \tag{7.20}$$

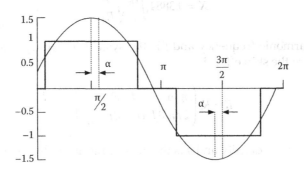

FIGURE 7.3
Switching function of a six-pulse converter.

$$I_c = I_{dc}S_c \tag{7.21}$$

$$V_{dc} = V_aS_a + V_bS_b + V_cS_c \tag{7.22}$$

The converter appears as a current source from the AC side and a voltage source from the DC side. The switching functions allow one to conduct studies for harmonic interaction between two or more converters in near proximity. The switching function assumes that the control system operates perfectly and delivers ignition pulses at regular intervals. Practically, there is interaction between the network harmonics and converter controls, which can be detected by modeling synchronizing loops.

7.2 Harmonic Modeling of System Components

7.2.1 Transmission Lines

The transmission line models are in Appendix A, Volume 1, which give impedance calculations for a mutually coupled three-phase line, with ground wires, bundled conductors, and symmetrical component transformations. The transmission line model to be used is determined by the wavelength of the highest frequency of interest. Long-line effects should be represented for lines of length $150/h$ miles, where h is the harmonic number. The effect of higher frequencies is to increase the skin effect and proximity effects. A frequency-dependent model of the resistive component becomes important, though the effect on the reactance is ignored. The resistance can be multiplied by a factor $g(h)$ [11]:

$$R(h) = R_{dc}g(h) \tag{7.23}$$

$$g(h) = 0.035X^2 + 0.938 > 2.4 \tag{7.24}$$

$$= 0.35X + 0.3 \le 2.4 \tag{7.25}$$

where

$$X = 0.3884\sqrt{\frac{f_h}{f}}\sqrt{\frac{h}{R_{dc}}} \tag{7.26}$$

where f_h is the harmonic frequency and f is the system frequency. Another equation for taking into account the skin effect is

$$R = R_e\left(\frac{j\mu\omega}{2\pi a}\frac{J_z(r)}{\partial J_z(r)/\partial r\,\big|_{r=a}}\right) \tag{7.27}$$

where $J_z(r)$ is the current density and a is the outside radius of the conductor.

7.2.1.1 Transmission Line Equations with Harmonics

See Volume 2, Appendix A for the calculation of transmission line parameters and the relevant equations. In the presence of harmonics,

$$V_{s(h)} = V_{r(h)} \cosh(\gamma l_{(h)}) + I_{r(h)} Z_{0(h)} \sinh(\gamma l_{(h)})$$

$$= V_{r(h)} \cosh\left(\sqrt{Z_{(h)} Y_{(h)}}\right) + I_{r(h)} \sqrt{\frac{Z_{(h)}}{Y_{(h)}}} \sinh\left(\sqrt{Z_{(h)} Y_{(h)}}\right) \tag{7.28}$$

$$I_{s(h)} = \frac{V_{r(h)}}{Z_{0(h)}} \sinh\left(\sqrt{Z_{(h)} Y_{(h)}}\right) + I_{r(h)} \cosh\left(\sqrt{Z_{(h)} Y_{(h)}}\right) \tag{7.29}$$

Similarly, from

$$\left| \begin{array}{c} V_r \\ I_r \end{array} \right| = \left| \begin{array}{cc} \cosh(\gamma l) & -Z_0 \sinh(\gamma l) \\ -\dfrac{\sinh(\gamma l)}{Z_0} & \cosh \gamma l \end{array} \right| \left| \begin{array}{c} V_s \\ I_s \end{array} \right| \tag{7.30}$$

equations of receiving end current and voltages can be written.

The variation of the impedance of the transmission line with respect to frequency is of much interest. We derived the equivalent π model in Volume 2.

$$Z_s = Z_0 \sinh \gamma l$$

$$Y_p = \frac{Y}{2}\left[\frac{\tanh \gamma l/2}{\gamma l/2}\right] \tag{7.31}$$

Here, we have denoted the series element of the π model with Z_s and the shunt element with Y_p. In presence of harmonics, we can write

$$Z_{s(h)} = Z_{0(h)} \sinh \gamma l_{(h)} = Z_{(h)} \frac{\sinh \gamma l_{(h)}}{\gamma l_{(h)}}$$

$$Y_{p(h)} = \frac{\tanh\left(\gamma l_{(h)}/2\right)}{Z_{0(h)}} = \frac{Y_{(h)}}{2} \frac{\tanh\left(\gamma l_{(h)}/2\right)}{\gamma l_{(h)}/2} \tag{7.32}$$

Also

$$\gamma_{(h)} = \sqrt{z_{(h)} y_{(h)}} \approx \frac{h\omega}{l}\sqrt{LC}$$

$$Z_{0(h)} = \sqrt{z_{(h)}/y_{(h)}} \approx \sqrt{\frac{L}{C}} \tag{7.33}$$

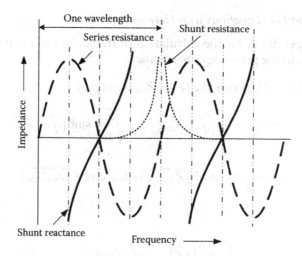

FIGURE 7.4
Impedance plot of a 230 kV line

Recall from Chapter 1 that

$$\lambda_{(h)} = 2\pi/\beta_{(h)} \approx \frac{1}{hf\sqrt{LC}}$$

$$V_{(h)} = f\lambda_h \approx \frac{1}{\sqrt{LC}} \tag{7.34}$$

$$f_{osc(h)} = \frac{V_{(h)}}{1} \approx \frac{1}{\sqrt{LC}}$$

Thus, the characteristic impedance, velocity of propagation, and frequency of oscillations are all independent of h, while wavelength varies inversely with h.

A specimen impedance plot of a 230 kV line is shown in Figure 7.4. The series and shunt reactances are predominant. The shunt resistance, normally considered zero in a π model, has a considerable effect on resonant frequency. The impedance variations are similar to a series and parallel resonating circuits.

Figure 7.5a shows the calculated impedance modulus, and Figure 7.5b shows the impedance angle versus frequency plots of a 400 kV line, consisting of four bundled conductors of 397.5 KCMIL ACSR per phase in horizontal configuration. The average spacing between bundles is 35 ft, and the height above the ground is 50 ft. Two ground wires at ground potential are considered in the calculations, and the earth resistivity is 100 Ω m. Each conductor has a diameter of 0.806 in. (0.317 cm) and bundle conductors are spaced 6 in. (0.15 m) center to center.

The plots show a number of resonant frequencies. The impedance angle changes abruptly at each resonant frequency.

7.2.2 Underground Cables

Calculation of cable constants is described in Appendix A, Volume 2.

Cables have higher capacitance than overhead lines; therefore, modeling of cable capacitances is of much significance for harmonic analysis. Much like transmission lines, the

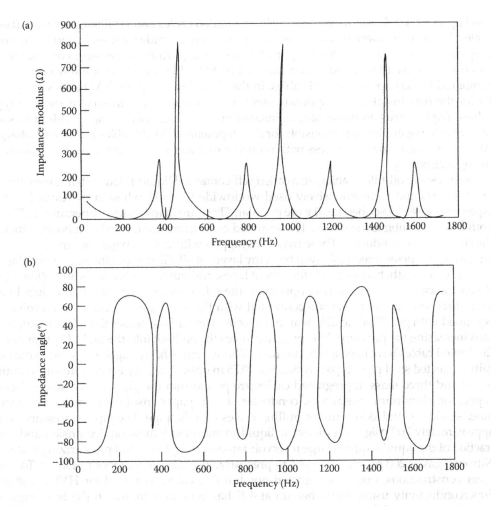

FIGURE 7.5
(a) Frequency scan of a 400 kV line with bundle conductors and (b) corresponding phase angle of the impedance modulus.

equivalent π model should be used, and not the nominal π model. A cable length can be considered short if the surge traveling time is lower than approximately 30% of the time constant of the voltage rise time in the system. The cross bonding of the sheaths of the cables presents another consideration, and each section of the cable length has sheath bonding and grounding connections.

A frequency-dependent model for armored power cables can be obtained using the finite element (FE) method, and the frequency response is obtained by coupling the cable FE domain model and external electrical circuits. The resulting vector fitting rational function can then be represented by an equivalent network. This frequency response is then fitted with rational function approximation. For harmonic frequencies to 3000 Hz, the resistance will increase. The slight decrease in inductance and the effect on shunt capacitance can be ignored. π models are considered adequate for frequency scans (though not for transient analysis).

Cables vary in constructional features, not discussed here. Extruded dielectric (XD), also called solid dielectric—principally cross-linked polyethylene (XLPE) cables—have been used up to 500 kV. These are replacing pipe-insulated cables and oil-filled, paper-insulated cables, because of lower costs and less maintenance. Extruded cables include ethylene–propylene rubber and low-density polyethylene, though XLPE is the most common insulation system used for transmission cables. The XD cables also have a lower capacitance compared to the paper-insulated cables. In the United States, though XD cables are popular for the new installations, approximately 70% of circuit miles in service are pipe-type cables. Triple extrusion (inner shield, insulation, and outer conducting shield) processed in one sealed operation is responsible for developments of XLPE cables for higher voltages. Also a dry vulcanization process with no steam or water is used in the modern manufacturing technology.

Low-pressure oil-filled cables, also called self-contained liquid-filled systems, were developed in 1920 and were extensively used worldwide till 1980, while in the United States, pipe-type cables (described next) were popular. There are many miles in operation at 525 kV, both land and submarine cables. The stranded conductors are formed as a hollow duct in the center of the conductor. These may be wrapped with carbon paper, nonmagnetic steel tape, and conductor screen followed by many layers of oil-filled paper insulation. Insulation screen, lead sheath, bedding, reinforcement tapes, and outer polymeric sheath follow next. When the cable heats up during operation, the oil flows through the hollow duct in an axial direction into oil reservoirs connected with the sealing ends. The oil reservoirs are equipped with gas-filled flexible-wall metal cells. The oil flow causes the cells to compress, thus increasing the pressure. This pressure forces the oil back into the cable. High-pressure fluid-filled cables have been a U.S. standard. The system is fairly rugged. A welded catholically protected steel pipe, typically 8.625 or 10.75 in optical density is pressure and vacuum tested and three mass-impregnated cables are pulled into the pipe. The cable consists of copper or aluminum conductors, conductor shield, paper insulation, insulation shield, outer shielding, and skid wire for pulling cables into the pipe. The pipe is pressurized to approximately 200 psig with dielectric liquid to suppress ionization. Expansion and contraction of the liquid requires large reservoir tanks and sophisticated pressurizing systems. Nitrogen gas at 200 psi may be used to pressurize the cable at voltages to 138 kV. To state other constructions, mass-impregnated nondraining cables are used for HVDC circuits. Superconductivity using liquid helium at 4 K has been known and in the last couple of years has produced high-temperature superconductors. These use liquid nitrogen temperatures (80 K). Special computer subroutines are required to correctly estimate the cable parameters for the harmonic analysis. Furthermore, the cable constants also depend on the geometry of cable installations, underground, overhead, in duct banks, and the like.

In EMTP simulations, the following cable models are available:

- Frequency-dependent cable model (FD)
- Constant parameter model (CP)
- Exact π model
- Wideband cable model

FD model provides an accurate representation of distributed nature of cable parameters R, L, G, and C, as well as their frequency dependence in modal quantities. It is assumed that the characteristic impedance and the propagation function matrices can be diagonalized by a modal transformation matrix. For harmonic frequencies, the frequency-dependent models are not necessary, though the effect on resistance is considerable and should be modeled.

The velocity of propagation on cables is much lower than on OH lines due to higher capacitance and, thus, smaller lengths of cables can be termed "long." Considering the velocity of propagation about 40%, the 40-mile-long cable can be termed "long." Long-line effects should be represented for cables of length $40/h$ miles, where h is the harmonic number. The distributed parameter model, akin to transmission lines, can be used.

With respect to frequency dependence, the rigorous modeling and mathematical treatment of calculations of cable constants becomes complex, not discussed here. Much cable data are required for each cable type, its geometry, and method of installation. Much work has been done in LINE CONSTANT routines for EMTP simulations.

7.2.3 Filter Reactors

The frequency-dependent Q of the filter reactors is especially important, as it affects the sharpness of tuning (Chapter 8). Resistance at high frequencies can be calculated by the following expressions:

$$R_h = \left[\frac{0.115h^2 + 1}{1.15} \right] R_f \text{ for aluminum reactors} \tag{7.35}$$

$$R_h = \left[\frac{0.055h^2 + 1}{1.055} \right] R_f \text{ for copper reactors} \tag{7.36}$$

7.2.4 Transformers

The single-phase and three-phase transformer models are discussed in load flow, Volume 2 and Appendix C, Volume 1. A linear conventional T-circuit model of the transformer is shown in Appendix C and fundamental frequency values of resistance and reactance can be found by a no-load and short-circuit test on the transformer. The resistance of the transformer can be modified with increase in frequency according to Figure 7.6. While the resistance increases with frequency, the leakage inductance reduces. The magnetizing branch in the transformer model is often omitted if the transformer is not considered a source of the harmonics. This simplified model may not be accurate, as it does not model the nonlinearity in the transformer on account of the following:

- Core and copper losses due to eddy current and hysteresis effects. The core loss is a summation of eddy current and hysteresis loss; both are frequency dependent:

$$P_c = P_e P_h = K_e B^2 f^2 + K_h B^s f \tag{7.37}$$

where B is the peak flux density, s is the Steinmetz constant (typically 1.5–2.5, depending on the core material), f is the frequency, and K_e and K_h are constants.
- Leakage fluxes about the windings, cores, and surrounding medium.
- Core magnetization characteristics, which are the primary source of transformer nonlinearity.

A number of approaches can be taken to model the nonlinearities. EMTP transformer models, *Satura* and *Hysdat*, are discussed in Appendix C, Volume 1. These models consider

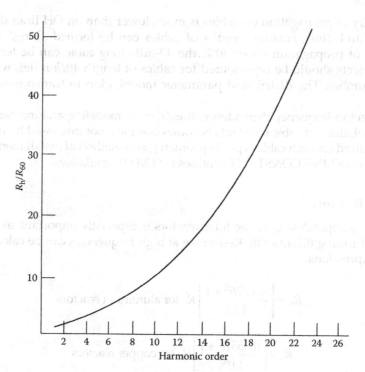

FIGURE 7.6
Increase in transformer resistance with frequency.

only core magnetization characteristics and neglect nonlinearities or frequency dependence of core losses and winding effects.

Capacitances of the transformer windings are generally omitted for harmonic analysis of distribution systems; however, for transmission systems, capacitances are included. Surge models of transformers are discussed in Appendix C, Volume 1. A simplified model with capacitances is shown in Figure 7.7; C_{bh} is the high-voltage bushing capacitance, C_b is the secondary bushing capacitance, C_{hg} and C_{lg} are distributed winding capacitances, and C_{hl} is the distributed winding capacitance between the windings. Typical capacitance values for core-type transformers are shown in Table 7.1.

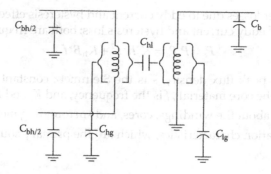

FIGURE 7.7
Simplified capacitance model of a two-winding transformer.

TABLE 7.1

Capacitance of Core-Type Transformers in nF

MVA Rating of Transformer	C_{hg}	C_{hl}	C_{lg}
1	1.2–14	1.2–17	3.1–16
2	1.4–16	1–18	3–16
5	1.2–14	1.1–20	5.5–17
10	4–7	4–11	8–18
25	2.8–4.2	2.5–18	5.2–20
50	4–6.8	3.4–11	3–24
75	3.5–7	5.5–13	2.8–30
100	3.3–7	5–13	4–40

Converter loads may draw DC and low-frequency currents through the transformers, i.e., a cycloconverter load. Geomagnetically induced currents flow on the earth's surface due to geomagnetic disturbance are typically at low frequencies (0.001–0.1 Hz), reaching a peak value of 200 A. These can enter the transformer windings through grounded wye neutrals and bias the cores to cause half-cycle saturation (see Appendix C, Volume 1). Also three-phase models of the transformers are discussed in detail in Volume 2.

7.2.5 Induction Motors

Equivalent circuits of induction motors are discussed in Volume 2, and Figure 7.8 is reproduced. The shunt elements g_m and b_m are relatively large compared to R_1, r_2, X_1, and x_2. Generally, the locked rotor current of the motor is known. At fundamental frequency, neglecting magnetizing and loss components, the motor reactance is

$$X_f = X_1 + x_2 \tag{7.38}$$

and the resistance is

$$R_f = R_1 + \frac{r_2}{s} \tag{7.39}$$

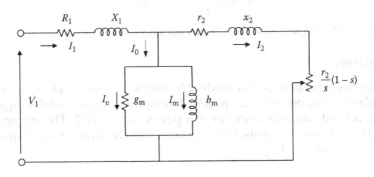

FIGURE 7.8
Equivalent circuit of an induction motor.

This resistance is not the same as used in short-circuit calculations. At harmonic frequencies, the reactance can be directly related to the frequency:

$$X_h = hX_f \tag{7.40}$$

Though this relation is only approximately correct. The reactance at higher frequencies is reduced due to saturation. The stator resistance can be assumed to vary as the square root of the frequency:

$$R_{1h} = \sqrt{h} \cdot (R_1) \tag{7.41}$$

The harmonic slip is given by

$$s_h = \frac{h-1}{h} \text{ for positive sequence harmonics} \tag{7.42}$$

$$s_h = \frac{h+1}{h} \text{ for negative sequence harmonics} \tag{7.43}$$

The rotor resistance at harmonic frequencies is

$$r_{2h} = \frac{\sqrt{(1 \pm h)}}{s_h} (r_2) \tag{7.44}$$

The motor impedance neglecting magnetizing resistance is infinite for triplen harmonics, as the motor windings are not grounded.

The motor impedance at harmonic h is (see Volume 2)

$$R_1 + jhX_1 + \frac{jhx_m \left(\dfrac{r_2}{s_h} + jhx_2 \right)}{(r_2 / s_h) + jh(x_m + x_2)} \tag{7.45}$$

where

$$s_h = 1 - \frac{n}{hn_s}, \quad h = 3n' + 1$$

$$n' = 1, 2, 3 \dots \tag{7.46}$$

$$s_h = 1 + \frac{n}{hn_s}, \quad h = 3n' - 1$$

$n_s = n$, Equation (7.45) ignores effect of harmonics on resistance, R_1 & r_2.

7.2.6 Generators

The synchronous generators do not produce harmonic voltages, and can be modeled with shunt impedance at the generator terminals. An empirical linear model is suggested which consists of a full subtransient reactance at a power factor of 0.2. The average inductance experienced by harmonic currents, which involve both the direct axis and quadrature axis reactances, is approximated by

$$\text{Average inductance} = \frac{L_d'' + L_q''}{2} \tag{7.47}$$

At harmonic frequencies, the fundamental frequency reactance can be directly proportioned. The resistance at harmonic frequencies is given by

$$R_h = R_{dc}\left[1 + 0.1(h_f / f)^{1.5}\right] \tag{7.48}$$

This expression can also be used for the calculation of harmonic resistance of transformers and cables having copper conductors. Model generators for harmonics with

$$Z_{0(h)} = R_h + jhX_0, \quad h = 3, 6, 9...$$

$$Z_{1(h)} = R_h + jhX_d'', \quad h = 1, 4, 7... \tag{7.49}$$

$$Z_{2(h)} = R_h + jhX_2, \quad h = 2, 5, 8...$$

where X_0, X_d'', and X_2 are the generator zero sequence, subtransient, and negative sequence reactances, respectively, at fundamental frequency.

7.3 Load Models

Figure 7.9a shows a parallel *RL* load model. It represents bulk power load as an *RL* circuit connected to ground. The resistance and reactance components are calculated from fundamental frequency voltage, reactive volt–ampère, and power factor:

$$R = \frac{V^2}{S \cos\phi} L = \frac{V^2}{2\pi fS \sin\phi} \tag{7.50}$$

The reactance is frequency dependent and resistance may be constant or it can also be frequency dependent. Alternatively, the resistance and reactance may remain constant at all frequencies.

Figure 7.9b shows a CIGRE (Conference Internationale des Grands Reseaux Electriques à Haute Tension) type-C load model [12], which represents bulk power, valid between 5th and 30th harmonics. Here, the following relations are applicable:

$$R_s = \frac{V^2}{P} X_s = 0.073 h R_s \; X_p = \frac{h R_s}{6.7 \dfrac{Q}{P} - 0.74} \tag{7.51}$$

This load model was derived by experimentation.

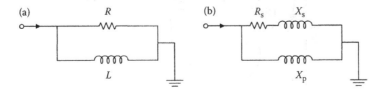

FIGURE 7.9
(a) Parallel *RL* load model and (b) CIGRE type-C model.

For the distribution systems modeling:

- Cables are represented as equivalent π model.
- For short lines model capacitance at the bus.
- Transformers represented by their equivalent circuits.
- Power factor correction capacitors modeled at their locations.
- Harmonic filters, generators represented as discussed above.
- All elements should be uncoupled using techniques discussed in this book.
- Loads modeled as per their composition and characteristics.

7.4 System Impedance

The system impedance to harmonics is not a constant number. Figure 7.10 shows the R–X plot of a system impedance. The fundamental frequency impedance is inductive, its value representing the *stiffness* of the system. The resonances in the system make the R–X plots a spiral shape, and the impedance may become capacitive. Such spiral-shaped impedances have been measured for high-voltage systems, and resonances at many frequencies are common. These frequencies at resonance points and also at a number of other points on the spiral-shaped curves are indicated as shown in Figure 7.10. At the resonance, the impedance reduces to a resistance. The system impedance can be ascertained by the following means:

- A computer solution can be used to calculate the harmonic impedances.
- In noninvasive measurements the harmonic impedance can be calculated directly from the ratio of harmonic voltage and current reading.

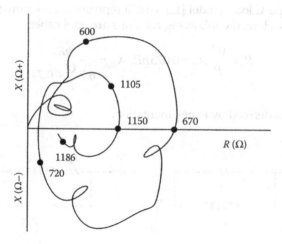

FIGURE 7.10
R–X plot of a supply system source impedance.

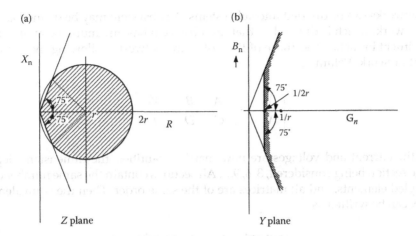

FIGURE 7.11
Generalized impedance plot (a) in the R–X plane and (b) in the Y plane.

- In another measurement method, shunt impedance is switched in the circuit and the harmonic impedance is calculated by comparing the harmonic voltages and currents before and after switching.

The spiral-shaped impedance plots can be bounded in the Z plane by a circle on the right side of the Y-axis and tangents to it at the origin at an angle of 75° (Figure 7.11a). This configuration can also be translated in the Y plane (Figure 7.11b) [13].

7.5 Three-Phase Models

The power system elements are not perfectly symmetrical. Asymmetry is involved in the circuit loading and mutual couplings and unbalanced self- and mutual impedances result. The problem is similar to the three-phase load flow, and is compounded by the nonlinearities of the harmonic loads [14]. Single-phase models are not adequate when

- Telephone interference is of concern. The influence of zero sequence harmonics is important, which gives rise to most of the interference with the communication circuits.
- There are single-phase capacitor banks in the system.
- There are single-phase or unbalanced harmonic sources.
- Triplen harmonics are to be considered, ground currents are important, and significant unbalanced loading is present.

The network asymmetries and mutual couplings should be included, which require a 3×3 impedance matrix at each harmonic. Three-phase models of transformers with neutral grounding impedances, mutually coupled lines, distributed parameter transposed, and untransposed transmission lines should be possible in the harmonic analysis program, also see Volume 2.

The networks can be divided into subsystems. A subsystem may be defined as any part of the network, which is divided so that no subsystem has any mutual coupling between its constituent branches and those of the rest of the network. Following the example of a two-port network, Volume 2:

$$\begin{vmatrix} \bar{V}_s \\ \bar{V}_r \end{vmatrix} = \begin{vmatrix} \bar{A} & \bar{B} \\ \bar{C} & \bar{D} \end{vmatrix} \begin{vmatrix} \bar{V}_r \\ \bar{I}_r \end{vmatrix} \tag{7.52}$$

That is the current and voltages are now matrix quantities, the dimensions depending upon the section being considered, 3, 6, 9…. All sections contain the same number of mutually coupled elements, and all matrices are of the same order. Then the equivalent model Y matrix can be written as

$$\bar{Y} = \begin{vmatrix} [\bar{D}][\bar{B}]^{-1} & \vdots & [\bar{C}]-[\bar{D}][\bar{B}]^{-1}[\bar{A}] \\ \cdots & \vdots & \cdots \\ [\bar{B}]^{-1} & \vdots & -[\bar{B}]^{-1}[\bar{A}] \end{vmatrix} \tag{7.53}$$

7.5.1 Noncharacteristic Harmonics

Noncharacteristic harmonics will be present. These may originate from variations in ignition delay angle. As a result, harmonics of the order $3q$ are produced in the direct voltage and $3q \pm 1$ are produced in the AC line current, where q is an odd integer. Third harmonic and its odd multiples are produced in the DC voltage and even harmonics in the AC line currents. The noncharacteristic harmonics may be amplified, as these may shift the times of voltage zeros and result in more unbalance of firing angles. A firing angle delay of 1° causes approximately 1% third harmonic. The modeling of any user-defined harmonic spectrum and their angles should be possible in a harmonic analysis program.

On a simplistic basis for a 12-pulse converter model noncharacteristic harmonics like 5, 7, 17, 19, … at 15% of the level computed for a 6-pulse converter and model triplen harmonics 3, 9, 15, … as 1% of the fundamental. But this does not account for unbalance. If we write the unbalance voltages as

$$V_a = V \sin \omega t$$

$$V_b = V(1+d)\sin(\omega t - 120°)$$

$$V_c = V \sin(\omega t + 120°) \tag{7.54}$$

where d is the unbalance; then conduction intervals for the current are

$$t_a = t_c = 120° - e$$

$$t_b = 120° + 2e \tag{7.55}$$

where

$$e = \tan^{-1} \frac{\sqrt{3}(1+d)}{3+d} - 30° \tag{7.56}$$

The instantaneous currents ignoring overlap are as follows:

$$i_a = \frac{4}{\pi} I_d \frac{(-1)^h}{h} \sin\left(\frac{ht_a}{2}\right) \sin(h\omega t)$$

$$i_b = \frac{4}{\pi} I_d \frac{(-1)^h}{h} \sin\left(\frac{ht_b}{2}\right) \sin(h\omega t - 120° - e/2) \qquad (7.57)$$

$$i_c = \frac{4}{\pi} I_d \frac{(-1)^h}{h} \sin\left(\frac{ht_c}{2}\right) \sin(h\omega t + 120° - e)$$

If there is 1% unbalance $d = 0.1$, then third harmonic currents are

$$i_{a3} = 0.01586k < 180°$$

$$i_{b3} = 0.03169k < -2.36° \qquad (7.58)$$

$$i_{c3} = 0.01586k < 175.27°$$

where

$$k = \frac{2\sqrt{3}I_d}{\pi} \qquad (7.59)$$

The positive and negative sequence components of these currents will be injected into the network even if a delta-connected transformer is used [15].

7.5.2 Converters

Chapter 1 discusses the harmonic generation from the converters. There is an interaction between the AC voltages and the DC current in weak AC/DC interconnections. Hatziadoniu and Galanos [5] describe voltage instabilities, control system instabilities, transient dynamic overvoltages, temporary and harmonic overvoltages, and low-order harmonic resonance. Tamby and John [16] describe the inadequacy of commercial harmonic analysis programs to model transient system harmonics and development of converter models and nonlinear resistor models to represent mercury arc and sodium vapor lamps. The computational algorithm to determine the harmonic components of the converter current and reactive power drawn considers overlap angle, calculated V_d and I_d, and reiterates with overlap angle till $V_d I_d = P$, calculated the Fourier components of the input current and the reactive power drawn by the converter. A power series model is adopted for the nonlinear resistor.

The harmonic currents produced by nonlinear loads are derived on the assumption of a perfectly sinusoidal source that is a strong sinusoidal voltage system. These harmonic currents are then injected into the AC system to determine levels of voltage distortion. However, when the injected harmonic frequency is close to the parallel resonant frequency, the calculation algorithm often diverges and a transient converter simulation is required. This requires AC system equivalents responding accurately to power and harmonic frequencies. An equivalent circuit consisting of number of tuned RLC branches has been proposed [17] as a solution to HVDC studies. Arrillaga et al. [10] describe frequency-dependent equivalent circuits suitable for integration in the time-domain solutions.

The converter and its AC system and the converter and the DC system can be reduced to Thévenin or Norton equivalent. Referring to Figure 7.12a, the harmonic relationships can be written as

$$\bar{V}_{dc} = \bar{A}\bar{V}_{ac} + \bar{B}\beta + \bar{C}\bar{I}_{dc}$$

$$\bar{I}_{ac} = \bar{D}\bar{I}_{dc} + \bar{E}\beta + \bar{F}\bar{V}_{ac}$$

$$\bar{V}_{ac} = Z_{ac}\bar{I}_{ac} + \bar{V}_{ac0}$$ (7.60)

$$\bar{I}_{dc} = Y_{dc}\bar{V}_{dc} + \bar{I}_{dc0}$$

where $A, B, C,...$ are matrices representing converter transfer functions from AC voltage to DC voltage, firing angle to DC voltage, DC current to AC current.... Z_{ac} and Y_{dc} are the diagonal matrices of AC-side harmonic impedance and DC-side harmonic admittances. I_{dc0} and V_{dc0} are the vectors of harmonic current sources on the DC side and harmonic voltage sources on the AC side. The elements of these matrices are derived by numerical techniques [18]. The frequency-dependent behavior of the converter may be defined as returned distortion (current or voltage) as a result of applied distortion at the same frequency. Manipulations of the Equation 7.60 lead to the interaction as shown in Figure 7.12b [10].

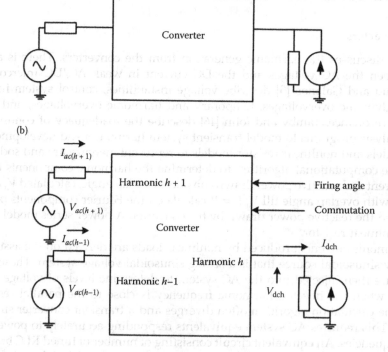

FIGURE 7.12
(a) Converter interconnecting AC and DC systems and (b) frequency interactions.

7.6 Modeling of Networks

The modeling is dependent on the network being studied. Networks vary in complexity and size and generally it is not possible to include the detailed model of every component in the study. A high-voltage grid system may incorporate hundreds of generators, transmission lines, and transformers. Thus, the extent to which a system should be modeled has to be decided. The system to be retained and deriving an equivalent of the rest of the system can be addressed properly only through a sensitivity study. An example of the effect of the extent of system modeling is shown in Figure 7.13 for a 200 MW DC tie in a 230 kV system [4]. A 20-bus model shows resonances at the 5th and 12th harmonics, rather than at the 6th and 13th when larger numbers of buses are modeled. This illustrates the risk of inadequate modeling.

7.6.1 Industrial Systems

Industrial systems vary in size and complexity, and some industrial plants may generate their own power and have operating loads of 100 MW or more. The utility ties may be at high voltages of 115, 138, or even 230 kV. It is usual to represent the utility source by its short-circuit impedance. There may be nearby harmonic sources or capacitors which will impact the extent of external system modeling. Generators and large rotating loads may be modeled individually, while an equivalent motor model can be derived connected through fictitious transformer impedance representing a number of transformers to which the motors are connected. This aggregating of loads is fairly accurate, provided that harmonic source buses and the buses having capacitor compensations are modeled in detail.

7.6.2 Distribution Systems

The primary distribution system voltage levels are 4–44 kV, while the secondary distribution systems are of low voltage (<600 V). A distribution system harmonic study is undertaken to investigate harmonic problems, resonances, and distortion in the existing system or to investigate the effect of adding another harmonic source. Single- or three-phase models can be used as demanded by the study. At low-voltage distribution level, the unbalance

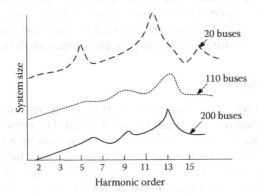

FIGURE 7.13
Errors introduced into a 200-bus system by inadequate modeling (G.D. Breuer et al., *IEEE Trans Power Appar Syst*, vol. PAS-101, no. 3, pp. 709–717 © 1982 IEEE.)

loads and harmonic sources are of particular importance, demanding a three-phase representation. The distribution systems are tied into the interconnected power network, and the system may be too large to be modeled. In most cases, it will be sufficiently accurate to represent a transmission network by its short-circuit impedance. This impedance may change, depending upon modifications to the system or system operation and is not a fixed entity. When capacitors or harmonic sources are present, a more detailed model will be necessary. The transmission system itself can be a significant source of harmonics in the distribution system and measurement at points of interconnection may be necessary to ascertain it. The load models are not straightforward, if these are lumped with harmonic-producing loads. The motor loads can be segregated from the nonrotating loads which can be modeled as an equivalent parallel R–L impedance. Again it is not feasible to model each load individually and feeder loads can be aggregated into large groups without loss of much accuracy.

The studies generally involve finite harmonic sources and the background harmonic levels are often ignored. These can be measured and analysis, generally, combines modeling and measurements for accuracy. A study shows that the harmonic currents at higher frequencies have widely varying phase angles, which result in their cancellation [19]. At lower frequencies up to the 13th, the cancellation is not complete. Unbalance loads on the feeder result in high harmonic currents in the neutral and ground paths. At fifth and seventh harmonics, the loads can be modeled as (1) harmonic sources, (2) a harmonic source with a parallel RL circuit, and (3) a harmonic source with a series RL circuit. Radial distribution systems will generally exhibit a resonance or cluster of resonances between fifth and seventh harmonics. See also Reference [20].

7.6.3 Transmission Systems

Transmission systems have higher X/R ratios and lower impedances and the harmonics can be propagated over much longer distances. The capacitances of transformers and lines are higher and these need to be included. The operating configuration range of a transmission system is much wider than that of a distribution system. A study may begin by identifying a local area which must be modeled in detail. The distant portions of the system are represented as lumped equivalents. Equivalent impedance based on short-circuit impedance is one approach, the second approach uses a frequency versus impedance curve of the system, and there is a third intermediate area, whose boundaries must be carefully selected for accuracy. These can be based on geographical distance from the source bus. Series line impedance and the number of buses distant from the source are some other criteria. Figure 7.14 shows the division of a network into areas of main study and external systems. Sensitivity methods provide a better analytical tool.

7.6.4 Sensitivity Methods

The purpose is to ascertain the sensitivity of the system response when a component parameter varies. Adjoint network analysis can be used. The network N consisting of linear passive elements is excited at the bus of primary interest by a unit current at the harmonic source bus and branch currents $I_1 I_2, ..., I_n$ are obtained. The transfer impedance T is defined as the voltage output across the bus of primary interest divided by the harmonic current of the input bus. The adjoint network $N*$, which has the same topology as the original network, is excited by a unit current source from the output to obtain adjoint network branch currents $I_1^*, I_2^*, ..., I_n^*$ are obtained. The sensitivity of a transfer impedance T is defined as

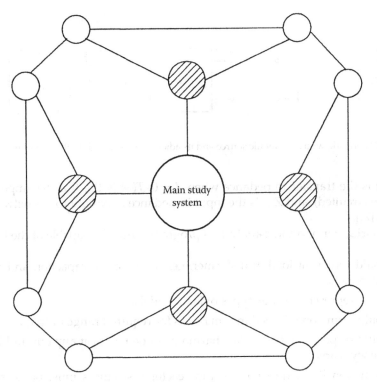

FIGURE 7.14
Division of a network into main study and external interconnected systems.

$$S_x^T = \frac{\partial T}{\partial x}\left(\frac{x}{T}\right) \tag{7.61}$$

where x is any parameter R, L, or C at frequency denoted by S_x^T. The sensitivities can be calculated using

$$\frac{\partial T}{\partial x} = I_x \cdot I_x^* \tag{7.62}$$

where I_x and I_x^* are x element branch currents from two-network analysis of N and $N*$, respectively (Figure 7.15a). The efficacy of the method is limited to small variations in the parameter. When large variations occur in external system equivalents, these can cause serious changes in the transfer function, and the bilinear theorem can be applied. These large changes in the transfer function of a two-port network to changes in an internal parameter are analyzed by pulling out Z of the network, effectively forming a three-port network (Figure 7.15b). For the transfer impedance, the following equation is obtained:

$$T = \frac{V_2}{I_1} = \frac{Z_{\text{xin}}T(0) + ZT(\infty)}{Z + Z_{\text{xin}}} \tag{7.63}$$

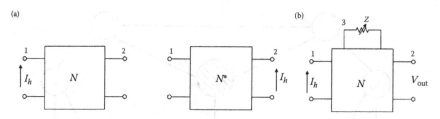

FIGURE 7.15
(a) Two-port N network, with a harmonic source and its adjoint network and (b) three-port network.

where $T(0)$ is the transfer impedance when $Z = 0$, $T(\infty)$ is the transfer impedance when $Z = \infty$ (open circuited), and Z_{xin} is the input impedance looking into the network from the nodes of Z [21].

To summarize, for harmonic analysis, a program should be capable of the following:

- It should represent load, transformer, generator, shunt capacitor, and induction motor models.
- It should represent various types of load models.
- It should form nodal impedance matrices for required range of frequencies.
- It should be possible to calculate harmonic impedances at any bus and harmonic frequency scans.
- Transmission lines, transformer and reactor resistances must be corrected for skin/proximity effects.
- It should be possible to calculate harmonic current spectrum for line flows.
- Harmonic current injection at each harmonic source and its reaction with system should be considered.
- Sensitivity analysis should be possible.
- A powerful plotting interface is required.
- The parameters like TIF, KVT, and IT should be possible to be calculated.
- When harmonic filters are applied, their loading and harmonic load flow throughout should be calculated, with distortion indices, according to IEEE 519. The system and filter resonant frequencies should be capable of being plotted.
- The three-phase models should be accommodated.
- The impact of transformer winding connections should be possible to be modeled.
- The harmonic derating factors of the transformers should be possible to be calculated.

7.7 Power Factor and Reactive Power

The power factor of a converter is given by the following expression:

$$\text{Total PF} = \frac{q}{\pi} \sin\left(\frac{\pi}{q}\right) \qquad (7.64)$$

where q is the number of converter pulses and πq is the angle in radians. This ignores commutation overlap, no-phase overlap, and neglects transformer magnetizing current. For a six-pulse converter, the power factor is $3/\pi = 0.955$. A 12-pulse converter has a theoretical power factor of 0.988. With commutation overlap and phase retard the power factor is given by Reference [22]:

$$\text{PF} = \frac{E'_d I_d}{\sqrt{3} E_L I_L} = \frac{3}{\pi} \frac{1}{\sqrt{1 - 3f(\mu, \alpha)}} \left(\cos\alpha - \frac{E_x}{E_{do}} \right) \qquad (7.65)$$

where
$E'_d = E_d + E_r + E_f$
E_d = average direct voltage under load
E_r = resistance drop
E_f = total forward drop per circuit element
I_d = DC load current in average amperes
E_L = primary line-to-line AC voltage
I_L = AC primary line current in rms ampères
α = phase retard angle
μ = angle of overlap or commutation angle
E_{do} = theoretical DC voltage
E_x = direct voltage drop due to commutation reactance

and

$$f(\mu, \alpha) = \frac{\sin\mu[2 + \cos(\mu + 2\alpha)] - \mu[1 + 2\cos\alpha\,\cos(\mu + \alpha)]}{[2\pi\cos\alpha - \cos(\mu + \alpha)]^2} \qquad (7.66)$$

The displacement power factor is

$$\cos\phi'_1 = \frac{\sin^2\mu}{\sqrt{\mu^2 + \sin^2\mu - 2\mu\sin\mu\cos\mu}} \qquad (7.67)$$

This relationship neglects transformer magnetizing current. The correction for magnetizing current is approximately given by

$$\cos\phi_1 = \cos[\text{arc }\cos\phi'_1 + \text{arc }\tan(I_{mag}/I_1)] \qquad (7.68)$$

where $\cos\phi'_1$ is the displacement power factor without transformer magnetizing current. The power factor of converters will vary with the type of converter and DC filter. In a PWM inverter, driven from a DC link voltage with a reactor and capacitor, the drive motor power factor is not truly reflected on the AC side, and is compensated by the filter capacitor and reactor.

Line-commutated inverters require reactive power from the supply system. The closer the operation to zero voltage DC, the more the reactive power required. Figure 7.16 shows the reactive power requirement of a fully controlled bridge circuit versus half-controlled bridge circuit. The maximum reactive power input for a half-controlled circuit is seen to be half of the fully controlled circuit. By sequential control, two converter sections are operated in series, one section fully phased on and the other section adding or subtracting from the voltage of the other section, see Figure 7.16.

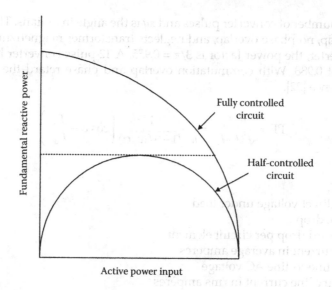

FIGURE 7.16
Reactive power requirements of fully controlled and half-controlled bridge circuits.

In the case of sinusoidal voltage and current, the following relationship holds:

$$S^2 = P^2 + Q^2 \tag{7.69}$$

where P is the active power, Q is the reactive volt–ampère and S is the volt–ampère. This relationship has been amply explored in load flow section: $S = V_f I_f$, $Q = V_f I_f \sin(\theta_f - \delta_f)$, and $PF = P/S$.

In the case of nonlinear load or when the source has nonsinusoidal waveform, the reactive power Q can be defined as

$$Q = \sum_{h=1}^{h=\infty} V_h I_h \sin(\theta_h - \delta_h) \tag{7.70}$$

and the apparent power can be defined as

$$S = \sqrt{P^2 + Q^2 + D^2} \tag{7.71}$$

where D is the distortion power. Consider D^2 up to the third harmonic:

$$D^2 = (V_0^2 + V_1^2 + V_2^2 + V_3^2)(I_0^2 + I_1^2 + I_2^2 + I_3^2)$$

$$-(V_0 I_0 + V_1 I_1 \cos\theta_1 + V_2 I_2 \cos\theta_2 + V_3 I_3 \cos\theta_3)^2$$

$$-(V_1 I_1 \sin\theta_1 + V_2 I_2 \sin\theta_2 + V_3 I_3 \sin\theta_3)^2 \tag{7.72}$$

An expression for distortion power factor can be arrived from current and voltage harmonic distortion factors. From the definition of these factors, rms harmonic voltages and currents can be written as

$$V_{\text{rms}(h)} = V_f \sqrt{1+(\text{THD}_V/100)^2} \qquad (7.73)$$

$$I_{\text{rms}(h)} = I_f \sqrt{1+(\text{THD}_I/100)^2} \qquad (7.74)$$

Therefore, the total power factor is

$$\text{PF}_{\text{tot}} = \frac{P}{V_f I_f \sqrt{1+(\text{THD}_V/100)^2} \sqrt{1+(\text{THD}_I/100)^2}} \qquad (7.75)$$

Neglecting the power contributed by harmonics and also voltage distortion, as it is generally small,

$$\text{PF}_{\text{tot}} = \cos(\theta_f - \delta_f) \cdot \frac{1}{\sqrt{1+(\text{THD}_I/100)^2}}$$

$$= \text{PF}_{\text{displacement}} \text{PF}_{\text{distortion}} \qquad (7.76)$$

The total power factor is the product of displacement power factor (which is the same as the fundamental power factor) and is multiplied by the distortion power factor as defined above.

7.8 Shunt Capacitor Bank Arrangements

Formation of shunt capacitor banks from small to large sizes and at various voltages is required for filter design and reactive power compensation. These can be connected in a variety of three-phase connections, which depend on the best utilization of the standard voltage ratings, fusing, and protective relaying. To meet certain kvar and voltage requirements, the banks are formed from standard unit power capacitors available in certain ratings and voltages. For high-voltage applications, these are outdoor rack mounted. For medium-voltage applications a bank may be provided in an indoor or outdoor metal enclosure or it can be rack mounted outdoors. Table 7.2 [23] shows the number of series groups for wye-connected capacitor banks required for line operating voltages from 12.47 to 500 kV, i.e., for 500 kV application, 14 series strings of 21.6 kV rated capacitors, or 38 strings of 7.62 kV rated capacitors are required in a formation as shown in Figure 7.17. The unit sizes are, generally, limited to 100, 200, 300, and 400 kvar, and in each string a number of units are connected in parallel to obtain the required kvar. There are limitations in forming the series and parallel strings. A minimum number of units should be placed in parallel per series group to limit voltage on the remaining units to 110%, if any one unit goes out of service, say due to operation of its fuse. This is because capacitors are rated to withstand 10% overvoltage; though for filter designs, other considerations enter into this picture. One is that larger the size of the capacitor unit lost (say due to operation of its fuse), the greater will be the detuning effect. This may overload the units remaining in service, shift the design frequencies, and increase the distortion at the point of common coupling (PCC), beyond acceptable limits.

Individual capacitor-can fusing is selected to protect the rupture/current withstand rating of the can. The maximum clearing time curve of the fuse and the case rupture curve are plotted together. The case rupture characteristics vary with the design and size and these

TABLE 7.2

Number of Series Groups in Y-Connected Capacitor Banks

V(kV)	V(kV)	Available Capacitor Voltage KV per Unit												
		21.6	7.92	14.4	13.8	13.28	12.47	9.96	9.54	8.32	7.96	7.62	7.2	6.64
500.0	288.7	14	15	20	21	22		29	30	35	36	38		
345.0	79.2		10			15	16	20	21	24	25	27		
230.0	132.8					10			14	16	17	18		20
161.0	92.9					7							13	14
138.0	79.7		4	6	6	6		8			10		11	12
115.0	66.4					5			7	8	9	9		10
69.0	39.8		2		3	3		4			5			6
46.0	26.56					2								4
34.5	7.92	1						2						3
24.9	14.4			1									2	
23.9	13.8				1									2
23.0	13.28					1								
14.4	8.32									1				
13.8	7.96										1			
13.2	7.62											1		
12.47	7.2												1	

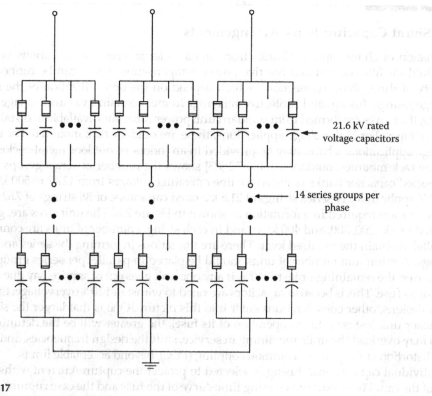

21.6 kV rated voltage capacitors

14 series groups per phase

FIGURE 7.17

Formation of a 500 kV shunt capacitor bank with series groups.

data should be obtained from a manufacturer. The curves are for 10% and 50% probability boundary. The probability of case rupture can be defined as opening of the case as a result of failure, from a mere cracked seam or bushing seal to a violent bursting of the case. Within the safe zone, no greater damage than a slight swelling of the case will occur, though a case rupture is possible for low levels of short-circuit current flowing for extended period of time. To avoid such ruptures the fuse link is coordinated so that it clears the fault in 300 s. The overvoltage duration that will occur on the capacitors remaining in service is of consideration. Also the fuse must be selected for continuous current, switching inrush current, and lightning surge current requirements. The available short-circuit and its clearing time at the point of application should be compared with the manufacturer's published data on short-circuit withstand. As the capacitor enclosures are grounded, a phase-to-ground fault can still occur, even if the capacitors are connected in ungrounded configuration. The maximum size of capacitor bank for group fusing is limited by the maximum size of the available fuse, current-limiting or T- and K-expulsion-type links. The expulsion-type fuses are generally used for outdoor rack mounted capacitor banks. In an ungrounded wye-connected bank, the maximum short-circuit current is only three times the full load current, and a group fuse selection to meet all the criteria and fast clearance time becomes difficult. For externally fused banks, fuses should be fast to coordinate with the fast unbalance relay settings, but should not operate during switching or external faults.

Table 7.3 shows the minimum units in a parallel group [23]. Also there is a limit to the number of parallel units connected in a series group when expulsion-type fuses are used for individual capacitor-can protection. The energy liberated and fed into the fault when a fuse operates is required to be limited, depending on the fuse characteristics (Figure 7.18). The energy release may be as follows:

$$E = 2.64 \text{ J per KVAC rated voltage} \tag{7.77}$$

$$E = 2.64(1.10)^2 \text{ J per KVAC 110\% voltage} \tag{7.78}$$

$$E = 2.64(1.20)^2 \text{ J per KVAC 120\% voltage} \tag{7.79}$$

TABLE 7.3

Minimum Number of Units in Parallel per Series Group to Limit Voltage on Remaining Units to 110% with One Unit Out

Number of Series Groups	Grounded Y or Δ	Ungrounded Y	Double Y, Equal Sections
1	–	4	2
2	6	8	7
3	8	9	8
4	9	10	9
5	9	10	10
6	10	10	10
7	10	10	10
8	10	11	10
9	10	11	10
10	10	11	11
11	10	11	11
12 and over	11	11	11

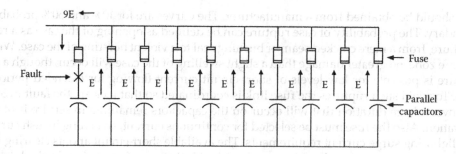

FIGURE 7.18
Energy fed into a fault from parallel capacitor units.

Normally, capacitors up to 3100 kvar can be connected in parallel when expulsion fuses are used. This limit can be exceeded if capacitor units are fused with current-limiting fuses (generally limited to indoor metal-enclosed installations).

Apart from the multiple series group grounded wye banks, the capacitors may be connected in

- Ungrounded wye connection
- Ungrounded double-wye neutrals
- Grounded double-wye neutrals
- Delta connection

These connections are shown in Figure 7.19. Delta connection is common for low-voltage application with one-series group rated for line-to-line voltage. A wye ungrounded group can be formed with one group per phase, when the required operating voltage corresponds to the standard capacitor unit rating. The wye neutral is left ungrounded. Grounded Y neutrals and multiple series groups are used for voltages above 34.5 kV. Multiple series groups limit the fault current. Grounded capacitors provide a low impedance path for lightning surge currents and give some protection from surge voltages; however, third-harmonic currents can circulate, and these may overheat the system grounding impedances. In the case of an ungrounded multiple series group wye-connected bank, third-harmonic currents do not flow, but the entire bank, including the neutral, should be insulated for the line voltage. Double-wye banks and multiple series groups are used when a capacitor bank becomes too large for the 3100 kvar per group for the expulsion type of fuses. The design of grounding grids and connection of neutral points of capacitor banks is of importance. Two methods of grounding are (1) single point grounding and (2) peninsula grounding. With single point grounding, the neutrals of capacitor banks of a given voltage are all connected through an insulated cable which is connected to the grid at one point. There will be substantial voltage of the order of tens of kV between the ends of the neutral bus and single point ground during switching. The use of shielded cable will help reduce the voltage stress. In the event of a fault high-frequency currents can flow back into power system via the substation ground grid.

With peninsula grounding, the grounding grid is built under capacitor banks and bus work in a form resembling a series of peninsulas. In this arrangement, one or more ground conductors may be carried under the capacitor rack of each phase of each group and ties to main station ground at one point at the edge of capacitor area. All capacitor neutral

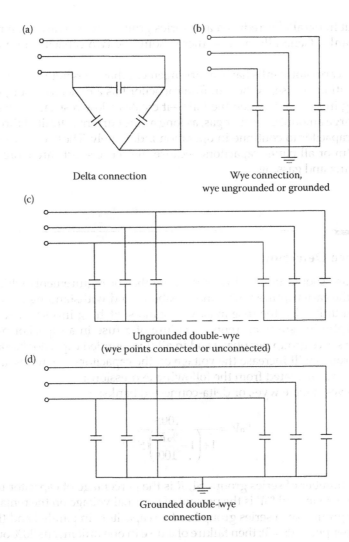

FIGURE 7.19
Three-phase connections of shunt capacitor banks.

connections are made to this isolated peninsula grounding grid conductors [23]. The capacitor bank and associated current transformer and voltage transformer potential will rise during capacitor switching, but the transients in the rest of the system will be reduced. Also with peninsula grounding, all equipment at the neutral ends tends to rise to the same potential and differential voltages can be avoided.

In the externally fused capacitors, the selection of the fuse is determined by the coordination with the capacitor case rupture withstand curve and the transient inrush currents. This fuse selection and coordination is not covered in this book.

The capacitor banks may be internally fused, where one fuse is connected in series with each capacitor element. On a puncture or short-circuit in a capacitor element, the current through the fuse increases in proportion to the number of units in parallel. This will melt the fuse in a short time of a few milliseconds. Generally, the number of units that can be put in parallel is reduced and series groups are increased compared to externally fused banks. Generally, at least there should be two capacitor units in parallel in each group.

The fuseless capacitor banks have an arrangement that does not use fuses; it is not simply the elimination of fuses. When all-film capacitor has a dielectric short, the film burns away resulting in a weld between the foils—it creates a low resistance short, which does not generate large amount of heat or gas, as long as the current is limited through the short, allowing the capacitor to continue in operation indefinitely. These designs were not used with paper/film or all paper capacitors because dielectric shorts are more likely to have localized heating and gassing.

7.9 Unbalance Detection

An external fuse should be sized to meet a number of requirements which include the protection of the case rupture withstand capability and withstanding the discharge current of a failed unit. Also the fuse must withstand switching inrush currents, restriking current, and lightning surge currents. A failure of a fuse in a capacitor bank involving more than one series group per phase and in all ungrounded capacitor banks irrespective of the series groups will increase the voltage on the capacitors remaining in service. This magnitude can be calculated from the following expressions:

Grounded wye-, double wye-, or delta-connected banks:

$$\%V = \frac{100S}{1+\left(1-\dfrac{\%R}{100}\right)(S-1)} \tag{7.80}$$

where S is the number of series groups >1, R is the percentage of capacitor units removed from one series group, and $\%V$ is the percent of nominal voltage on the remaining units in affected series group. Say a series group has five capacitors in parallel and there are three series groups per phase ($S = 3$); then failure of a fuse in one unit means 20% outage, ($R = 20$), and $\%V = 115\%$. A capacitor cannot withstand higher than 10% voltage continuously, and depending upon the specific application, this may not be acceptable.

Ungrounded wye or ungrounded double-wye neutrals are isolated:

$$\%V = \frac{100S}{S\left(1-\dfrac{\%R}{100}\right)+\dfrac{2}{3}\left(\dfrac{\%R}{100}\right)} \tag{7.81}$$

For the same configuration as for grounded bank, the $\%V$ will be 118.4%. The failure of a unit results in detuning in case of a filter bank, further discussed in Chapter 8. Generally, the voltage rise is limited to a maximum of 10% and, sometimes, even lower. A bank can be designed so that on loss of a single unit, the capacitor bank continues to operate in service and tripped immediately on loss of two units in the same group. The protection schemes to detect unbalance are not discussed; however, with modern microprocessor-based relays, it is possible to detect an unbalance as low as 1%–2% on failure of a capacitor unit in banks of parallel and series groups.

7.10 Study Cases

At each step of the calculations in these examples, we will analyze the results and draw inferences as we proceed. Modifications to the original problem and its solution reveal the nature of the study and harmonic resonance problems. We will then continue with the same examples in Chapter 8 to address the resolution of harmonic problems that we discover in this section.

Example 7.1: A Substation with Single Harmonic Source

Consider a six-pulse drive system load and a 2000 hp induction motor connected as shown in Figure 7.20a; 10 MVA and 5 MVA transformers have a percentage impedance of 6%. A 4.16 kV bus 2 is defined as the PCC. It is required to calculate the harmonic voltages and currents into the 2000 hp motor and 10 MVA transformer T1, and also harmonic distortion factors starting with hand calculations. The source impedance of the supply system is first neglected in the hand calculations; it is relatively small compared to the transformer impedance. We will itemize the steps of the calculations and make observations as we proceed.

$$I_{\text{rms}} = \left[\sum_{h=1}^{h=35} I_h^2\right]^{1/2} = 806.6\,\text{A}$$

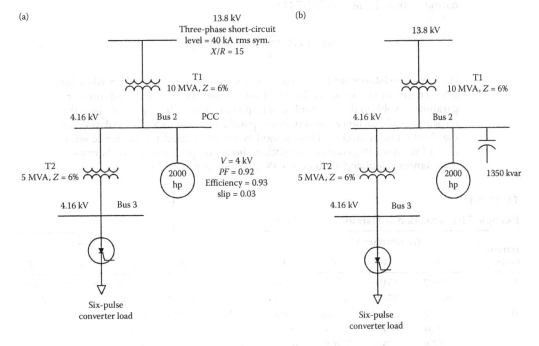

(a)

(b)

FIGURE 7.20

(a) System with harmonic load for Example 7.1 and (b) with a shunt capacitor of 1.35 Mvar added at bus 2.

1. *Derive a harmonic current injection model*: The converter operates at a firing angle α of 15°, and a 5 MVA transformer T2 is fully loaded. The hand calculations are limited to the 17th harmonic. This is to demonstrate the procedure of calculations, rather than a rigorous solution of the harmonic flow problem. Harmonic flow calculations are generally carried to the 40th harmonic and preferably to the 50th. By limiting the calculation to lower frequencies, a possible higher frequency resonance can be missed and serious errors in calculations of distortion factors can occur.

   The commutating reactance is 0.06 + 0.03 = 0.09%, approximately, on a 5 MVA base. The overlap angle is calculated using Equation 1.66 and it is ≈ 14.5°. From Chapter 5, the percentage harmonic currents are 18%, 13%, 6.5%, 4.8%, and 2.8%, respectively, for the 5th, 7th, 11th, 13th, and 17th harmonics. In terms of the fundamental current of transformer T2 = 695 A, the injected harmonic currents are 125, 90, 76, 33, and 19 A, respectively, for the 5th, 7th, 11th, 13th, and 17th harmonics. The transformer must be capable of supplying these nonlinear loads.

2. *Calculate bus harmonic impedances*: Table 7.4 shows the calculation of the transformer T1 resistance and reactance at harmonic frequencies. The resistance at fundamental frequency is multiplied by appropriate values read from the curve shown in Figure 7.6. The reactance at harmonic frequencies is directly proportional to the harmonic frequency, though this is an approximation.

   To calculate the harmonic impedances of the induction motor, its locked rotor reactance at the rated voltage is taken as 16.7% (the same as in the short-circuit calculations). For a 2000 hp motor, this gives an $(R_1 + r_2) + j(X_1 + x_2)$ of 1.658 Ω. The fundamental frequency resistance is calculated from the X/R ratio of 30. This gives a resistance of 0.0522 Ω. Assume that 45% of the resistance is the stator resistance and 55% is the rotor resistance. The stator resistance is then 0.02487 Ω and the rotor resistance is 0.03033 Ω. The slip at the fifth harmonic (negative sequence harmonic) is 1.2. The resistance at the fifth harmonic from Equations 7.41–7.44 is

$$0.02487\sqrt{5} + \frac{\sqrt{6}(0.03033)}{1.2} = 0.1186\,\Omega$$

Harmonic resistance and reactance for the motor, calculated in a likewise manner, are shown in Table 7.4. The transformer and motor impedances are paralleled to obtain the harmonic bus impedances at each frequency. Here, the resistance and reactance are separately paralleled, akin to X/R calculation for the short-circuit currents. This method is not to be used for harmonic studies and complex impedance calculation should be performed. The harmonic impedances calculated at bus 2, the PCC, are shown in Table 7.4.

TABLE 7.4

Example 7.1: Calculation of Harmonic Impedances

Harmonic Order	Transformer T1			Motor			Bus 2		
	R	X	Z	R	X	Z	R	X	Z
5	0.0277	0.5166	0.5173	0.1186	8.290	8.291	0.0225	0.4863	0.4868
7	0.0416	0.7231	0.7243	0.1525	11.606	11.607	0.0368	0.6807	0.6817
11	0.0832	1.1363	1.1393	0.1789	18.238	18.239	0.0568	1.0696	1.0711
13	0.0901	1.3420	1.3450	0.787	21.554	21.555	0.0620	1.2614	1.2629
17	0.1730	1.7560	1.7645	0.2241	28.186	28.187	0.0976	1.653	1.6558

Note: All values are in ohms.

3. *Calculate harmonic currents and voltages*: As the harmonic currents are known, the bus harmonic voltages can be calculated, by *IZ* multiplication. The harmonic currents in the motor and transformer T1 are then calculated, based on the harmonic voltages. Note that these do not exactly sum to the harmonic injected currents (Table 7.5).
4. *Calculate harmonic current and voltage distortion factors*: The limits of current and voltage distortion factors are discussed in Chapter 5. The voltage harmonic distortion factor is given by

$$\frac{\sqrt{(60.85)^2 + (61.40)^2 + (81.40)^2 + (41.67)^2 + (31.46)^2}}{(4160/\sqrt{3})}$$

This gives 5.36% distortion. The calculation is not accurate as the higher order harmonics are ignored. Based on the current injected into the main transformer T1 it is

$$\frac{\sqrt{(117.36)^2 + (84.71)^2 + (71.45)^2 + (30.98)^2 + (17.80)^2}}{830}$$

Note that the TDD is based upon the total load demand, which includes the nonlinear load and the load current of the 2000 hp motor. This gives a TDD of 19.9%. This exceeds the permissible levels at the PCC. This can be expected as the ratio of the nonlinear load to the total load is approximately 83%. When the nonlinear load exceeds about 35% of the total load, a careful analysis is required.

5. *Reactive power compensation*: Consider now that a power factor improvement shunt capacitor bank at bus 2 shown in Figure 7.20b is added to improve the overall power factor of the supply system to 90%.

    The operating power factor of the converter is 0.828. A load flow calculation shows that 5.62 MW and 3.92 Mvar must be supplied from the 13.8 kV system to transformer T1. This gives a power factor of 0.82 at 13.8 kV, and there is 0.58 Mvar of loss in the transformers.

    A shunt capacitor bank of 1.4 Mvar at bus 2 will improve the power factor from the supply system to approximately 90%.

6. *Form a capacitor bank and decide its connections*: A capacitor rating of 2.77 kV and 300 kvar is standard. The reasons for using a voltage rating higher than the system voltage are discussed in Chapter 8. At the voltage of use, two units per phase, connected in ungrounded wye configuration will give

$$3(600)\left[\frac{2.4}{2.77}\right]^2 = 1351.25 \text{ kvar}$$

TABLE 7.5

Example 7.1: Harmonic Current Flow and Harmonic Voltages

Harmonic Order	Harmonic Current Injected (A)	Current into Transformer T1 (A)	Current into 2000 hp Motor (A)	Harmonic Voltage (V)
5	125	117.63	7.33	60.85
7	90	84.71	5.28	61.35
11	76	71.45	4.46	81.40
13	33	30.98	1.93	41.67
17	7	17.80	1.12	31.46

Note: Harmonic voltage distortion factor = 5.36%.

and

$$X_c = \frac{(kV)^2 \cdot 10^3}{\text{Kvar}} = \frac{2.77^2 \cdot 10^3}{600} = 12.788 \ \Omega$$

or $C = 2.074E{-}4F$.

7. *Estimate resonance with capacitors*: The short-circuit level at bus 2 shown in Figure 7.20a is approximately 150 MVA and parallel resonance can be estimated to coincide with the 11th harmonic. To ascertain the resonant frequency correctly, a frequency scan is made at small increments of frequency. To capture the resonance correctly, this increment of frequency should be as small as practical. A scan at five-cycle intervals will miss the resonance peak by four cycles.

8. *Limitations of hand calculations*: This brings a break point in the hand calculations. Assuming that the impedances are calculated at five-cycle intervals, to cover the spectrum up to even the 17th harmonic, 192 complex calculations must be made in the first iteration. The hand calculations are impractical even for a small system.

9. *Frequency scan with capacitors*: Resorting to a computer calculation, the current injection is extended to the 35th harmonic and the system short-circuit level of 40 kA is inputted.

   The results of frequency scan at two-cycle intervals show that the resonant frequency is between 630 and 632 (a two-cycle frequency step is used in the calculations). This result was expected. The maximum angle is 88.59° and the minimum −89.93°; the impedance modulus is 122.70 Ω. The impedance modulus is shown in Figure 7.21a and phase angle in 7.21b. If a frequency scan is made without the capacitors, it shows that the maximum impedance is 3 Ω and it is a straight line of uniform slope (Figure 7.21a, dotted lines). The impedance at parallel resonance increases manyfold. Figure 7.21b shows an abrupt change in impedance phase angle with the addition of capacitor.

10. *Harmonic study with capacitors*: The distribution of harmonic currents at each frequency is shown in Table 7.6. It shows as follows:

    The harmonic currents throughout the spectrum are amplified. This amplification is high at the 11th harmonic (close to the resonant frequency of 630–632 Hz). While the injected current is 76 A, the current in the capacitor bank is 884 A, in the supply transformer 753 A, and in the motor 55.7 A.

    The 11th harmonic voltage is 1030 V. The harmonic voltage distortion is 43.3%, and the voltage–time wave shape shown in Figure 7.22 shows large distortion due to the 11th harmonic. The total harmonic current demand distortion factor is 94.52%. Chapter 5 shows that total TDD for this system should not exceed 8%.

    The harmonic loading on the capacitor banks is calculated as follows:

$$I_{\text{rms}} = \left[\sum_{h=1}^{h=35} I_h^2 \right]^{1/2} = 806.6\,\text{A}$$

The permissible rms current in the capacitors including harmonics is 1.35 times the rated current = 337.56 A. The current loading is exceeded. The rms voltage is calculated in a similar way to current and it is 2614 V, V_{rms} rated = 2770 V, V_{rms} permissible = 3047 V. The voltage rating is not exceeded. The kvar loading is calculated from

$$\text{Kvar} = \sum_{h=1}^{h=35} I_h V_h = 2283$$

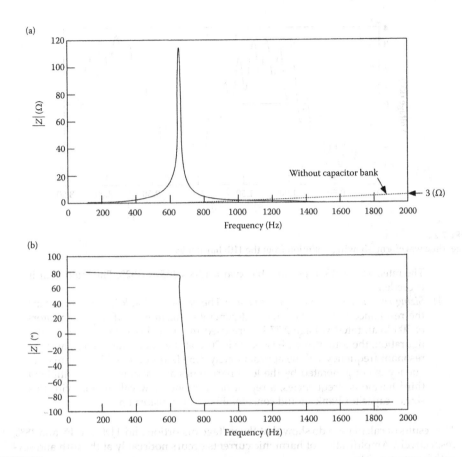

FIGURE 7.21
(a) Impedance modulus showing resonance at the 11th harmonic with addition of 1.35 Mvar capacitor bank (Example 7.1) and (b) impedance phase angle plot.

TABLE 7.6

Example 7.1: Harmonic Currents and Voltages with a Shunt Capacitor Bank

Order of the Harmonic	Injected Current (A)	Current in the Capacitor Bank (A)	Current in the Transformer T1, at PCC (A)	Current in the 2000 hp Motor (A)	Harmonic Voltage, Bus 2 (V)
5	125	36.4	150	11.1	93.5
7	90	71.5	150	11.1	13.9
11	76	884	753	55.7	1030
13	33	95	58	4.3	94.2
17	7	30	11	0.81	23.2
7	10.5	15.9	4.5	0.33	10.70
23	3.5	4.4	0.86	0.064	2.47
25	2.8	3.4	0.56	0.042	1.74
29	2	2.3	0.28	0.021	1.02
31	1.4	1.6	0.17	0.013	0.65
35	1.2	1.4	0.14	0.009	0.65

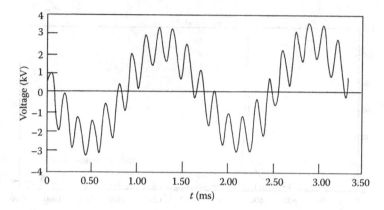

FIGURE 7.22
Voltage–time waveform, showing pollution from the 11th harmonic.

The rated kvar = 1351, permissible kvar = 1.35 × 1351 = 1823. The rated kvar is exceeded.

11. *Sizing capacitor bank to escape resonance*: The capacitor bank is sized to escape the resonance with any of the load-generated harmonics. If three capacitors of 300 kvar, rated voltage 2.77 kV are used in parallel per phase in wye configuration, the total three-phase kvar is 2026 at the operating voltage, and the resonant frequency will be approximately 515 Hz (Equation 7.20). As this frequency is not generated by the load harmonics and also is not a multiple of third harmonic frequencies, a repeat of harmonic flow calculation with this size of capacitor bank should considerably reduce distortion.

The results of calculation do show that the voltage distortion and TDD are 4% and 18%, respectively. Amplification of harmonic currents occurs noticeably at the fifth and seventh harmonics. Thus,

- The mitigation of the resonance problem by selecting the capacitor size to escape resonance may not minimize the distortion to acceptable levels.
- Current amplifications occur at frequencies adjacent to the resonant frequencies, though these may not exactly coincide with the resonant frequency of the system.
- The size of the capacitor bank has to be configured based on series parallel combinations of standard unit sizes and this may not always give the desired size. Of necessity, the capacitor banks have to be sized a step larger or a step smaller to adhere to configurations with available capacitor ratings and over- or undercompensation results.
- The resonant frequency swings with the change in system operating conditions and this may bring about a resonant condition, however carefully the capacitors were sized in the initial phase. The resonant frequency is especially sensitive to change in the utility's short-circuit impedance.

The example is carried further in Chapter 8 for the design of passive filters.

Example 7.2: Large Industrial System

This example portrays a relatively large industrial distribution system with plant generation and approximately 18% of the total load consisting of six-pulse converters. The system configuration is shown in Figure 7.23. The loads are lumped on equivalent

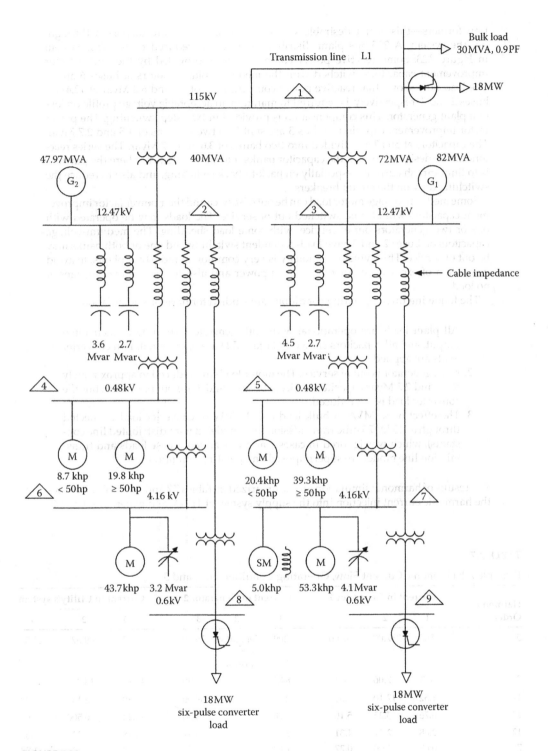

FIGURE 7.23
Single-line diagram of a large industrial plant, loads aggregated for harmonic study (Example 7.2).

transformers—this is not desirable when harmonic sources are dispersed throughout the system. A 225-bus plant distribution system is reduced to an 8-bus system in Figure 7.23. Some reactive power compensation is provided by the power factor improvement capacitors switched with the medium-voltage motors at buses 6 and 7. The load flow shows that reactive power compensation of 7.2 and 6.3 Mvar at 12.47 kV buses 2 and 3, respectively, is required to maintain an acceptable voltage profile on loss of a plant generator. This compensation is provided in two-step switching. The power factor improvement capacitors at bus 3 are split into two sections of 4.5 and 2.7 Mvar. The capacitors at bus 2 are divided into two banks of 3.6 and 2.7 Mvar. The series reactors can be designed to turn the capacitor banks into shunt filters, but here the purpose is to limit inrush currents, especially on back-to-back switching, and also to reduce the switching duty on the circuit breakers.

Some medium-voltage motor load can be out of service and their power factor improvement capacitors will also be switched out of service. The loads may be operated with one or two generators out of service, with some load shedding. The medium-voltage capacitors on buses 2 and 3 have load-dependent switching and one or both banks may be out of service. This switching strategy is very common in industrial plants, to avoid generation of excessive capacitive reactive power and also to prevent overvoltages at no load.

The following three operating conditions are studied for harmonic simulation:

1. All plant loads are operational with both generators running at their rated output, and all capacitors shown in Figure 7.23 are operational. Full converter loads are applied.
2. No. 2 generator is out of service. The motor loads are reduced to approximately 50%, and 2.7 Mvar capacitor banks at buses 2 and 3 are out of service, but the converter load is not reduced.
3. The effect of 30 MVA of bulk load and 18 MW of converter load connected through a 115 kV 75 mile transmission line (modeled with distributed line constants), which was ignored in cases 1 and 2, is added. These loads and transmission line model are superimposed on operating condition 1.

The results of harmonic simulation are summarized in Tables 7.7 and 7.8. Table 7.7 shows the harmonic current injection into the supply system at 115 kV and generators 1 and 2.

TABLE 7.7

Example 7.2: Harmonic Current Flow, Operating Conditions 1, 2, and 3

Harmonic Order	Current in Generator 1			Current in Generator 2			Current in Utility's system		
	1	2	3	1	2	3	1	2	3
5	187	257.00	167.00	209	Generator out of service	76.00	3.22	69.02	23.3
7	353	7.06	340	84.2		82.90	44.9	18.31	44.60
11	1.52	17.10	1.83	102		109.00	3.80	2.13	4.32
13	6.76	3.25	5.10	21.40		18.50	0.22	0.500	3.67
17	2.88	2.71	4.31	2.74		3.34	0.48	0.33	1.55
7	0.68	7.00	0.27	1.03		0.74	1.28	2.23	1.88
23	0.06	0.93	0.09	0.12		0.15	0.003	0.18	1.64
25	0.7	0.37	0.16	0.11		0.09	0.03	0.08	1.63

TABLE 7.8

Example 7.2: Resonant Frequencies and Impedance Modulus, Operating Conditions 1, 2, and 3

Bus ID	1			2			3		
	Parallel Resonance	Series Resonance	Impedance Modulus	Parallel Resonance	Series Resonance	Impedance Modulus	Parallel Resonance	Series Resonance	Impedance Modulus
Utility	403	424	239.6	373	385	318.6	400	425	213.6
				496	530				
Bus 2	586	1436	17.68	367	647	41.5	589	1436	28.26
				905					
Bus 3	400	635	33.76	493	860	26.24	401	635	12.89
	893	1343		1153	1699		892	1342	

Table 7.8 shows the parallel and series resonant frequencies. The impedance modulus versus frequency plots of the 115 kV bus and 12.47 kV buses 2 and 3 for all three cases of study are shown in Figures 7.24 through 7.26. The R/X plots of the utility's supply system and impedance modulus versus frequency plots are shown in Figure 7.27.

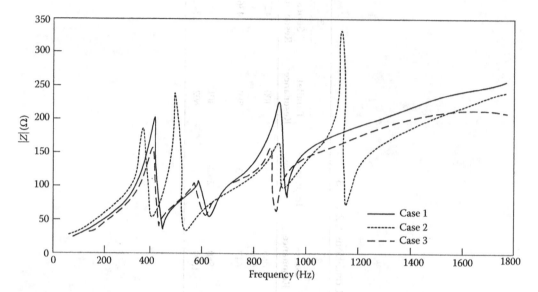

FIGURE 7.24
Impedance modulus versus frequency plot for three conditions of operation, 115 kV utility's bus (Example 7.2).

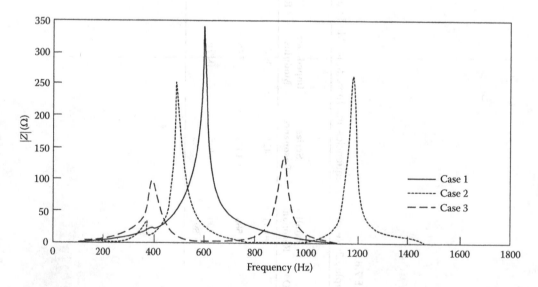

FIGURE 7.25
Impedance modulus versus frequency plot for three conditions of operation, 12.47 kV bus 2 (Example 7.2).

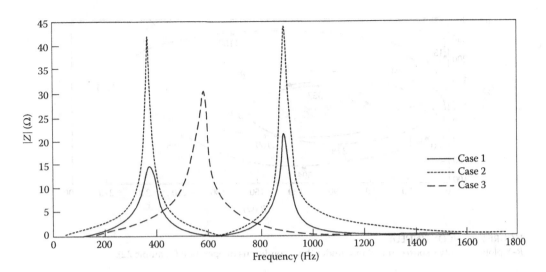

FIGURE 7.26
Impedance modulus versus frequency plot for three conditions of operation, 12.47 kV bus 3 (Example 7.2).

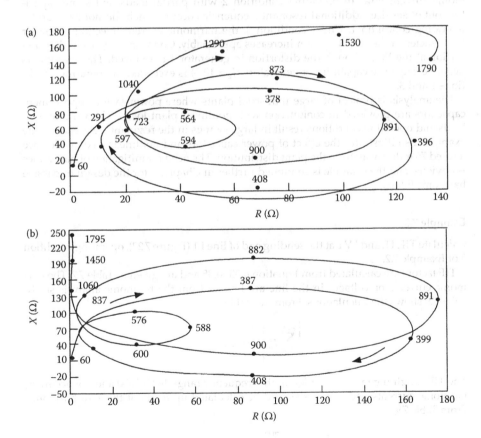

FIGURE 7.27
R–X plots of utility's source impedance under three conditions of operation (Example 7.2).

(*Continued*)

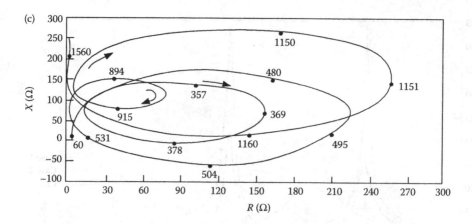

FIGURE 7.27   (CONTINUED)
R–X plots of utility's source impedance under three conditions of operation (Example 7.2).

It is seen that the resonant frequencies vary over wide limits and so does the harmonic current flow. In operating condition 2 with partial loads and some capacitors out of service, additional resonant frequencies occur, which did not exist under operating condition 1. Under condition 3, the harmonic current injection at higher frequencies in the utility system increases appreciably. Condition 2 gives higher distortion at the PCC, though the distortion in generator 1 is reduced. The continuous negative sequence capability of generators ($I_2 = 10\%$) is exceeded in operating conditions 1 and 3.

The analysis is typical of large industrial plants where power factor improvement capacitors are provided in conjunction with nonlinear plant loads. The variations in loads and operating conditions result in large swings in the resonant frequencies. The example also illustrates the effect of power capacitors and nonlinear loads which are located 75 miles away from the plant distribution. The need for mitigation of harmonics is obvious, and the example is continued further in Chapter 8 for the design of passive harmonic filters.

Example 7.3

Calculate TIF, IT, and kVT at the sending end of line L1 (Figure 7.23), operating condition 3 of Example 7.2.

TIF factors are calculated from Equation 6.33–6.35 and are given in Table 7.7. The harmonic currents or voltages in the line are known from the harmonic analysis study. Table 7.9 shows the calculations. From this table,

$$\left[\sum_{h=1}^{h=49} I_h^2\right]^{1/2} = 0.159 \text{ kA}$$

The TIF weighting factors are high in the frequency range 1620–3000 and, for accuracy, harmonic currents and voltages should be calculated up to about the 49th harmonic. From Table 7.9,

$$\left[\sum_{h=1}^{h=49} W_f^2 I_f^2\right]^{1/2} = \sqrt{4700.73} = 68.56$$

TABLE 7.9

Example 7.3: Calculation of IT Product, Sending End, Distributed Parameter Line L1 (Figure 7.22)

Harmonic	Frequency	TIF Weighting (W_f)	Harmonic Current in kA (I_f)	$W_f I_f$	$(W_f I_f)^2$
1	60	0.5	0.158	0.079	0.006
5	300	225	0.0127	2.857	8.165
7	420	650	0.00269	1.749	3.057
11	660	2260	0.0110	24.860	618.020
13	780	3360	0.00438	14.717	216.584
17	1020	5100	0.00431	21.980	483.164
7	1140	5630	0.00381	21.450	460.115
23	1380	6370	0.00257	16.371	268.006
25	1500	6680	0.00324	21.643	468.428
27	1740	7320	0.00292	21.374	456.865
31	1860	7820	0.00247	7.315	373.085
35	2100	8830	0.00175	15.453	238.780
37	2220	9330	0.00157	14.648	214.567
41	2460	10,360	0.00146	15.126	228.784
43	2580	10,600	0.00150	15.900	252.810
47	2820	10,210	0.00150	15.315	234.549
49	2940	9820	0.00135	13.256	175.748

Therefore, IT = 68.56/0.159 = 360.85; kVT can be similarly calculated using harmonic voltages in kV. TIF factors for industrial distribution systems are, normally, not a concern.

Example 7.4: Transmission System

Harmonic load flow in a transmission system network is illustrated in this example. The system configuration is shown in Figure 7.28. There are eight transmission lines and the line constants are shown in Table 7.10. These are modeled with distributed parameters or π networks. Also, the 6-pulse and 12-pulse converters are modeled as ideal converters.

The frequency scan with *only harmonic injections* and without bulk linear loads shows numerous resonant frequencies with varying impedance modulus; for example, Figure 7.29 shows these at the 230 kV bus 1. The resonance at higher frequencies is more predominant. The harmonic current flows in lines as shown in Table 7.11.

Figure 7.30 is a frequency scan with bulk loads applied. The resonant frequencies and a number of smaller resonance points are eliminated. The impedance modulus is reduced. Table 7.12 shows the cluster of resonant frequencies and impedance modulus in these two cases.

The effect on harmonic current flows in line Sections 1–5 and 1–2 under harmonic injections only and under harmonic injections and loads is shown in Figures 7.31 and 7.32.

Example 7.5: Three-Phase Modeling

Consider a 2.0 MVA, 2.4–0.48 kV delta–wye-connected transformer serving a mixed three-phase load and single-phase load, as shown in Figure 7.33. A single-phase 480–240 V, 500 kVA transformer serves lighting and switch-mode supply loads. The

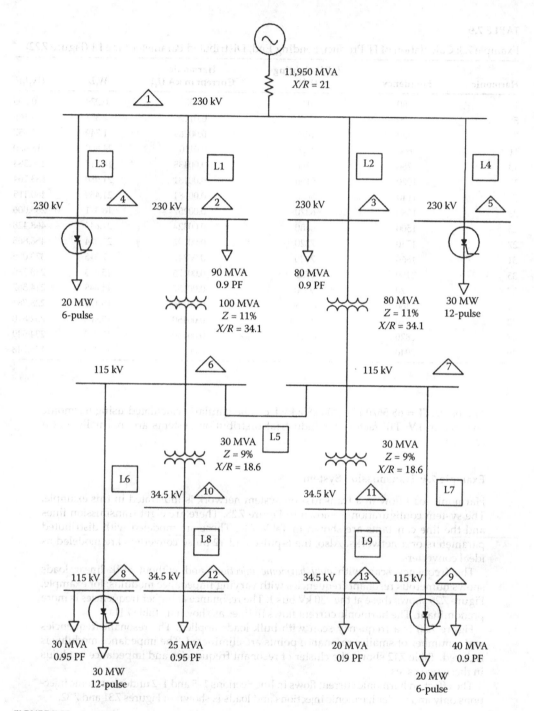

FIGURE 7.28
A transmission system for harmonic analysis study (Example 7.4).

TABLE 7.10

Example 7.4, Transmission Line Data and Models

Line	Voltage (kV)	Conductor ACSR KCMIL	Length (Miles)	GMD	R (per Unit, 100 MVA Base)	X_L (per Unit, 100 MVA Base)	X_c (per Unit 1 100 MVA Base)	Model Type
L1	230	636	100	34	0.00003	0.08585	1.8290	Distributed
L2	230	636	120	34	0.00004	0.10302	1.5240	Distributed
L3	230	636	50	34	0.00002	0.04293	3.6580	Distributed
L4	230	636	40	34	0.00001	0.03434	4.5720	Distributed
L5	115	715	30	21	0.00004	0.08828	20.347	π model
L6	115	715	40	21	0.00005	0.11771	15.260	Distributed
L7	115	715	50	21	0.00006	0.14713	12.208	Distributed
L8	34.5	397.5	30	5	0.00072	0.62473	131.228	π model
L9	34.5	397.5	30	5	0.00072	0.62473	131.228	π model

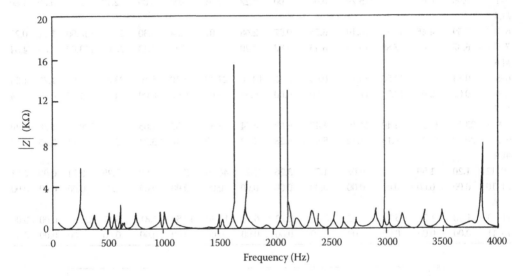

FIGURE 7.29

Cluster of resonant frequencies at 230 kV bus 1 without bulk load applied (Example 7.4).

harmonic spectrum is modeled according to Chapter 1. The results of harmonic flow study are shown in Table 7.13 and the harmonic currents injected into the wye secondary of a 2 MVA, 2.4–0.48 kV transformer are shown. The capacitors at bus B amplify third harmonic currents. As the single-phase transformer is connected between phases *a* and *b*, only these phases carry harmonic currents. The harmonic currents in phase *c* are zero. If the 350 kvar capacitors are removed, the harmonic injection pattern changes. The currents at the second and third harmonics are reduced, while those at the fourth, fifth, and seventh harmonics are increased, that is, capacitors amplify lower order harmonics and attenuate higher order harmonics.

TABLE 7.11

Harmonic Current Flow, with Harmonic Current Injection, Bulk Loads Not Applied

Line	\multicolumn{14}{c}{Harmonic Current Flow (A)}												
	5th	7th	11th	13th	17th	7th	23rd	25th	29th	31st	35th	37th	41st	43rd
#L1														
1–2	5.71	7.16	7.73	2.55	6.17	0.90	0.52	18.5	7.92	2.71	0.41	42.81	0.04	1.31
2–1	4.39	1.50	0.06	1.39	2.79	0.95	0.47	0.32	7.91	2.24	0.40	46.20	0.08	0.89
#L2														
1–3	6.70	21.60	16.03	4.06	10.70	21.50	1.18	0.87	21.60	3.68	1.40	17.20	1.7	0.11
3–1	13.62	2.56	13.01	5.7	0.53	3.89	0.20	2.99	6.58	0.42	3.28	29.30	0.44	0.04
#L3														
1–4	4.30	12.63	8.31	8.39	24.70	7.54	0.38	2.39	1.71	1.92	9.14	72.80	1.39	3.38
4–1	10.00	7.17	4.56	3.86	2.95	2.64	2.18	2.01	1.73	1.62	1.43	1.36	1.22	1.07
#L4														
1–5	5.63	1.95	0.03	1.73	20.20	30.00	0.29	15.10	15.30	0.95	7.30	12.60	0.16	2.17
5–1	0.00	0.00	6.85	5.79	0.00	0.00	3.27	3.01	0.00	0.00	2.15	2.04	0.00	1.60
#L5														
6–7	7.59	4.45	11.31	16.10	6.75	0.07	2.68	9.01	1.24	0.30	23.6	148.00	1.61	0.25
7–6	6.37	5.67	7.88	15.10	6.45	0.87	3.20	18.50	7.40	3.43	22.0	173.00	0.88	4.34
#L6														
6–8	0.84	1.00	12.50	13.10	0.60	0.86	10.30	22.40	24.10	5.10	33.2	250.00	1.79	1.24
8–6	0.00	0.00	13.70	11.60	0.00	0.00	6.55	6.02	0.00	0.00	4.30	4.70	0.00	3.24
#L7														
7–9	33.20	11.10	32.40	24.40	4.47	6.09	4.64	8.46	2.52	2.05	30.5	239.00	1.78	4.40
9–7	20.10	14.30	9.13	7.72	5.91	5.28	4.37	4.02	3.46	3.24	2.87	2.71	2.45	2.14
#L8														
10–12	1.20	1.50	3.62	0.07	1.73	3.26	22.2	46.70	23.5	3.09	5.25	3.31	0.00	2.61
21–10	0.00	0.00	0.00	0.00	0.00	0.00	0.00	0.00	0.00	0.00	0.00	0.00	0.00	0.00
#L9														
11–13	1.40	1.7	5.29	3.63	3.03	2,72	6.13	13.50	12.50	1.81	6.17	23.8	0.00	0.08
13–11	0.00	0.00	0.00	0.00	0.00	0.00	0.00	0.00	0.00	0.00	0.00	0.00	0.00	0.00

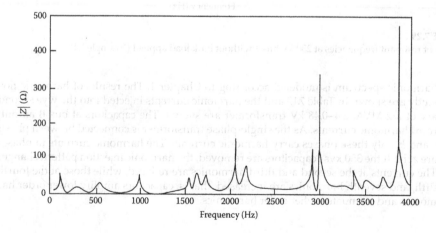

FIGURE 7.30
Shift in resonant frequencies at 230 kV bus 1 with bulk load applied (Example 7.4).

TABLE 7.12

Example 7.4: Resonant Frequencies and Maximum Impedance Modulus

	Without Load			With Load		
Bus ID	Parallel Resonance (Hz)	Series Resonance (Hz)	Impedance Modulus (Ω)	Parallel Resonance	Series Resonance (Hz)	Impedance Modulus (Ω)
1	242,362,524, 581,739,958, 997,1123, 1186,1513, 1646,1735, 1832,2053, 2126,2212, 2381,2569, 2881,2979	187,340,464, 572,733,872, 979,1114, 1165,1189, 1520,1687, 1823,1856, 2084,279, 2370,2555, 2612,2909	27,410	545,974, 1742,2074, 2153,2582, 2902,2990	870,1166, 1874,2114, 2574,2610, 2934	3749
6	362,580,739, 958,997, 1123,1187, 1294,1315, 1514,1646, 1735,1832, 2054,2126, 2213,2381, 2638	469,643,757, 967,1022, 1129,1279, 1312,1346, 1538,1685, 187,2008, 2056,2149, 2372,2636, 2872	46,300	1542,1662, 1762,2150, 2298,2442, 2665	1558,1702, 2014,2166, 2394,2657	1135
10	362,580,739, 958,997, 1123,1186, 1295,1315, 1513,1645, 2212	392,589,742, 962,1004, 1126,1243, 1306,1502, 1639,1738, 2224	32,820	2306,2402, 2622	2357,2466, 2860	448.6

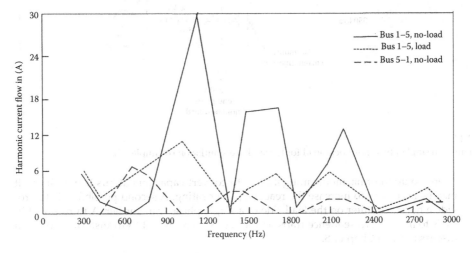

FIGURE 7.31
Harmonic current flow in bus ties 1–5 (Example 7.4).

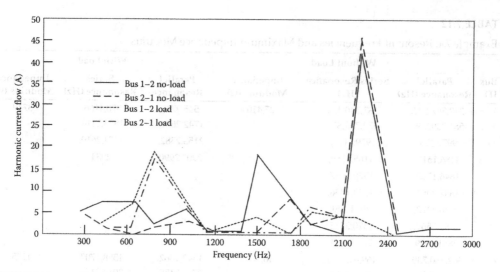

FIGURE 7.32
Harmonic current flow in bus ties 1–2 (Example 7.4).

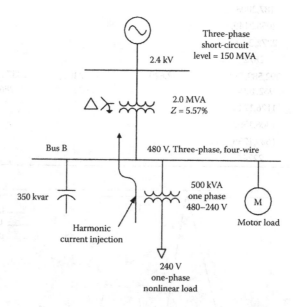

FIGURE 7.33
System with single-phase nonlinear load for three-phase simulation (Example 7.5).

One solution of harmonic mitigation is to convert capacitors at bus B into filters. It may be necessary to increase the reactive power rating. A second possibility is to provide a harmonic filter on the load side of a single-phase transformer. A third possibility is to install zero sequence traps on the wye side of the 2.4 kV transformer. We will discuss these in Chapter 8.

TABLE 7.13

Example 7.5: Harmonic Current Injected into Wye Secondary of Transformer T1 (in A), Figure 7.29

	With 350 kvar Capacitors on Bus B			350 kvar Capacitors on Bus B removed		
h	Phase *a*	Phase *b*	Phase *c*	Phase *a*	Phase *b*	Phase *c*
2	141	141	0	46.3	46.3	0
3	456	456	(Single- phase	241	241	(Single- phase load)
4	8.8	8.8	load)	150	150	
5	7.82	7.82		24	24	
6	6.30	6.30		18	18	
7	8.26	8.26		60	60	

Problems

7.1 Calculate overlap angle and harmonic spectrum of a six-pulse converter operating in the following configuration:

Delay angle = 30°, supply source short-circuit level at 13.8 kV = 875 MVA, converter transformer 3000 kVA, percentage impedance = 5.5%, converter load = 75% of the transformer kVA rating.

7.2 A 10 MVA transformer has a percentage impedance of 10% and X/R ratio of 10. Tabulate its resistance and reactance for harmonic spectrum of a six-pulse converter up to the 25th harmonic.

7.3 Repeat Problem 7.2 for an induction motor of a 2.3 kV, four-pole, 2500 hp, full-load power factor = 0.92, full load efficiency = 0.93%, locked rotor current = six times the full-load current at 20% power factor.

7.4 Calculate PF, DF, distortion power, and active and reactive power for the converter in Problem 7.1.

7.5 Harmonic propagation in a transmission system is to be studied. Describe how the extent of the system to be modeled will be decided. What models of lines and transformers will be used?

7.6 Form a capacitor bank of 15 Mvar, operating voltage 44 kV in wye configuration from unit capacitor sizes in Table 7.2. Expulsion fuses are to be used for fusing of individual capacitor units.

7.7 Explain why IT and kVT will be different at the sending and receiving end of a transmission line.

7.8 Plot the impedance versus frequency across a transmission line using equivalent π model for a long line. Consider line parameters as follows:

$$Y_{sr} = 0.464 - j2.45$$

$$Y = j0.0538$$

References

1. G.D. Breuer, J.H. Chow, T.J. Gentile, C.B. Lindh, F.H. Numrich, G. Addis, R.H. Lasseter, J.J. Vithayathil. HVDC–AC interaction. Part-1 Development of harmonic measurement system hardware and software, *IEEE Trans Power Appar Syst*, vol. PAS-101, no. 3, pp. 701–706, 1982.

2. G.D. Breuer, J.H. Chow, T.J. Gentile, C.B. Lindh, G. Addis, R.H. Lasseter, J.J. Vithayathil. HVDC–AC interaction. Part-II AC system harmonic model with comparison of calculated and measured data, *IEEE Trans Power Appar Syst*, vol. PAS-101, no. 3, pp. 709–717, 1982.

3. IEEE. Task force on harmonic modeling and simulation, modeling and simulation of propagation of harmonics in electrical power systems—Part I: Concepts, models and simulation techniques, *IEEE Trans Power Deliv*, vol. 11, pp. 452–465, 1996.

4. IEEE. Task force on harmonic modeling and simulation, modeling and simulation of propagation of harmonics in electrical power systems—Part II: Sample systems and examples, *IEEE Trans Power Deliv*, vol. 11, pp. 466–474, 1996.

5. C. Hatziadoniu, G.D. Galanos. Interaction between the AC voltages and DC current in weak AC/DC interconnections, *IEEE Trans Power Deliv*, vol. 3, pp. 1297–1304, 1988.

6. D. Xia, G.T. Heydt. Harmonic power flow studies Part 1—Formulation and solution, *IEEE Trans Power Appar Syst*, vol. 101, pp. 1257–1265, 1982.

7. D. Xia, G.T. Heydt. Harmonic power flow studies—Part II, Implementation and practical applications, *IEEE Trans Power Appar Syst*, vol. 101, pp. 1266–1270, 1982.

8. A.A. Mohmoud, R.D. Shultz. A method of analyzing harmonic distribution in AC power systems, *IEEE Trans Power Appar Syst*, vol. 101, pp. 1815–1824, 1982.

9. W. Xu, J.R. Marti, H.W. Dommel. A multiphase harmonic load flow solution technique, *IEEE Trans Power Deliv*, vol. 6, no. 1, pp. 174–244, 1991.

10. J. Arrillaga, B.C. Smith, N.R. Watson, A.R. Wood. *Power System Harmonic Analysis*. John Wiley & Sons, Chichester, West Sussex, 1997.

11. EPRI. HVDC-AC system interaction for AC harmonics. EPRI Report 1: 7.2-7.3, 1983.

12. CIGRE. Working Group 36–05. Harmonic characteristic parameters, methods of study, estimating of existing values in the network, *Electra* pp. 35–54, 1977.

13. E.W. Kimbark. *Direct Current Transmission*. John Wiley & Sons, New York, 1971.

14. M. Valcarcel, J.G. Mayordomo. Harmonic power flow for unbalance systems, *IEEE Trans Power Deliv*, vol. 8, pp. 2052–2059, 1993.

15. D.J. Pileggi, N.H. Chandra, A.E. Emanuel. Prediction of harmonic voltages in distribution systems, *IEEE Trans on Power Appar Syst*, vol. 100, no. 3, pp. 1307–1315, 1981.

16. J.P. Tamby, V.I. John. Q'Harm—A harmonic power flow program for small power systems, *IEEE Trans Power Syst*, vol. 3, no. 3, pp. 945–955, 1988.

17. N.R. Watson, J. Arrillaga. Frequency dependent AC system equivalents for harmonic studies and transient converter simulation, *IEEE Trans Power Deliv*, vol. 3, no. 3, pp. 1196–1203, 1988.

18. E.V. Larsen, D.H. Baker, J.C. Mclver. Low order harmonic interaction on AC/DC systems, *IEEE Trans Power Deliv*, vol. 4, no. 1, pp. 493–501, 1989.

19. T. Hiyama, M.S.A.A. Hammam, T.H. Ortmeyer. Distribution system modeling with distributed harmonic sources, *IEEE Trans Power Deliv*, vol. 4, pp. 1297–1304, 1989.

20. M.F. McGranaghan, R.C. Dugan, W.L. Sponsler. Digital simulation of distribution system frequency-response characteristics, *IEEE Trans Power Appar Syst*, vol. 100, pp. 1362–1369, 1981.

21. M.F. Akram, T.H. Ortmeyer, J.A. Svoboda. An improved harmonic modeling technique for transmission network, *IEEE Trans Power Deliv*, vol. 9, pp. 1510–1516, 1994.

22. IEEE Standard 519. IEEE recommended practice and requirements for harmonic control in electrical systems, 1992.

23. ANSI/IEEE Standard C37.09. Guide for protection of shunt capacitor banks, 2005.

8

Harmonic Mitigation and Filters

With the increase in consumer nonlinear load, the harmonics injected into the power supply system and their consequent effects are becoming of greater concern. Harmonic currents seeking a low impedance path or a resonant condition can travel through the power system and create problems for the consumers who do not have their own source of harmonic generation. IEEE standard 519 [1] stipulates who is responsible for what. A consumer can inject only a certain amount of harmonic current into the supply system, depending on his/her load, short-circuit current, and supply system voltage. The utility supply companies must ensure a certain voltage quality at the point of common coupling (PCC) with the consumer apparatus. These harmonic current and voltage limits [1, 2] are discussed in Chapter 5.

8.1 Mitigation of Harmonics

There are four major methodologies for mitigation of harmonics:

1. The equipment can be designed to withstand the effect of harmonics, e.g., transformers, cables, and motors can be derated. Motors for pulse-width modulation (PWM) inverters can be provided with special insulation to withstand high du/dt, and the relays can be rms sensing.

2. Passive filters at suitable locations, preferably close to the source of harmonic generation, can be provided so that the harmonic currents are trapped at the source and the harmonics propagated in the system are reduced.

3. Active filtering techniques, generally, incorporated with the harmonic-producing equipment itself can reduce the harmonic generation at source. Hybrid combinations of active and passive filters are also a possibility.

4. Alternative technologies can be adopted to limit the harmonics at source, e.g., phase multiplication, operation with higher pulse numbers, converters with interphase reactors, active wave-shaping techniques, multilevel converters, and harmonic compensation built into the harmonic-producing equipment itself to reduce harmonic generation.

The most useful strategy in a given situation largely depends on the currents and voltages involved, the nature of loads, and the specific system parameters, e.g., short-circuit level at the PCC.

8.2 Single-Tuned Filters

The operation of a single-tuned (ST) shunt filter is explained with reference to Figure 8.1. (Any other type of filter connected in the shunt can be termed a shunt filter.) Figure 8.1a shows a system configuration with nonlinear load and Figure 8.1b shows the equivalent circuit. Harmonic current injected from the source, through impedance Z_c, divides into the filter and the system. The system impedance for this case, shown as Z_s, consists of the source impedance Z_u in series with the transformer impedance Z_t and paralleled with the motor impedance:

$$I_h = I_f + I_s \qquad (8.1)$$

where I_h is the harmonic current injected into the system, I_f is the current through the filter, and I_s is the current through the system impedance. Also

$$I_f Z_f = I_s Z_s \qquad (8.2)$$

i.e., the harmonic voltage across the filter impedance (Z_f) equals the harmonic voltage across the equivalent power system impedance (Z_s).

$$I_f = \left[\frac{Z_s}{Z_f + Z_s} \right] I_h = \rho_f I_h \qquad (8.3)$$

$$I_s = \left[\frac{Z_f}{Z_f + Z_s} \right] I_h = \rho_s I_h \qquad (8.4)$$

where ρ_f and ρ_s are complex quantities which determine the distribution of harmonic current in the filter and system impedance. These equations can also be written in terms of admittances.

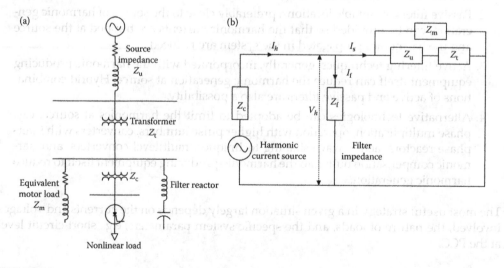

FIGURE 8.1
(a) Diagram of connections with a single-tuned (ST) shunt filter and harmonic source and (b) equivalent circuit looking from harmonic injection as the source.

A properly designed filter will have ρ_f close to unity, typically 0.995, and the corresponding ρ_s for the system will be 0.05. The impedance angles of ρ_f and ρ_s may be of the order of $-81°$ and $-2.6°$, respectively.

The harmonic voltages should be as low as possible. The equivalent circuit of Figure 8.1b shows that system impedance plays an important role in the harmonic current distribution. For infinite system impedance, the filtration is perfect, as no harmonic current flows through the system impedance. Conversely, for a system of zero harmonic impedance, all the harmonic current will flow into the system and none in the filter. In the case where there is no filtration, all the harmonic current passes on to the system. The lower the system impedance, i.e., the higher the short-circuit current, the smaller the voltage distortion, provided the filter impedance is lowered so that it takes most of the harmonic current.

In an ST filter, as the inductive and capacitive impedances are equal at the resonant frequency, the impedance is given by the resistance R:

$$Z = R + j\omega L + \frac{1}{j\omega C} = R \tag{8.5}$$

The following parameters can be defined:

ω_n is the tuned angular frequency in radians and is given by

$$\omega_n = \frac{1}{\sqrt{LC}} \tag{8.6}$$

X_0 is reactance of the inductor or capacitor at the tuned angular frequency. Here, $n = f_n/f$, where f_n is the filter-tuned frequency and f is the power system frequency.

$$X_0 = \omega_n L = \frac{1}{\omega_n C} = \sqrt{\frac{L}{C}} \quad \text{and} \quad \omega_n = \sqrt{\frac{1}{LC}} \tag{8.7}$$

The quality factor of the tuning reactor is defined as

$$Q = \frac{X_0}{R} = \frac{\sqrt{L/C}}{R} \tag{8.8}$$

It determines the sharpness of tuning. The pass band is bounded by frequencies at which

$$|Z_f| = \sqrt{2}R \tag{8.9}$$

$$\delta = \frac{\omega - \omega_n}{\omega_n} \tag{8.10}$$

At these frequencies, the net reactance equals resistance, capacitive on one side and inductive on the other. If it is defined as the deviation per unit from the tuned frequency, then for small frequency deviations, the impedance is approximately given by

$$|Z_f| = R\sqrt{1 + 4\delta^2 Q^2} = X_0\sqrt{Q^{-2} + 4\delta^2} \tag{8.11}$$

To minimize the harmonic voltage, Z_f should be reduced or the filter admittance should be high as compared to the system admittance.

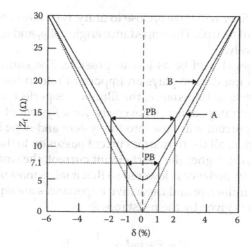

FIGURE 8.2
Response of an ST shunt filter, pass band, and asymptotes with varying Q factors.

The plot of the impedance is shown in Figure 8.2. The sharpness of tuning is dependent on R as well as on X_0 and the impedance of the filter at its resonant frequency can be reduced by reducing these. The asymptotes are at

$$|X_f| = \pm 2X_0|\delta| \tag{8.12}$$

The edges of the pass band are at $\delta = \pm 1/2Q$ and width $= 1/Q$. In Figure 8.2, curve A is for $R = 5\,\Omega$, $X_0 = 500\,\Omega$, and $Q = 100$, with asymptotes and pass band, as shown. Curve B is for $R = 10\,\Omega$, $X_0 = 500\,\Omega$, and $Q = 50$. These two curves have the same asymptotes. The resistance, therefore, affects the sharpness of tuning.

8.2.1 Tuning Frequency

ST filter is not tuned exactly to the frequency of the harmonic it is intended to suppress. The system frequency may change, causing harmonic frequency to change. The tolerance on filter reactors and capacitors may change due to aging or temperature effects. The tolerance on commercial capacitor units is ±20% and on reactors ±5%. For filter applications, it is necessary to adhere to closer tolerances on capacitors and reactors. Where a number of capacitor units are connected in series or parallel, these are carefully formed with tested values of the capacitance so that large-phase unbalances do not occur. Any such unbalances between the phases will result in overvoltage stress; in addition, the neutral will not be at ground potential in ungrounded wye-connected banks. A tolerance of ±2.0% on reactors and a plus tolerance of +5% on capacitors (no negative tolerance) in industrial environment are practical. Closer tolerances may be required for high-voltage DC (HVDC) applications.

A change in L or C of 2% causes the same detuning as a change of system frequency by 1% [3]:

$$\delta = \frac{\Delta_f}{f_n} + \frac{1}{2}\left(\frac{\Delta L}{L_n} + \frac{\Delta C}{C_n}\right) \tag{8.13}$$

Figure 8.3 shows the circuits and R–X and Z–ω plots of ST filters in isolation and in parallel.

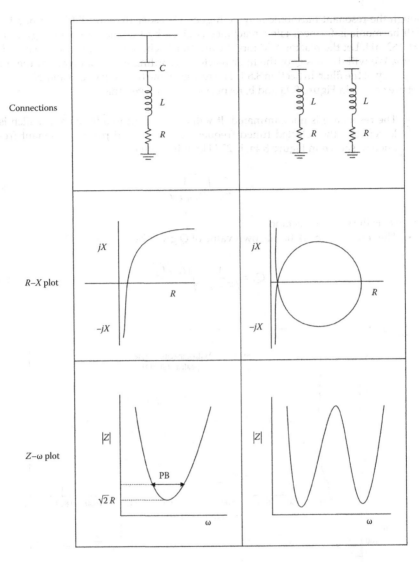

FIGURE 8.3
Circuit connections, and *R–X* and *Z–ω* plots of an ST shunt filter in isolation and two ST filters in parallel.

8.3 Practical Filter Design

Example 8.1

We will continue with Example 7.1 and go through the iterations of designing ST filters, so that the harmonic distortions are at acceptable limits.

Form an ST Shunt Filter

In Example 7.1, with the addition of a 1350 kvar capacitor bank, the resonant frequency is close to the 11th harmonic, which gives large magnification of harmonic currents and distortion factors. An ST filter design is generally started with the lowest harmonic, though the resonant harmonic may be higher. Consider that an ST filter tuned to the 47th harmonic is formed. Filter reactance is given by Equation 8.7, which gives filter $L = 0.682$ mH. Let the reactor X/R ratio be arbitrarily chosen as 40 *at the fundamental frequency*. We will discuss Q of the filter reactors according to Equation 8.8, at the tuned frequency of the filter, in Section 8.8.1. The results of the frequency scan and phase angle plot are shown in Figure 8.4a and b, respectively. It is noted that:

- The resonance is not eliminated. It will always shift to a frequency, which is lower than the selected tuned frequency. This shifted parallel resonant frequency, shown in Figure 8.4a, is 257 Hz. It is given by

$$f_{11} = \frac{1}{2\pi} \sqrt{\frac{1}{(L_s + L)C}} \tag{8.14}$$

where L_s is the system reactance.
- The resonance peak has its own value of Q given by

$$Q_1 = \frac{1}{(R_s + R)} \sqrt{\frac{(L_s + L)}{C}} \tag{8.15}$$

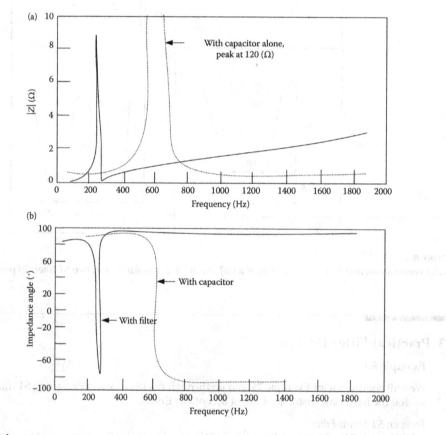

FIGURE 8.4
Impedance modulus and phase angle plots with application of an ST fifth harmonic filter (Example 8.1); response compared with capacitors alone.

where R_s is the system resistance. Figure 8.4a shows that the impedance modulus is considerably reduced. It is 9.17 Ω as compared to 122.7 Ω with capacitor alone.

- The harmonic currents flows are shown in Table 8.1. To calculate total demand distortion (TDD) according to the IEEE [1], short-circuit current and load-demand current at the PCC are required. From the data given in Figure 7.20a, short-circuit current is 20.82 kA. The average load current is 830 A. Thus, the ratio $I_{sc}/I_L = 25.08$. The permissible and calculated harmonic current distortion factors with the fifth harmonic ST filter are shown in Table 8.2. The TDD for 7th and 11th harmonics is above permissible limits and the total TDD is 11.87%. Though the total TDD is reduced from 94.52% (without filter) to 11.87%, it is still above permissive limits.

Add Another ST Filter for the Seventh Harmonic

As the distortion limits are not met, we can try splitting the 1350 kvar capacitor bank used to form the fifth harmonic ST filter into two equal parallel ST filters, one tuned to $n = 4.7$ as before and the other tuned to $n = 6.7$. We had used two units of 300 kvar, 2.77 kV per phase, giving an installed kvar of 1800 at 2.77 kV (= 1350 kvar at 2.4 kV). This is split into two equal banks, 900 kvar, each at 2.77 kV, and ST filters are formed. The results of TDD calculation are again shown in Table 8.2. TDD for the fifth harmonic worsens. The fifth harmonic current in the supply system is increased, while the seventh harmonic current is reduced. The harmonic current flow is shown in Table 8.3, Case 1.

Effect of Tuning Frequencies

As the fifth harmonic current is increased, a sharper tuning of the fifth ST filter is attempted by shifting the resonant frequency to $n = 4.85$. TDD calculations are shown in Table 8.2. The harmonic current distributions are shown in Table 8.3, Case 2. A closer tuning of the fifth ST filter increases the fifth harmonic loading of the filter, resulting in reduced TDD at these frequencies.

With the fifth ST filter tuned to $n = 4.85$ and the seventh to $n = 6.7$, the TDD for harmonics <11 meets the requirements; however, TDD for harmonics $11 < h < 17$ is still high. These point to the necessity of adding another ST filter for the 11th harmonic.

Add Another ST Filter for the 11th Harmonic

Another equal section of the capacitor bank is added and $n = 10.6$. The results of the simulation are shown in Table 8.4 and the TDD calculations are in Table 8.2. The TDD for each band of frequencies as well as total TDD now meets the requirements. A total of three ST filters of equal sections, each consisting of 900 kvar (300 kvar per phase) rated

TABLE 8.1

Example 8.1: Harmonic Currents and Voltages with Fifth Harmonic ST Filter

Order of Harmonic	Current in ST Filter (A)	Current in Transformer T1 (A)	Current in 2000 hp Motor (A)	Harmonic Voltage, Bus 2 (V)
5	79.56	42.30	3.13	26.3
7	24.07	61.40	4.54	53.4
11	14.95	56.8	4.20	77.76
13	6.18	25.00	1.85	40.38
17	3.38	14.50	1.08	30.7
19	1.93	8.44	0.63	19.9
23	0.60	2.70	0.20	7.71
25	0.48	2.16	0.16	6.71
29	0.34	1.54	0.11	5.57
31	0.24	1.08	0.08	4.17
35	0.20	0.95	0.06	3.26

TABLE 8.2

Example 8.1: Summary of TDD under Various Conditions Studied

Harmonic order →	<11		11≤h<17		17≤h<23		23≤h<35				35≤h	Total TDD
Operating condition ↓												
IEEE limits	7.0		3.5		2.5		1.0				0.5	8.0
Harmonic order →	5	7	11	13	17	19	23	25	29	31	35	
No capacitor	14.172[a]	10.208[a]	8.608[a]	3.733[a]	2.144[a]	Not calculated (total TDD not accurate)						19.91%[a]
1350 kvar capacitors	18.07[a]	18.07[a]	90.723[a]	6.987[a]	1.325	0.542	0.103	0.067	0.033	0.020	0.017	94.52[a]
Fifth harmonic ST filter, n=4.7	5.096	7.398[a]	6.843[a]	3.012	1.747	1.017	0.325	0.260	0.186	0.130	0.114	11.87[a]
Fifth and seventh harmonic ST filters, n=4.7 and 6.7	8.72[a]	2.783	5.892[a]	2.662	1.578	0.922	0.296	0.237	0.169	0.119	0.100	11.03[a]
Fifth and seventh harmonic ST filters n=4.85 and 6.7	5.422	2.759	5.855[a]	2.638	1.566	0.916	0.293	0.237	0.169	0.119	0.100	9.04[a]
5th, 7th, and 11th harmonic ST filters, n=4.85, 6.7, and 10.6	3.572	2.700	1.658	2.110	2.082	1.42	0.571	0.504	0.425	0.320	0.238	6.85
5th, 7th, and 11th harmonic ST filters, with tolerances	8.735[a]	4.614	1.205	1.313	0.988	0.600	0.200	0.163	0.118	0.083	–	10.10[a]
One large fifth harmonic ST filter, n=4.85	0.984	3.398	3.289	1.675	0.998	0.583	0.188	0.151	0.107	0.075	0.050	5.24

Note: $I_{sc}/I_L = 25$, $I_L = 830$ A = base current, six-pulse converter, 2.4 kV.

[a] TDD limits are exceeded.

TABLE 8.3

Example 8.1: Harmonic Simulation, Fifth and Seventh ST Filters, Effect of Tuning Frequency

Harmonic order h ↓	Fifth ST filter	Seventh ST filter	2000 hp motor	Supply System PCC
5	67.1[a]	30.2	5.36	72.4
	89.00[b]	12.23	3.33	45.00
7	45.2	60.7	1.71	23.1
	50.59	60.33	1.70	22.90
11	6.43	17.10	3.62	48.90
	6.91	16.89	3.60	48.6
13	2.73	6.56	1.63	22.1
	2.92	6.53	1.62	21.9
17	1.53	3.39	1.31	13.10
	1.62	3.37	0.96	13.00
19	0.87	1.91	0.57	7.65
	0.93	1.89	0.56	7.61
23	0.27	0.58	0.18	2.46
	0.29	0.58	0.18	2.44
25	0.22	0.46	0.15	1.97
	0.23	0.46	0.14	1.96
29	0.16	0.33	0.11	1.41
	0.16	0.33	0.10	1.41
31	0.11	0.23	0.07	0.99
	0.11	0.23	0.07	0.99
35	0.09	0.19	0.06	0.83
	0.09	0.19	0.06	0.83

Note: Harmonic current flow in ampères.

[a] Case 1: top row, fifth ST filter $n=4.7$, seventh ST, filter $n=6.7$.
[b] Case 2: bottom row, fifth ST filter $n=4.85$, seventh ST filter $n=6.7$.

voltage of 2.77 kV, connected in ungrounded wye and provided with series tuning reactors $n=4.85$, 6.7, and 10.6, respectively, meet the requirements. The frequency scan of the impedance and its phase angles, with all three filters in place, are plotted in Figure 8.5.

Minimum Filter
In this example, though additional reactive power is not required from fundamental frequency load flow conditions, yet to meet the TDD requirements an additional 11th ST filter has become necessary. The converse may be also true, depending on the load power factor and percentage of nonlinear load to total load. Compromises are often required when filters must meet the requirements of certain reactive power compensation as well as TDD. A filter designed to control only the harmonic distortion, without the limitation of meeting a certain reactive power demand, is termed the *minimum filter.*

One Large Fifth Harmonic ST Filter to Meet the TDD Requirements
As the relative harmonic loading of the filter and the system depends on the filter reactance, if the requirements of TDD are not met, the simplest artifice is to increase the size of the filter, i.e., increase the fundamental frequency Mvar rating and redesign the filter. Often this becomes a limitation of the passive filters: to achieve the required TDD, the size of the filter is too large, resulting in overcompensation of the reactive power and giving rise to overvoltage when in service and reduced voltage when out of service.

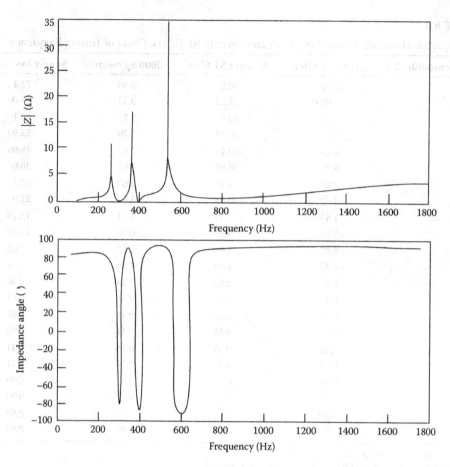

FIGURE 8.5
Example 8.1: impedance modulus and phase angle plots with 5th, 7th, and 11th harmonic filters.

TABLE 8.4

Example 8.1: Harmonic Simulation 5th, 7th, and 11th ST Filters

h ↓	5th ST Filter	7th ST Filter	11th ST Filter	2000 hp Motor	Supply System PCC	Harmonic Voltage at PCC (bus 2)
5	94.3	13.06	7.49	3.53	47.7	29.65
7	5.67	67.7	11.04	1.90	25.7	22.41
11	1.42	3.47	60.38	0.74	9.99	13.66
13	1.44	3.22	16.71	0.80	10.8	17.51
17	1.02	2.12	7.09	0.61	8.17	17.28
19	0.61	1.24	3.82	0.37	4.96	11.72
23	0.20	0.39	1.12	0.12	1.66	4.74
25	0.16	0.32	0.88	0.10	1.35	4.19
29	0.12	0.23	0.61	0.07	0.98	3.53
31	0.08	0.16	0.41	0.05	0.69	2.66
35	0.06	0.12	0.36	0.03	0.57	1.98

Note: Harmonic current flow in ampères; harmonic voltage in volts.

TABLE 8.5

Example 8.1: Current into PCC with One Large Fifth Harmonic ST Filter

$h \rightarrow$	5	7	11	13	17	19	23	25	29	31	35
Harmonic current $\rightarrow$	8.17	28.2	27.3	13.9	8.29	4.84	1.56	1.25	0.89	0.63	0.42

Note: Harmonic current flow in ampères.

In this example, we can iterate by trying increasing sizes of the fifth ST filter, without parallel 7th and 11th filters, until the TDD requirements are met. Calculations show that a single fifth ST filter requires 9000 kvar of capacitors (3000 kvar per phase, i.e., 10 units of 300 kvar, rated voltage 2.77 kV, connected in parallel) to satisfy the TDD requirements. The harmonic currents in the PCC with this filter are shown in Table 8.5 and TDD in Table 8.2.

The single large filter with 9.00 Mvar of capacitors is impractical. The example illustrates that *simply increasing the size of the filter may not give an optimum solution*.

8.3.1 Shifted Resonant Frequencies

Each parallel ST filter gives rise to a shifted resonant frequency, below its own tuned frequency. If the shifted resonance frequency coincides with one of the characteristics, non-characteristic, or triplen harmonics present in the system, current magnification at these frequencies will occur Reference [4]. The switching inrush current of a transformer is rich in even and third harmonics. As the transformers are switched in and out, harmonic current injections into the system and filters will increase, though this will last for the switching duration of the transformers. It is possible that these currents are sufficiently magnified to give rise to large harmonic voltages. High overvoltages can occur if the system is sharply tuned to the harmonic that is being excited by the transformer inrush current, (second, third, fourth, and even harmonics). Capacitor banks could also fail prematurely. This places a constraint in the design of ST filters. The shifted resonance frequencies should have at least 30 cycles difference between the adjacent and odd or even harmonics. Even then, some amplification of the transformer switching inrush current will occur. In this example, with three ST filters, the shifted frequencies are 271, 367, and 549 Hz. Reference [4] illustrates switching transients when capacitors and transformers are switched together. These increase the decay time of the switching transient and harmonic resonance can occur.

The last two frequencies are close to the sixth and ninth harmonics. A wider band can be attempted by slightly lowering the tuning frequency of the 7th and 11th ST filters. The shifted frequencies can also be calculated from Equation 8.14.

8.3.2 Effect of Tolerances on Filter Components

The tolerances on capacitors and reactors will result in detuning. Consider that components of the following tolerances are selected:

- Capacitors: +5%
- Reactors: ±2%

Let the capacitance of the fifth and seventh filters increase by 5% and the inductance by 2%. This is quite a conservative assumption for checking the detuning effect and resulting current distribution. The series-tuned frequencies of the fifth and seventh filters will shift to a lower value.

TABLE 8.6

Harmonic Simulation with Tolerances on Filters (See Text)

$h\downarrow$	5th ST filter	7th ST filter	11th ST filter	2000 hp motor	Supply system PCC
5	81.4	22.9	11.4	5.36	72.5
7	8.11	57.1	16.4	2.84	38.3
11	1.41	3.28	60.55	0.74	10.00
13	1.44	3.10	16.78	0.80	10.9
17	1.02	2.06	7.11	0.61	8.20
19	0.61	1.21	3.84	0.37	4.98
23	0.20	0.39	1.12	0.12	1.66
25	0.16	0.32	0.88	0.10	1.35
29	0.12	0.23	0.61	0.07	0.98
31	0.08	0.16	0.41	0.05	0.69

Note: Harmonic current flow in ampères.

The results of harmonic current flow are shown in Table 8.6. The harmonic distortion increases above the acceptable limits.

This points to the necessity of iterating the design with required tolerances and fine-tuning the selected tuning frequencies. Closer tolerances on the components are an option but that may not be practical and economically justifiable. Capacitors with metalized film construction lose capacitance as they age, resulting in a gradual increase in the tuning frequency. Nonmetalized electrode capacitors have a fairly stable capacitance. Tuning a harmonic filter more sharply than required to attain the desired performance unnecessarily stresses the components and generally makes the filter more prone to overload from other harmonic sources. Considerations should also be applied to the increase of the loads and consequent harmonics. Transformer energizing and clearing of nearby faults will lead to temporary surge of the filter harmonic current. The faults and energizing of transformer may result in saturation of transformers, which gives additional harmonic loading. If a harmonic filter is not going to be removed automatically during a system outage, it is desirable to do a system study to determine the filter performance during the outage. Switching transients are of concern and are discussed in Volume 1, though a more rigorous analysis will be required depending upon the application.

8.3.3 Outage of One of the Parallel Filters

Outage of one of the parallel ST filters should be considered. It will have the following effects:

- The current loading of remaining filters in service may increase substantially and the capacitors and reactors may be overloaded.
- The resonant frequencies will shift and may result in harmonic current amplification.
- The harmonic distortion will increase.

Table 8.7 shows the effect of outage of one of the three filters at a time. The harmonic distortion at the PCC increases in every case, though the filter components are not overloaded. According to IEEE standard 519 [1], it is permissible to operate the system on a short-term

TABLE 8.7

Example 8.1: Effect of Outage of One of the Parallel Filters

Harmonic Order $h \downarrow$	5th ST Out			7th ST Out			11th ST Out		
	7th ST Filter	11th ST Filter	Utility (PCC)	5th ST Filter	11th ST Filter	Utility (PCC)	5th ST Filter	7th ST Filter	Utility (PCC)
5	53.28	30.57	194	85.48	6.79	43.2	89.00	12.23	45.00
7	72.2	11.78	27.5	22.9	44.57	104.00	50.59	60.33	22.90
11	3.5	61.53	10.2	1.48	63.27	10.5	6.91	16.89	48.60
13	3.37	17.47	11.3	1.59	1.85	1.20	2.92	6.53	21.90
17	2.24	7.49	8.64	1.14	7.97	9.20	1.62	3.37	13.00
19	1.31	4.05	5.25	0.68	4.31	5.59	0.93	1.89	7.61
23	0.42	1.19	1.76	0.22	1.27	1.87	0.29	0.58	2.44
25	0.34	0.93	1.43	0.18	0.98	1.52	0.23	0.46	1.96
29	0.24	0.64	1.04	0.13	0.68	1.11	0.16	0.33	1.41
31	0.19	0.44	0.73	0.09	0.47	0.78	0.11	0.23	0.99
35	0.14	0.40	0.46	0.07	0.41	0.48	0.07	0.14	0.46

Note: Harmonic current in ampères.

basis with higher distortion limits at the PCC, provided that the faulty unit is placed back in service quickly after rectification.

Sometimes, the outage of a filter may result in overloading of the remaining filters in service. It then becomes necessary that parallel filters are also removed from service. The filter protection and switching scheme are designed so that, with the outage of a unit, the complete system is shut down. This brings another consideration, that is, redundancy in the filter applications so that the harmonic emission at PCC is controlled within IEEE limits. Alternatively, enough spare parts and services should be available to bring the faulty unit in service in a short time.

8.3.4 Operation with Varying Loads

When load-dependent switching is required for reactive power compensation, multiple capacitor banks are switched in an ascending order, i.e., 5th, 7th, and 11th. Generally, this will occur during start-up conditions; however, if sustained operation at reduced loads is required, it is necessary to control the harmonic distortion at each of the operating loads and switching steps. The harmonic loads may or may not decrease in proportion to the overall plant load. This adds another step in designing an appropriate passive filtering scheme to meet the TDD requirements.

8.3.5 Division of Reactive kvar between Parallel Filter Banks

When multiple parallel filters are required and the total kvar requirements are also known, it remains to find out the most useful distribution of kvar among the parallel filters. In the above example, the 5th, 7th, and 11th filters are based on equal kvar. This is too simplistic an approach, rarely implemented. As filters should be sized to handle the harmonic loading, one approach would be to divide the required kvar based on the percentage of harmonic current that each filter will carry. This will not be known in advance. The other method is to proportionate the filters with respect to harmonic current generation, i.e., the lower order harmonics are higher in magnitude, so more kvar are allocated to a lower order filter. Again some iteration will be required to optimize the sizes initially chosen, based on the actual fundamental and harmonic current loadings and the desired reactive power compensation.

8.3.6 Losses in the Capacitors

The power capacitors have some active power loss component, though small. Figure 8.6 shows the average losses versus ambient temperature for capacitors. At an operating temperature of 40°C, the loss is approximately 0.10 W/kvar and increases to 0.28 W/kvar at −40°C. This loss should be considered in the filter design, by an equivalent series resistance inserted in the circuit.

8.3.7 Harmonic Filter Detuning and Unbalance

The operation of an internal or external fuse or shorting of elements of a fuseless capacitor bank changes the capacitance of the filter and subjects it to higher overvoltage. Unbalance detection systems are applied and some considerations are as follows:

- The resonant frequency will change. Fuse operation in a parallel-connected capacitor bank will decrease the capacitance and increase the resonant frequency. There is a possibility that shorting of elements of an externally fused capacitor bank increases the capacitance and decreases the resonant frequency. It is desirable to ascertain the maximum plus minus capacitance change that can be tolerated. The detuning may be a more stringent condition than the overvoltage on remaining units.

- Ambiguous indications are a possibility. For example, a negligible current will flow through a CT connecting the neutrals of a balanced ungrounded double-wye bank, and this will not change if equal number of fuse failure or elements short-outs occurs in the same phases of the wye banks. Where such a possibility exists, an alarm on first failure of a fuse or first failure of an element in fuseless bank is desirable to be provided.

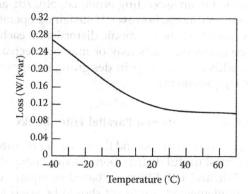

FIGURE 8.6
Average losses in film–foil capacitor units, with variation of temperature.

- An arcing fault external to a capacitor unit can result in large change in filter capacitance and detuning. The unbalance protection may not always operate, depending upon filter configurations. A phase overcurrent relaying scheme can be designed according to IEEE Std. C37.99-2000 [5]. Furthermore, harmonic currents should not adversely affect trip or alarm level relays and filters are required so that the operation is based upon fundamental currents and voltages.
- To simulate the fuse failure in one capacitor unit in a phase and its effect on detuning, three-phase harmonic load flow modeling and analyses are required.

8.4 Relations in an ST Filter

The reactive power output of a capacitor at fundamental frequency is V^2/X_c. In the presence of a filter reactor, it is given by

$$S_f = \frac{V^2}{X_L - X_c}$$

$$= \frac{V^2}{X_c / n^2 - X_c}$$

$$= \frac{n^2}{n^2 - 1} \times (\text{reactive power without reactor}) \tag{8.16}$$

The reactive power output with a filter reactor tuned to, say, $4.85 f$ is approximately 4% higher than without the reactor. This is so because the voltage drop in the reactor is added to the capacitor voltage and its operating voltage is

$$V_c = V + V_L = V + j\omega L(V / j - 1 / j\omega C))$$

$$= \frac{n^2}{n^2 - 1} \tag{8.17}$$

The capacitors in a fifth harmonic filter tuned to $4.85 f$ operate at approximately 4% higher than the system voltage. While selecting the voltage rating on the filter capacitors, the considerations are as follows:

- Higher operating voltage due to the presence of filter reactors.
- Sustained upward operating voltage of the utility power supply system. This may be due to location, e.g., close to generating stations, or may be due to voltage adjustment tap changing on transformers.
- The higher voltages that will be imposed when one or two capacitor units in a parallel group go out of service. The neutral unbalance detection schemes may be set to alarm if a fuse operates on one of the parallel capacitor units and trip if the fuse operates on two units in the same phase.

The steady state fundamental frequency voltage can be

$$V_r = V\left(\frac{n^2}{n^2-1}\right) + \sum_{h=2}^{\infty} I_h X_{ch}$$ (8.18)

where V_r is the rated voltage, I_h is the harmonic current, V is the maximum system voltage across capacitor excluding the voltage rise across the reactor, and X_{ch} is the capacitive reactance at the harmonic order.

For transient events, such as capacitor bank switching, circuit breakers restrikes, etc., the voltage rating is calculated from

$$V_r = \frac{V_{tr}}{\sqrt{2}k}$$ (8.19)

where V_{tr} is the peak transient voltage and k factor is read from IEEE standard 1531-2003 (Figure 8.7), not reproduced here. If there are 100 transient events in a year, $k=2.6$. For a 13.8 kV rated voltage $V_{tr}=50.73$ kV peak, which is approximately 2.6 times the rated voltage. This implies that the capacitors can withstand 100 surges in a year of magnitude approximately 2.6 times the rated voltage. Higher voltage surges will be applicable if the frequency of surges is decreased.

For dynamic events generally lasting for a few fundamental cycles to several seconds, such as transformer energization, bus fault clearance, the voltage rating of the capacitor is calculated from the equation:

$$V_r = \frac{V_d}{\sqrt{2}}$$ (8.20)

where V_d is the voltage across the capacitor reached during dynamic event. The power frequency short-time overvoltage capability is specified in Figure 8.8 of IEEE standard 1531-2003 [6], not reproduced here. A capacitor unit can withstand 2.2 pu of rated voltage for 0.01–0.1 s and at about 15 s the overvoltage withstand capability is reduced to 1.4 pu.

Power system studies, for example, EMTP switching transient studies, are required to establish the transient and dynamic overvoltages in a system. The rated voltage selected should be the highest of the voltages given by Equations 8.18, 8.19, or 8.20. Generally, an operating voltage 8%–10% higher than the nominal system voltage is selected. This reduces the reactive capability at the voltage of application as the square of the voltages.

Example 8.2

A 3 Mvar capacitor bank is required to be formed for a ST filter application at 13.8 kV. The ST filter is wye connected and ungrounded, and has the harmonic loading $I_1=124$ A, $I_5=40$ A, $I_7=20$ A, $I_{11}=5$ A, higher order harmonics neglected for this example. The switching overvoltages studies show that these overvoltages are not more than 2.5 times the rated system voltage and the estimated switching operations per year are 50. Calculate the rated voltage of the capacitor units required to form the bank.

$$X_{ch} = \left(\frac{4.7^2}{4.7^2-1}\right)\left(\frac{13.8^2}{3}\right) = 66.49\ \Omega \text{ at fundamental frequency}$$

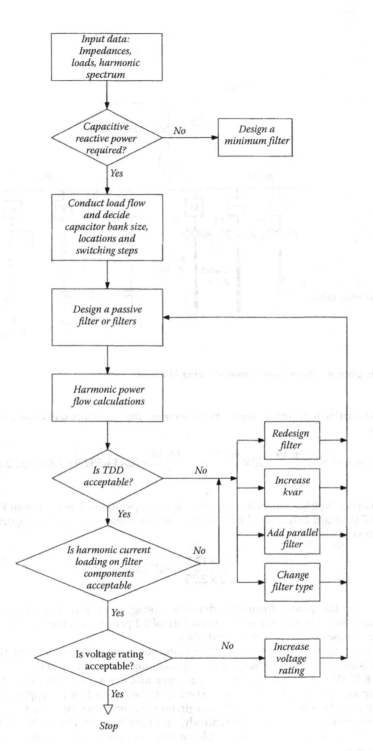

FIGURE 8.7
Flowchart for design of ST filters.

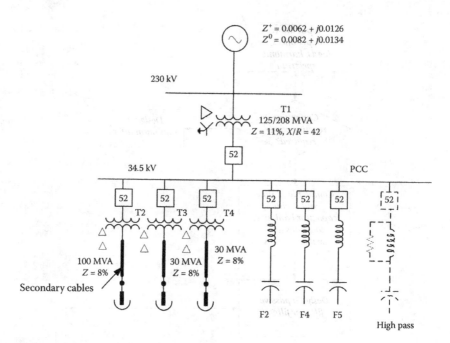

FIGURE 8.8
Single-line diagram of a furnace installation, showing ST filters.

From Equation 8.18, under steady-state operation, the capacitors will experience a voltage of

$$124 \times 66.49 + 40 \times \frac{66.49}{5} + 20 \times \frac{66.49}{7} + 5 \times \frac{66.49}{11} = 8244 + 531.9 + 190 + 30.22 = 8.90 \text{ kV}$$

For transient voltage capability and 50 switching operations, k as read from Figure 8.7 of IEEE standard 1531-2003 [6] is 2.75. Then the rms voltage rating of the capacitor from Equation 8.19 is

$$\frac{2.5 \times 13.8}{\sqrt{2} \times 2.75} = 8.872 \text{ kV}$$

Check for the power frequency dynamic voltage capability from Figure 8.8 of Reference [6]. This shows a voltage capability of 2.2 per unit of rated voltage, which is adequate based upon the switching studies.

In the above example, if the switching frequency is raised to 500 per year, then the k factor is 2.4 and the rated voltage of the capacitor unit should be 11.09 kV.

An 8.32 kV standard voltage rating capacitor unit has a continuous operating voltage capability of 9.152 kV. This voltage rating can be selected for the application; however, it is preferable to reserve 10% margin on the continuous rating of the capacitors for contingency operations. Alternatively, an immediate shutdown should be taken for replacement of failed unit or a blown fuse, as detected by unbalance detection equipment.

8.4.1 Number of Series and Parallel Groups

Table 7.3 gives the minimum number of units in parallel per series group to limit the voltage on remaining unit to 110% with one unit out of service. Equations 7.80 and 7.81 can be used for accurate calculations. The shift in tuning frequency and consequent overloading will be a more stringent condition, as compared to overvoltage limitation. Consider the 3000 kvar bank of Example 8.2; per phase 1000 kvar is required. The reactive power output of 8.32 kV capacitor will reduce by a factor of 0.917 when applied at 13.8 kV ungrounded wye configuration and a minimum of four units per phase are required. A reference to the manufacturer's standard available unit sizes shows that four units of 300 kvar will give 1100 kvar per phase at the operating voltage of 13.8 kV, when connected in wye configuration and a three-phase bank size of 3.3 Mvar. Thus, an adjustment in the required size has to be made based upon the standard voltage and capacitor kvar size ratings. IEEE standard 519 allows that harmonic limits may be exceeded by 50% for shorter periods during start-ups or unusual conditions (Chapter 5).

To meet the requirement of a certain reactive power output, a larger number of units are then required.

The fundamental loading of the capacitors is given by

$$\frac{V_c^2}{X_c} = \frac{V^2}{X_c}\left[\frac{n^2}{n^2-1}\right]^2 = s_f\left[\frac{n^2}{n^2-1}\right] \tag{8.21}$$

and the harmonic loading is

$$\frac{I_h^2 X_c}{h} = \frac{I_h^2 V^2}{S_f}\frac{n^2}{n^2-1} \tag{8.22}$$

When harmonic voltages and current flows are known from harmonic simulation, the harmonic loading can be found from

$$\sum_{h=2}^{h=\infty} V_h I_h \tag{8.23}$$

The fundamental frequency loading of the filter reactor is

$$\frac{V_L^2}{X_L} = \left[\frac{V_c}{n^2}\right]^2\left[\frac{n^2}{X_c}\right] = \frac{V_c^2}{n^2 X_c} = \frac{S_f}{n^2}\left[\frac{n^2}{n^2-1}\right] \tag{8.24}$$

The harmonic loading for the reactor is the same as for the capacitor.

The increase in bus voltage on switching a capacitor at a transformer secondary bus is approximately given by

$$\%\Delta V = \frac{\text{kvar}_{\text{capacitor}} Z_t}{\text{kVA}_t} \tag{8.25}$$

A flowchart of the design of ST filters is thus shown in Figure 8.7.

8.5 Filters for an Arc Furnace

Example 8.3

Figure 8.8 shows an arc furnace installation. The total operating load is 150 MVA. A 34.5 kV bus is considered a PCC. The furnace loads are connected through step-down transformers of 34.5–7.2 kV as shown in this figure. A reactive power compensation of 98 Mvar is required, which is provided by three ST filters formed with 49, 24.5, and 24.5 Mvar capacitors, respectively, for second, fourth, and fifth harmonics. These are connected at the main 34.5 kV bus. This reactive power compensation is calculated at full load. As stated in Chapter 2, arc furnace loads give rise to flicker due to variations in reactive power demand and erratic load patterns and normally a thyristor controlled reactor (TCR), SVC (static var compensator), or similar fast response device is required.

1. Model with current injection

   Harmonic current spectrum during melting and refining, from IEEE 519, is shown in Chapter 1. The melting period gives higher magnitudes of harmonics, which are considered in this simulation.

   The third harmonic is generated; however, no filter is provided for it because the delta windings of the transformers will filter out the third harmonics and no third harmonics appear in the lines. The three ST filters are formed as follows:

   • Second harmonic ST filter: *double wye*, ungrounded, three series groups, each group containing eight units of 400 kvar capacitors of rated voltage 7.2 kV. This gives a three-phase Mvar of 49 (Figure 8.9).

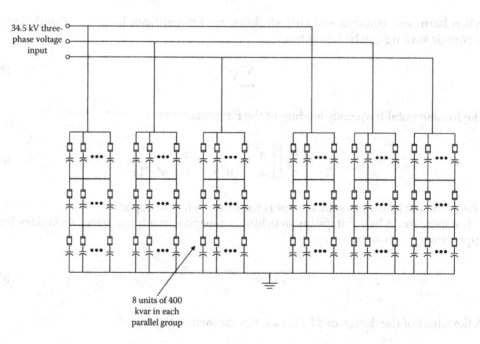

8 units of 400
kvar in each
parallel group

FIGURE 8.9
Formation of a 49 Mvar double wye, 34.5 kV capacitor bank for second harmonic filter (Example 8.3).

- Fourth and fifth ST filters: single wye, ungrounded, three series groups, each group containing eight units of 400 kvar capacitors of rated voltage 7.2 kV. This gives a three-phase Mvar of 24.5.

The tuning frequencies for second, fourth, and fifth ST filters are 1.95, 3.95, and 4.95 times the fundamental frequency, respectively. This gives series reactors of 1.694E–2, 8.259E–3, and 5.260E–3 for the second, fourth, and fifth filters, respectively. Q factors are as follows: second harmonic filter=60, fourth and fifth harmonic filters=40.

The system is impacted with a harmonic spectrum during the melting cycle of the furnaces. We observe that each ST filter operates effectively providing a low impedance path for the harmonic it is intended to shunt away.

The load demand is 319.7 A at 230 kV, the short-circuit current at PCC = 20 kA, ratio I_s/I_r=62.5. The permissible total TDD = 3.5% at 230 kV bus. And at 34.5 kV, it is 12%. Also the limits of each of the harmonics as per IEEE 519 should be met. The results of harmonic flow calculations and TDD at the PCC are shown in Table 8.8, which shows that when 34.5 kV bus is declared as PCC, all the harmonic emission limits are met, *but not so if we declare 230 kV bus as the PCC*. The permissible second harmonic limit at 230 kV is 0.75%, while the actual calculated is 2.20%. *To control even harmonics in arc furnace installations as per permissive limits requires much larger second harmonic filter*. The second harmonic filter size will require to be approximately doubled to reduce it to 0.75%. This will be indeed a very large filter. The voltage distortion at 230 kV is 0.16% and at 34.5 kV bus is 1.27%. Figures 8.10 and 8.11 illustrate the impedance modulus and impedance angle.

2. Model with typical voltage harmonics

Generally, it is not the current injection model but harmonic voltage injection model that is used for the arc furnace harmonic studies. The voltage harmonics are shown in Table 8.9, normal and worst case scenarios. The voltage harmonics are applied through the furnace transformer secondary leads, which should be properly modeled.

3. Model with maximum voltage harmonics

If the worst case harmonics shown in Table 8.9 are modeled, it is seen that the second harmonic ST filter is not even adequate to control the second harmonic

TABLE 8.8

Calculations of Harmonic Emissions, Arc Furnace, Three ST Filters (49, 24.5, 24.5 Mvar), IEEE 519 Current Injection Model

	Current through Filters			TDD Calculations			Harmonic Voltage Distortion	
Harmonic	Second ST Filter	Fourth ST Filter	Fifth ST Filter	TDD At 230 kV or 34.5 kV bus	TDD IEEE Limits 34.5 kV	TDD IEEE Limits 230 kV	34.5 kV PCC	230 kV Bus
Fundamental A or V	1117.5 A	439.9 A	429.3 A	310.76 A–23 V 2071.7 A, 34.5 kV-	2071.7	310.76	34.5 kV	230 KV
$h=2$	15.39	1.38	1.25	**2.20**	2.5	0.75	0.54	0.07
$h=4$	0.09	15.11	1.07	0.20	2.5	0.75	0.10	0.01
$h=5$	0.06	0.75	24.32	0.15	10	3	0.09	0.01
$h=7$	0.51	3.58	7.86	1.34	10	3	1.14	0.14
Total TDD, THDV				2.54	12	3.75	1.27	0.16

Note: All harmonic currents and voltages are in % of the fundamental currents and voltages.

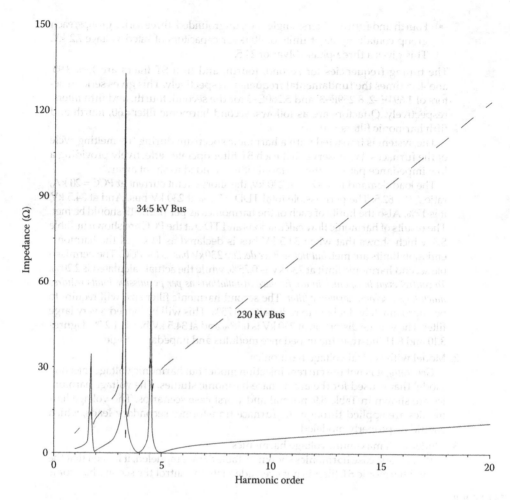

FIGURE 8.10
Impedance modulus with second, fourth, and fifth harmonic filters and typical current injection model.

to acceptable level of 2.5% for 34.5 kV bus as the PCC. The distortion at the second harmonic becomes 3.5%, though all the other harmonic indices meet IEEE
requirements at 34.5 kV bus. It is necessary to raise the size of this filter from
49 Mvar to 72 Mvar to control the second harmonic. It will be a very large filter,
double-wye ungrounded connection, three series groups of 7.2 kV capacitors,
each rated for 400 kvar and 12 units in parallel per series group. This will give
73.5 Mvar at 34.5 kV.

The results of a study with this configuration are shown in Table 8.10. Even
with increased filter size, the TDD and voltage distortion are slightly more
than those in Case 2.

4. Study with second harmonic filter as an ST filter and fourth and fifth harmonic
   filters as second-order high-pass damped filters.

   Damped filters for arc furnace installations are shown in Figure 8.15. As we
   discussed in Chapter 2, arc furnaces loads give rise to interharmonics and with
   ST filters there is a possibility of their magnification (though the interharmonics

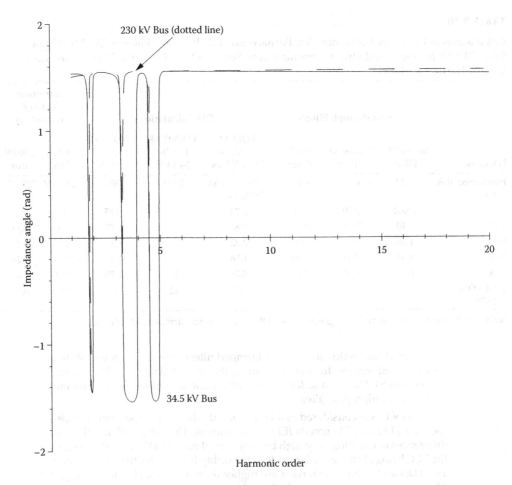

FIGURE 8.11
Impedance angle with second, fourth, and fifth harmonic filters and typical current injection model.

TABLE 8.9

Voltage Harmonics for Harmonic Analysis of Arc Furnace

Harmonic Order	Maximum Voltage Distortion	Typical Voltage Distortion (% of Fundamental Voltage)
2	17	5.0
3	29	20
4	7.5	3.0
5	10	10
6	3.5	1.5
7	8	6
8	2.5	1
9	5	3

TABLE 8.10

Calculations of Harmonic Emissions, Arc Furnace, Second Harmonic Filter of 73.5 Mvar Type C Filter, 24.5 Mvar Fourth and Fifth Harmonic Filters, Second-Order Damped Filters, Worst Case Voltage Harmonic Model

	Current through Filters			TDD Calculations			Harmonic Voltage Distortion	
Harmonic	Second ST Filter	Fourth ST Filter	Fifth ST Filter	TDD At 230 kV or 34.5 kV bus	TDD IEEE Limits 34.5 kV	TDD IEEE Limits 230 kV	34.5 kV PCC	230 kV Bus
Fundamental A or V	1714	462.6	451.3	388.9 A–230 kV 2593A, 34.5 kV	2593	388.9	34.5 kV	230 KV
$h=2$	25.62	2.10	1.91	2.72	2.5	0.75	0.85	0.11
$h=4$	1.40	21.10	9.01	1.86	2.5	0.75	1.16	0.14
$h=5$	1.86	13.20	19.57	2.32	10	3	1.81	0.23
$h=7$	1.55	5.47	11.48	1.58	10	3	1.73	0.21
$h=8$	0.52	1.52	3.19	0.48	2.5	0.75	0.59	0.07
Total TDD, THDV				4.35	12	3.5	2.91	0.37

Note: All harmonic currents and voltages are in % of the fundamental currents and voltages.

are not modeled in this study case). Damped filters do not give rise to shifted resonant frequencies. In this case study, the second harmonic filter is still retained as ST filter, while fourth and fifth harmonic filters are turned into second-order high-pass filter.

If 230 kV bus is considered as a PCC, not only the second harmonic but also the total TDD of 4.73 exceeds IEEE requirements. This amply shows the vast differences in decaling very high bus voltages above 161 kV as the PCC versus the PCC being at the secondary side of the utility interconnecting transformer. The TDD and voltage distortion are higher than in case (c) because damped filters are not as effective as ST filters. If we increase the size of the damped filters, the distortions can be reduced.

5. As case (c), except that the second harmonic filter is turned into a type C filter.

C-type filter for arc furnace installations will control the transients. The type C filter is further described in this chapter.

- $C_1 = 161\,\mu F$
- $C = 485\,\mu F$
- $L = 14.62\,mH$
- $R = 18\,\Omega$

The study results of this case are almost identical results to case (c).

We may conclude that all the designs meet the requirement of IEEE harmonic limits when the 34.5 kV bus is declared as the PCC, but not so when the 230 kV bus is declared as PCC. Limiting the second harmonic distortion to 0.75% at the 230 kV bus will require an almost 150 Mvar filter at 34.5 kV, which will bring down the total TDD also to less than 3.5% in all the cases. However, this is a large filter that will give rise to overvoltages and two double-wye-connected filters in parallel will need to be provided. The overcompensation of reactive power will not be acceptable and a STATCOM, SVC, or TCR should be provided.

8.6 Filters for an Industrial Distribution System

Example 8.4

Example 7.2 for application of power capacitors in an industrial plant showed that depending on the operating condition, resonant frequencies swing over a wide spectrum and the harmonic distortion at the PCC is high. The equivalent negative sequence current loading of generators is exceeded. Even without harmonics, a part of the generator negative sequence capability may be utilized due to unbalance loads and voltages and system asymmetries. The harmonic loading on the generators and harmonic distortion is reduced by turning capacitor banks at buses 2 and 3 into parallel fifth and seventh ST filters.

The following details are applicable for the final filter designs:

Bus 3, fifth harmonic ST filter: five units of 300 kvar, 7.2 kV, per phase, total capacitor Mvar=4.5, connected in ungrounded wye configuration; $n=4.85$, $C=76.75\,\mu F$, $L=3.897\,mH$.

Buses 2 and 3, seventh harmonic ST filter: three units of 300 kvar, 7.2 kV, per phase, giving 2.7 Mvar total, connected in ungrounded wye configuration; $n=6.75$, $C=46.05\,\mu F$, $L=3.353\,mH$.

Bus 2, fifth harmonic ST filter: four units of 300 kvar, 7.2 kV, per phase, total capacitor Mvar=3.6, $n=4.85$, $C=61.40\,\mu F$, $L=4.872\,mH$, $Q=100$.

X/R ratio of reactors=100 at fundamental frequency. The harmonic current flow is studied under the same three conditions as in Example 7.2 and TDD at the PCC is calculated. The normal load demand current $I_r=250$ A in cases 1 and 3, and 410 A in case 2, when No. 2 generator is out of service. The three-phase short-circuit current is 30.2 kA sym. The results of calculation are shown in Table 8.11. The following observations are of interest.

When the 75-mile 115 kV line and its harmonic load are modeled, TDD increases over normal operating condition 1. (See Example 7.3 for a description of the operating conditions for this system.) This shows the impact of harmonic loads that may be located at

TABLE 8.11

Example 8.4: Harmonic Simulation with Filters

Harmonic Order ↓	Current in Generator 1			Current in Generator 2			Current into the Utilities system (PCC)		
	1[a]	2	3	1	2	3	1	2	3
5	37	35.6	31.7	51.7	Generator	48.7	6.78	7.05	12.8
7	24.4	104	27.3	24.9	out of	26.3	4.13	25.81	5.44
11	2.03	42.3	2.3	25.7	service	27.2	1.24	5.45	3.70
13	3.76	3.99	2.87	17.6		15.2	1.17	0.21	4.15
17	2.83	1.76	4.0	10.0		12.1	0.75	0.56	0.90
19	2.03	1.60	8.20	7.13		4.96	0.54	0.45	2.13
23	1.94	1.81	2.89	7.71		9.53	0.54	0.47	0.95
25	1.75	1.73	1.30	7.44		6.19	0.48	0.44	1.80
Calculated TDD at PCC→							3.28	6.66	6.12
Permissible TDD at PCC→							7.5	6	7.5

Note: Harmonic current flow in ampères.

[a] 1, 2, 3 refer to study cases, see Example 7.2.

considerable distance from the consumer. These should invariably be considered in a harmonic analysis study.

In operating condition 2, seventh harmonic ST filters on buses 2 and 3 are out of service. TDD is slightly above the limits. This is acceptable for short-term operation.

The smaller generator G2 rated at 47.97 MVA has a higher harmonic loading as compared to the larger generator G1 of 82 MVA. The impedance modulus shows some interaction between ST filters and capacitors at motors. Generally, it is desirable to observe one strategy of reactive power compensation in a distribution system, due to the problem of secondary resonance, discussed in Section 4.4.

The resonant frequency varies by a maximum of 1.2% in the three cases. This can be compared to the much wider swings shown in Table 7.8, without filters. The resulting harmonic current flows through the system and filters change with switching operation, yet the TDD at the PCC remains within acceptable limits (Table 8.11). Also, the negative sequence loadings of the generators are at safe levels.

8.7 Band Pass Filters

Band pass filters are a new breed of filters for harmonics. Referring to Figure 8.12a, a simple LC circuit can act as a band pass filter but requires large components. It is free of resonance problems, but at no load the output voltage can be high and PF is leading at all loads [7].

An improved LLCL filter is shown in Figure 8.12b. Filter capacitors C_f are delta connected and damping resistors are connected to C_f. The filter output terminals V_0 are connected to the rectifier terminals and Lo is relatively small, 3–5%. At dominant rectifier harmonics, the large filter input reactor L_1 provides high impedance with respect to shunt filter impedance over a wide frequency range. Thus, it impedes the flow of harmonic currents generated by the rectifier into the AC lines and also minimizes the effect of line voltage harmonics on the rectifier. The filter parallel resonance frequency is given by

$$f_p = \frac{1}{2\pi\sqrt{(L_1+L_f)C_f}} \tag{8.26}$$

The parallel resonant frequency is selected between the fundamental frequency and the first dominant frequency of the rectifier circuit, which is equal to fifth harmonic. The voltages

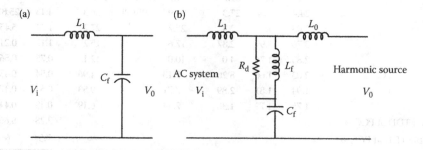

FIGURE 8.12
(a) An RC circuit for a band pass filter and (b) an improved band pass filter for harmonic mitigation.

are maintained in a narrow band on load variations. The components C_f and L_f provide a low impedance path to the dominant rectifier harmonic, like an ST filter.

In the design process, L_0 (Figure 8.12b) is selected as 4% reactor:

$$L_0 = 0.04 \frac{V^2}{\omega P} \tag{8.27}$$

where P is the rated power and V is the line-to-line voltage.

The other filter components are selected based on total harmonic distortion (THD), PF, and voltage excursions, no load to full load. This is rather a complex mathematical problem. GA (genetic algorithm—not discussed in this book, see Reference [8]) for optimization of these parameters, using the fitness function, is applied:

$$\text{Fitness} = \text{THD} + \Delta V_0 + \left(\frac{1}{PF} \right) \tag{8.28}$$

The parameters thus calculated for a 5.5 kW ASD (adjustable speed drive) are as follows:

$$L_1 (mH) = 10.1$$

$$L_f (mH) = 8.1$$

$$C_f (\mu F) = 21$$

The THD is limited to 6–6.5%, PF 0.98–0.99, and =3.0–3.2% for the 5.5 kW ASD. The parameters must be calculated for each rating of ASD. The filter has limitation when a number of different nonlinear sources are present.

8.8 Filter Reactors

The rated steady-state voltage is calculated as arithmetic sum of fundamental and harmonic voltages, similar to the capacitor.

$$V_r = \sum_{h=1}^{h=\infty} I_h X_{R(h)} \tag{8.29}$$

where $X_{R(h)}$ is the reactance at the given harmonic order.

At medium-voltage levels, the harmonic filter is generally connected in ungrounded configuration and when the reactor is located on the source side of the capacitor, the available fault current is limited, while an iron core reactor in the same situation may saturate, and will not decrease the short-circuit current. For high-voltage grounded banks, the filter reactor may be located on the neutral side. This may allow basic insulation level (BIL) of the reactor to be less than that of the system.

8.8.1 Q Factor

Apart from its impact on the filter performance, the Q factor determines the fundamental frequency losses and this could be an overriding consideration, especially when the reactors at medium-voltage level are required to be located indoors in metal or fiberglass enclosures and space is at a premium. Consider a second harmonic filter for the furnace installation shown in Figure 8.8. The capacitor is 0.00128 μF, the inductor is 0.01371 mH, i.e., inductive reactance is 5.1687 Ω. An X/R of 50 gives a reactor resistance of 0.1032 Ω. The fundamental frequency current is 1280 A. This gives a loss of approximately 507 kW/h (at fundamental), which is very substantial.

Equation 8.8 defines the filter Q based on the inductive or capacitive reactance at the tuned frequency (these are equal). The fundamental frequency losses and heat dissipation are of major consideration. This does not mean that the effect on filter performance can be ignored. The higher the value of Q, the more pronounced the valley at the tuned frequency. For industrial systems, the value of R can be limited to the resistance built in the reactor itself.

Example 8.5

The effect of change in Q of the filter is examined in this example. We have an X/R of 100 at fundamental frequency for the filter reactors in Example 8.4. Thus, the resistance values of the reactors are as follows:

Fifth harmonic filter at bus 2=0.01469 Ω
Fifth harmonic filter at bus 3=0.01836 Ω
Seventh harmonic filter at buses 2 and 3=0.012642 Ω.

Figure 8.2 shows that the sharpness of tuning is dependent upon the resistance. Harmonic load flow of Example 8.3 is repeated with tuning reactors of $X/R=10$ and the results are shown in Table 8.12. There is hardly an appreciable difference in the harmonic current flow. In industrial systems, the performance of ST filters will be, generally, indistinguishable for Q (Equation 8.8)=20 to $Q=100$.

The X/R of tuning reactors at 60 Hz is given by $3.07K^{0.377}$, where K is the three-phase kVA= (I is the rated current in ampères and X the reactance in ohms). X/R of a 1500 kVA reactor

TABLE 8.12

Effect of Change of Q: Harmonic Simulation, Condition 1 of Example 8.5

Harmonic Order	Current in Generator 1		Current in Generator 2		Current into Utilities System (PCC)	
	$X/R=100$	$X/R=7$–10	$X/R=100$	$X/R=7$–10	$X/R=100$	$X/R=7$–10
5	37	37.5	51.7	54.0	6.78	6.90
7	24.4	25.0	24.9	25.7	4.13	4.24
11	2.03	2.03	25.7	25.7	1.24	1.24
13	3.76	3.76	17.6	17.6	1.17	1.17
17	2.83	2.83	10.0	10.0	0.75	0.75
19	2.03	2.03	7.13	7.13	0.54	0.54
23	1.94	1.94	7.71	7.71	0.54	0.54
25	1.75	1.75	7.44	7.44	0.48	0.48

Note: Harmonic current flow in ampères.

will be 50 while that of a 10 MVA reactor will be 100. High X/R reactors can be purchased at a cost premium. Thus, selection of X/R of the reactor depends upon the following:

- Initial capital investment
- Active energy losses
- Effectiveness of the filtering

The optimization of filter admittance and Q for the impedance angle of the network and δ *are required* for the transmission systems. The following optimum value of Q is given by Reference [3]:

$$Q = \frac{1 + \cos\phi_m}{2\delta_m + \sin\phi_m} \tag{8.30}$$

where φ_m is the network impedance angle. Consider a frequency variation of ±1%, a temperature coefficient of 0.02% per degree Celsius, and a temperature variation of ±30°C on the inductors and capacitors; then from Equation 8.13, $\delta = 0.006$. For an impedance angle $\varphi_m = 80°$, the optimum Q from Equation 8.24 is 99.31. The higher the tolerances on components and frequency deviation, the lower the value of Q.

8.9 Double-Tuned Filter

A double-tuned filter is derived from two ST filters and is shown in Figure 8.13. Its R–X plot and Z–ω plots are identical to those of two ST filters in parallel, as shown in Figure 8.3. The advantage with respect to two ST filters is that the power loss at fundamental frequency is less and one inductor instead of two is subjected to full impulse voltage. In Figure 8.13,

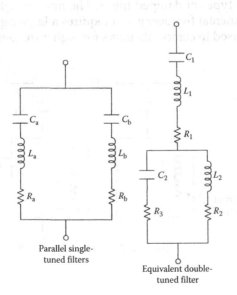

Parallel single-
tuned filters

Equivalent double-
tuned filter

FIGURE 8.13
Equivalent circuits of two ST parallel filters and a single double-tuned filter.

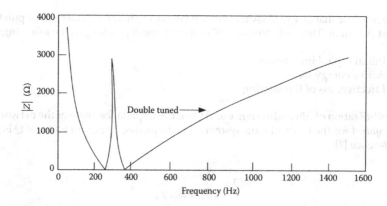

FIGURE 8.14
Z–ω plot of double-tuned filter.

the BIL on reactor L2 is reduced, while reactor L1 sees the full impulse voltage. This is an advantage in high-voltage applications, see Reference [9].

Generally, R1 is omitted and R2 and R3 are modified so that the impedance near resonance is practically the same. Figure 8.14 shows the response of a double-tuned filter.

The equations for conversion of parameters for two ST filters to those for double-tuned filters are provided in References [8, 9]; however, these are not shown here, as these do not seem to be accurate.

8.10 Damped Filter

Figure 8.15 shows four types of damped filters. The first-order filter is not used as it has excessive loss at fundamental frequency and requires a large capacitor. The second-order high pass is generally used in composite filters for higher frequencies.

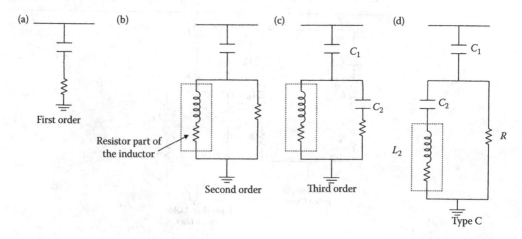

FIGURE 8.15
Circuits of damped filters: (a) first-order filter, (b) second-order filter, (c) third-order filter, and (d) type C filter.

If it were to be used for the full spectrum of harmonics, the capacitor size would become large and fundamental frequency losses in the resistor would be of consideration. This will be illustrated with an example. The filter is more commonly described as a second-order high-pass filter. The third-order filter has a substantial reduction in fundamental frequency losses, due to the presence of C_2 which increases the filter impedance; C_2 is very small as compared to C_1. The filtering performance of type-C filters lies between that of second- and third-order filters. C_2 and L_2 are series tuned at fundamental frequency and the fundamental frequency loss is reduced. Also see References [10].

The band pass filters give rise to a shifted resonance frequency, while damped filters do not. This advantage of damped filters can be exploited and possible resonances at shifted frequencies can be avoided. Unlike ST parallel multiple filters, there are no parallel branches, yet the component sizing becomes comparatively large and it may not possible to exploit this advantage in every system design. The performance and loading is less sensitive to tolerances. The behavior of damped filters can be described by the following two parameters [9]:

$$m = \frac{L}{R^2 C} \tag{8.31}$$

$$f_0 = \frac{1}{2\pi CR} \tag{8.32}$$

The impedance can be expressed in the parallel equivalent form:

$$Y_f = G_f + jB_f \tag{8.33}$$

where

$$G_f = \frac{m^2 x^4}{R_1 \left[(1 - mx^2)^2 + m^2 x^2 \right]} \tag{8.34}$$

$$B_f = \frac{x}{R_1} \left[\frac{1 - mx^2 + m^2 x^2}{(1 - mx^2)^2 + m^2 x^2} \right] \tag{8.35}$$

where

$$x = \frac{f}{f_0} \tag{8.36}$$

Considering that the filter is in parallel with an AC system of admittance $Y_a < \pm\varphi_a$ (max), the minimum total admittance as φ_a and Y vary is as follows:

$$Y = B_f \cos\phi_a + G_f \sin\phi_a \tag{8.37}$$

provided that the sign of each term is taken as positive and χ is less than the value that gives

$$|\cot\phi_f| = \left| \frac{G_f}{B_f} \right| = |\tan\phi_a| \tag{8.38}$$

For a given C, select parameters f_0 and m to obtain a sufficiently high admittance (low impedance) over the required frequency range. Values of m are generally between 0.5 and 2.

8.10.1 Second-Order High-Pass Filter

The characteristics of a second-order high-pass filter are shown in Figure 8.16, with its R–X and Z–ω plots. It has a low impedance above a corner frequency; thus, it will shunt a large percentage of harmonics at or above the corner frequency. The sharpness of tuning in high-pass filters is the reciprocal of tuned filters:

$$Q = \frac{R}{(L/C)^{1/2}} = \frac{R}{X_L} = \frac{R}{X_c} \tag{8.39}$$

With high Q, the filtering action is more pronounced. Filter impedance is given by

$$Z = \frac{1}{j\omega C} + \left(\frac{1}{R} + \frac{1}{j\omega L} \right)^{-1} \tag{8.40}$$

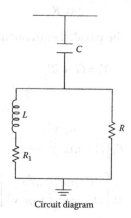

Circuit diagram

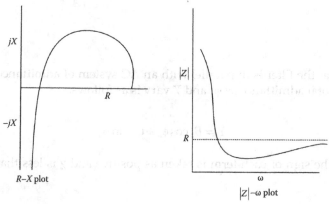

R–X plot

$|Z|$–ω plot

FIGURE 8.16
Circuit and R–X and Z–ω plots of a second-order high-pass filter.

The higher the resistance, the greater the sharpness of tuning. The Q value may vary from 0.5 to 2 and there is no optimum Q, unlike with band pass filters.

The reactive power of the capacitor at fundamental frequency is the same as for an ST filter. The loading at harmonic h is

$$I_h^2 \frac{X_c}{h} = \frac{1}{S_f} \frac{I_h^2}{h} V^2 \left[\frac{n^2}{n^2-1} \right] \qquad (8.41)$$

Thus, the total harmonic loading is

$$V^2 \frac{n^2}{S_f(n^2-1)} \sum_{h=min}^{h=max} \frac{I_h^2}{h} \qquad (8.42)$$

The reactor loading at fundamental frequency can be calculated by assuming that current through the parallel resistor is zero, i.e., current through the inductor is the same as through the capacitor; then, fundamental frequency loading is

$$I_L^2 X_L = I_c^2 \frac{X_c}{n^2} = \frac{S_f}{n^2} \left[\frac{n^2}{2} - 1 \right] \qquad (8.43)$$

At harmonic h, the harmonic current I_h divides into the resistance and inductance. The inductive component of the current is

$$I_{hL} = I_h \frac{R}{R+j\omega L} = I_h \frac{Q}{[Q^2 + (h/n)^2]^{1/2}} \qquad (8.44)$$

The total harmonic loading is, therefore,

$$= Q^2 \frac{V^2}{S_f} \left[\frac{n^2}{n^2-1} \right] \sum_{n=min}^{h=max} \left[n \frac{I_h^2}{Q^2 n^2 + h^2} \right] \qquad (8.45)$$

The loss in the resistor can be calculated as follows:

$$R = QhX_L \qquad (8.46)$$

$$|I_R| = \frac{|I_L| X_L}{R} = \frac{I_L}{Qn} \qquad (8.47)$$

Thus, the power loss is

$$I_R^2 R = \frac{1}{Qn} I_L^2 X_L \qquad (8.48)$$

$$I_R^2 R = \frac{1}{Qn} (\text{Mvar loading}) \qquad (8.49)$$

$$I_R^2 R = \frac{S_f}{Qn^3} \left[\frac{n^2}{n^2-1} \right] \qquad (8.50)$$

8.10.2 Type C Filter

Type C filter was first introduced in the France–England HVCD interconnection project [11–12] and then in Intermountain and Quebec–New England HVDC projects. It can replace conventional ST filters effectively and finds its use in arc furnace and ladle furnace installations [13]. Figure 8.15d shows the equivalent circuit of a type C filter. Neglecting the resistance of the reactor, the impedance of a C type filter is given by

$$Z(\omega) = \left(\frac{1}{R} + \frac{1}{j\omega L + 1/(j\omega C)} \right)^{-1} + \frac{1}{j\omega C_1}$$

$$= \frac{R(\omega^2 LC - 1)^2 + jR^2 \omega C(\omega^2 LC - 1)}{(R\omega C)^2 + (\omega^2 LC - 1)^2} - j\frac{1}{\omega C_1} \tag{8.51}$$

The impedance varies with the frequency. To avoid power loss at fundamental frequency, f, in damping resistor R, the components L and C are tuned to fundamental frequency:

$$\omega_f^2 LC = 1 \tag{8.52}$$

Therefore, impedance of a filter at fundamental frequency is determined by C_1:

$$Z(\omega_f) = \frac{-j}{\omega_f C_1} = -j\frac{V_s}{Q_f} \tag{8.53}$$

where Q_f is the reactive power requirement at fundamental frequency and V_s is the nominal system voltage. This allows a straightway calculation of C_1.

8.11 Design of a Second-Order High-Pass Filter

Example 8.6

In Example 8.1, we concluded that three ST filters for the 5th, 7th, and 11th harmonics designed around 900 kvar capacitors, rated voltage 2.77 kV (equivalent to 675 kvar at a system voltage of 2.4 kV) for each filter leg, provided adequate filtration, and controlled TDD to acceptable values. A total of 2700 kvar was required. We also observed that if an ST filter for the fifth harmonic is used it has to be of impractically large size, requiring

TABLE 8.13

Example 8.6: High-Pass Filter

Harmonic Order h ↓	Case 1	Case 2	Case 3	Case 4
5	322	145	79	46.2
7	173	60.8	33.5	21.5
11	31.6	36.7	20.3	25.6
13	9.21	14.3	7.81	13.0
17	3.05	6.80	3.65	7.96
19	1.44	3.61	1.93	4.66
23	0.33	0.98	0.52	1.49
25	0.23	0.73	0.39	1.19
29	0.13	0.46	0.24	0.84
31	0.08	0.30	0.16	0.58
35	0.06	0.21	0.09	0.43

Note: Harmonic currents in PCC, in ampères.

9000 kvar of similarly rated capacitors. *Capacitor units rated at 2.4 kV could be used, depending on the system overvoltage profile.*

If a single second-order high-pass filter is designed to control TDD to acceptable levels in Example 8.1, its size will be still larger than the single fifth ST filter. This is so because a high-pass filter has a higher impedance at notch frequency, as compared to an ST filter. The application of a high-pass filter is, generally, for higher frequencies and notch reduction. Four cases of study are presented in Table 8.13.

Case 1: Capacitor size 1.8 Mvar per phase; $C=622\,\mu F$, reactor $L=0.4805\,mH$ (reactor $X/R=100$), $R=0.181\,\Omega$.

Case 2: As case 1, except that filter $Q=4.85$. For $n=4.85$, $C=622\,\mu F$, $L=0.4805$, $R=0.875\,\Omega$. A marked difference in the harmonic currents injected at the PCC occurs with higher Q. The TDD is still high.

Case 3: The size of capacitors is doubled, i.e., 3.6 Mvar per phase, 10.8 Mvar total. The TDD at the PCC is still not in control, especially at lower harmonics.

The frequency scan of these three cases is shown in Figure 8.17. For larger resistance, the filter reverts to an ST filter.

Case 4: An ST fifth harmonic filter of 300 kvar per phase ($C=10.37\,\mu F$, $L=2.88\,mH$, $Q=100$, $n=4.85$) is paralleled with a high-pass filter; $C=1.2$ Mvar per phase (3.6 Mvar total), $C=414.9\,\mu F$, $L=0.346\,mH$, filter $Q=4.5$, $n=7$. The result of harmonic current flow into the PCC almost meets the TDD requirements.

This shows that three ST filters are the best design choice for Example 8.1. The circuit diagram of a high-pass filter in conjunction with parallel ST filters and R–X and Z–ω plots are shown in Figure 8.18.

The second-order high-pass filter is often applied for higher order harmonics and notch reduction. Reference [14] describes unique application for two 8000 hp ID fan drives to mitigate a wide spectrum of harmonics.

8.12 Zero Sequence Traps

Zigzag transformers and delta–wye transformers will act as zero sequence traps when connected in the neutral circuit of a three-phase four-wire system. Figure 8.19 shows a delta–wye transformer serving single-phase nonlinear loads of switched-mode power supplies, PCs, printers, and fluorescent lighting. As discussed in Chapter 1, the neutral can carry excessive harmonic currents. A zigzag or delta–wye transformer connected as shown in Figure 8.19 will reduce harmonic currents and voltage.

As we discussed in Volume 1, the zero sequence impedance of the core-type delta–wye transformer is low as the zero sequence flux seeks a high reluctance path through air or the transformer tank. The delta winding carries the zero sequence currents to balance the primary ampère turns. In an unbalanced system, the positive and negative sequence components will also be present, and these will not be suppressed. In a zigzag transformer, all windings have the same number of turns, but each pair of windings on a leg is wound in the opposite direction. A zigzag transformer has low zero sequence impedance and works in the same manner as a delta–wye transformer.

Figure 8.19 shows that the three-phase four-wire system, with neutral solidly grounded, serves single-phase loads. The neutral currents have two parallel paths, both of low impedance, through the delta–wye or zigzag transformer and also through the grounded neutral. The neutral voltage rise will be much less, though it will not be completely stable.

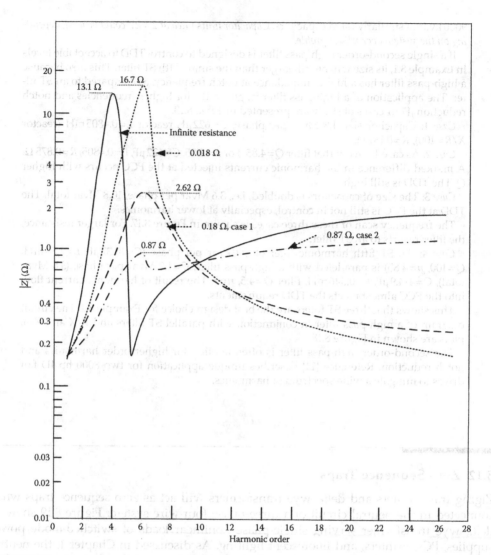

FIGURE 8.17
$Z–\omega$ plots of a second-order high-pass filter with varying Q and capacitor bank size (Example 8.6).

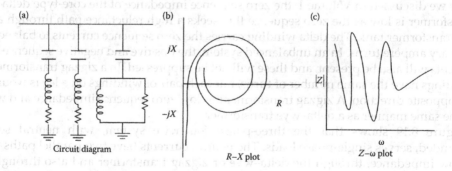

FIGURE 8.18
High-pass filter for higher frequencies in parallel with two ST filters. $R–X$ and $Z–\omega$ characteristics.

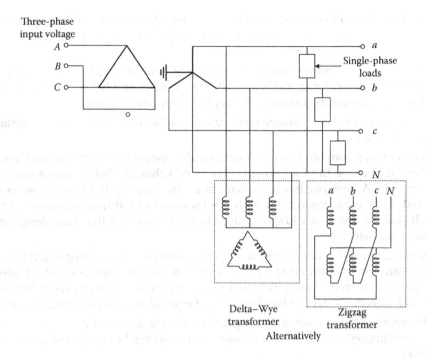

FIGURE 8.19
Delta–wye or zigzag transformer used as neutral trap in a three-phase four-wire system serving nonlinear loads.

8.13 Limitations of Passive Filters

Passive filters have been widely applied to limit harmonic propagation, improve power quality, reduce harmonic distortion, and provide reactive power compensation simultaneously. These can be designed for large current applications and high voltages. Many such filters are in operation for HVDC links, and passive filters are still the only choice when high voltages and currents are involved.

Some of the limitations of the passive filters are apparent from the Examples [15, 16]. These can be summarized as follows:

- Passive filters are not adaptable to the changing system conditions and, once installed, are *rigidly* in place. Neither the tuned frequency nor the size of the filter can be changed so easily. The passive elements in the filters are close tolerance components.

- A change in the system or operating condition can result in detuning and increased distortion. This can go undetected, unless there is online monitoring equipment in place.

- The design is largely affected by the system impedance. To be effective, the filter impedance must be less than the system impedance, and the design can become a problem for stiff systems. In such cases, a very large filter will be required. This may give rise to overcompensation of reactive power, and overvoltages on switching and undervoltages when out of service.

- Often, passive filters will require a number of parallel shunt branches. Outage of a parallel unit totally alters the resonant frequencies and harmonic current flows. This may increase distortion levels beyond permissible limits.

- Power losses in the resistance elements of passive filters can be very substantial for large filters.

- The parallel resonance between filter and the system (for single- or double-tuned filters) may cause amplification of currents of a characteristic or noncharacteristic harmonic. A designer has a limited choice in selecting the tuned frequency to avoid all possible resonances with the background harmonics. System changes will alter this frequency to some extent, however carefully the initial design might have been selected.

- Damped filters do not give rise to a system parallel resonant frequency; however, these are not so effective as a group of ST filters. The impedance of a high-pass filter at its notch frequency is higher than the corresponding ST filter. The size of the filter becomes large to handle the fundamental and the harmonic frequencies.

- The aging, deterioration, and temperature effects detune the filter in a random manner (though the effect of maximum variations can be considered in the design stage).

- If the converters feed back DC current into the system (with even harmonics), it can cause saturation of the filter reactor with resulting increase in distortion.

- Definite-purpose breakers are required. To control switching surges, special synchronous closing devices or resistor closing is required (see Volume 1).

- The grounded neutrals of wye-connected banks provide a low impedance path for the third harmonics. Third-harmonic amplification can occur in some cases.

- Special protective and monitoring devices (not discussed) are required.

8.14 Active Filters

By injecting harmonic distortion into the system, which is equal to the distortion caused by the nonlinear load, but of opposite polarity, the waveform can be corrected to a sinusoid. The voltage distortion is caused by the harmonic currents flowing in the system impedance. If a nonlinear current with opposite polarity is fed into the system, the voltage will revert to a sinusoid.

Active filters can be classified according to the following way these are connected in the circuit [17, 18]:

- In series connection
- In parallel shunt connection
- Hybrid connections of active and passive filters

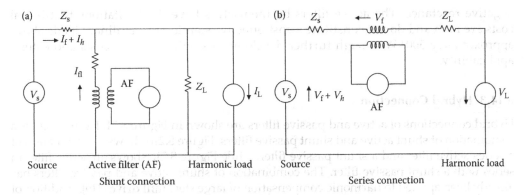

FIGURE 8.20
(a) Shunt connection of an active filter and (b) series connection of an active filter.

8.14.1 Shunt Connection

As we have seen, the voltage in a weak system is very much dependent on current, while a stiff system of zero impedance will have no voltage distortion. Thus, provided that the system is not too stiff, a nonsinusoidal voltage can be corrected by injecting proper current. A harmonic current source is represented as a Norton equivalent circuit, and it may be implemented with a voltage-fed PWM inverter to inject a harmonic current of the same magnitude as that of the load into the system, but of harmonics of opposite polarity. A shunt connection is shown in Figure 8.20a. The load current will be sinusoidal, so long as the load impedance is higher than the source impedance:

$$I_\mathrm{fl} = I_h \text{ for } |Z_\mathrm{L}| \gg |Z_\mathrm{s}| \quad I_h = 0 \tag{8.54}$$

In Chapter 1, we studied two basic types of converters: current and voltage. A converter with DC output reactor and constant DC current is a current harmonic source. A converter with a diode front end and DC capacitor has a highly distorted current depending on the AC source impedance, but the voltage at rectifier input is less dependent on AC impedance. This is a voltage harmonic source. It presents a low impedance and a shunt connection will not be effective. A shunt connection is more suitable for current-source controllers where the output reactor resists the change of current.

8.14.2 Series Connection

Figure 8.20b shows a series connection. A voltage V_f is injected in series with the line and it compensates the voltage distortion produced by a nonlinear load. A series active filter is more suitable for harmonic compensation of diode rectifiers where the DC voltage for the inverter is derived from a capacitor, which opposes the change of the voltage.

Thus, the compensation characteristics of the active filters are influenced by the system impedance and load. This is very much akin to passive filters; however, active filters have better harmonic compensation characteristics against the impedance variation and frequency variation of harmonic currents. The control systems of the active filters have a profound effect on the performance and a converter can have even a negative reactance. The active filters by themselves have the limitations that initial costs are high and do not constitute a cost-effective solution for nonlinear loads above approximately 500 kW, though further developments will lower the costs and extend applicability.

8.14.3 Hybrid Connection

Hybrid connections of active and passive filters are shown in Figure 8.21. Figure 8.21a is a combination of shunt active and shunt passive filters. Figure 8.21b shows a combination of a series active filter and a shunt passive filter, while Figure 8.21c shows an active filter in series with a shunt passive filter. The combination of shunt active and passive filters has already been applied to harmonic compensation of large steel mill drives [19]. Addition of a large shunt capacitor will reduce the load resistance and Equation 8.54 is no longer valid. The shunt passive filter will draw a large source current from a stiff system and may act as a sink to the upstream harmonics. It is required that in a hybrid combination, the filters share compensation properly in the frequency domain.

In a series connection, the active filter is connected in series with the passive filter, both being in parallel with the load, as shown in Figure 8.21c. With suitable control of the active filter, it is possible to avoid resonance and improve filter performance. The active filter can

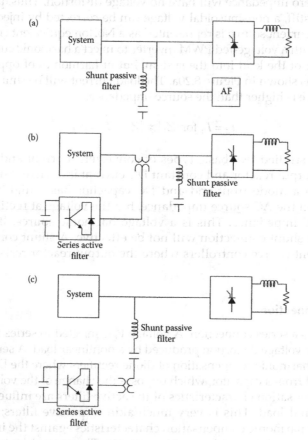

FIGURE 8.21
(a)–(c) Hybrid connections of active and passive filters.

be either voltage or current controlled. In current-mode control, the inverter is a voltage source to compensate for current harmonics. In voltage-mode control, the converter is a voltage-source inverter controlled to compensate for the voltage harmonics. The advantage is that the converter itself is far smaller, only about 5% of the load power. The active filter in such schemes regulates the effective source impedance as experienced by the passive filter, and the currents are forced to flow in the passive filter rather than in the system. This makes the passive filter characteristics independent of the actual source impedance and a consistent performance can be obtained.

8.14.4 Combination of Active Filters

A combination of series and shunt active filters is shown in Figure 8.22. This looks similar to the unified power controller discussed in Volume 2, but its operation is different [19]. A series filter blocks harmonic currents flowing in and out of the distribution feeders. It detects the supply current and is controlled to present a zero impedance to the fundamental frequency and high resistance to the harmonics. The shunt filter absorbs the harmonics from the supply feeders and detects the bus voltage at the point of connection. It is controlled to present infinite impedance to the fundamental frequency and low impedance to the harmonics. The harmonic currents and voltages are extracted from the supply system in the time domain.

The electronics and power devices used in both types of converters for filters are quite similar; Figure 8.23 shows three-phase voltage-source and current-source PWM converters. The current-source active filter has a DC reactor with a constant dc current while the voltage-source active filter has a capacitor on the DC side with constant DC voltage. An output filter is provided to attenuate the inverter switching effects. In a current-source type, LC filters are necessary (Figure 8.23b). Transient oscillations can appear because of resonance between filter capacitors and inductors. The controls are implemented so that the inverter outputs a harmonic current equivalent but opposite to that of the load. The source side current is therefore sinusoidal, but the voltage will be sinusoidal only if the source does not generate any harmonics. Bipolar junction transistors are used with switching frequencies up to 50 kHz for modest ratings. Silicon-controlled rectifiers and gate turn-off thyristors are used for higher power outputs.

A further classification is based on the control system, i.e., time-domain and frequency-domain corrections.

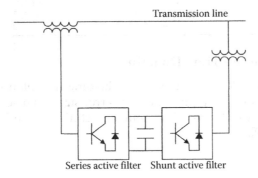

FIGURE 8.22
Connections of a unified power quality conditioner.

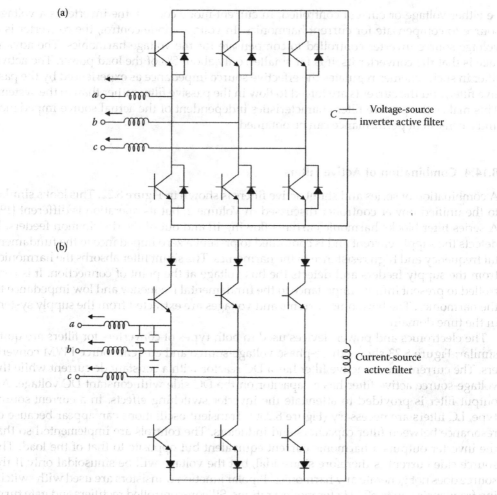

FIGURE 8.23
(a) Voltage-source inverter active filter and (b) current-source active inverter filter.

8.15 Corrections in the Time Domain

Corrections in the time domain are based on holding instantaneous voltage or current within reasonable tolerance of a sine wave. The error signal can be the difference between actual and reference waveforms. Time-domain techniques can be classified into three main categories [18, 19]:

- Triangular wave
- Hysteresis
- Deadbeat

The error function can be instantaneous reactive power (IRP, described in Section 8.17) or EXT (extraction of fundamental frequency component). For EXT (Extraction of fundamental frequency component), the fundamental component of the distorted waveform is extracted through a 60 Hz filter and then the error function is $e(t) = f60(t) - f(t)$. For IRP, the error function is given by the difference between the instantaneous orthogonal transformation of *actual* and 60 Hz components of voltages and currents.

The triangular-wave method is easiest to implement and can be used to generate two-state or three-state switching functions. A two-state function can be connected positively or negatively, while a three-state function can be positive, negative, or zero (Figure 8.24).

In the two-state system, the inverter is always *on* (Figure 8.24a). The extracted error signal is compared to a high-frequency triangular carrier wave, and the inverter switches

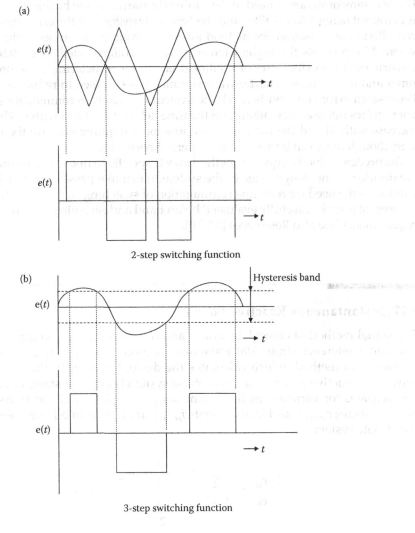

FIGURE 8.24
(a) Two-step switching function and (b) three-step switching function.

each time the waves cross. The result is an injected signal that produces equal and oppo-site distortion.

In a three-state system (hysteresis method), preset upper and lower limits are compared to an error signal (Figure 8.24b). So long as the error is within a tolerable band, there is no switching and the inverter is off.

The advantage of time-domain methods is fast response, though these are limited to one-node application, to which these are connected and take measurement from.

8.16 Corrections in the Frequency Domain

Fourier transformation is used to determine the harmonics to be injected. The error signal is extracted using a 60 Hz filter and the Fourier transform of the error signal is taken. The cancellation of M harmonics method allows for compensation up to the Mth harmonic, where M represents the highest harmonic to be compensated. A switching function is constructed by solving a set of nonlinear equations to determine the precise switching times and magnitudes. Quarter-wave symmetry is assumed to reduce the computations. Because an error function is used, the system can easily accommodate system changes but requires intense calculations and the time delays associated with it. The computations increase with M and the increased computational requirements are the main disadvantage, though these can be applied in dispersed networks.

The predetermined frequency method injects specific frequencies into the system, which are decided in the design stage of the system, much like passive harmonic filtering. This eliminates the need for real-time commutation of switching signals, but the harmonic levels present must be carefully evaluated beforehand and each filter designed for the specific requirements (see also References [20, 21]).

8.17 Instantaneous Reactive Power

The signal method of control generates an error signal based on input voltage or current and a reference sinusoidal waveform. A more elaborate function is the instantaneous power method, which calculates the desired current so that the instantaneous active and reactive power in a three-phase system is kept constant, i.e., the active filter compensates for variation in instantaneous power [22]. By linear transformation, the phase voltages e_a, e_b, e_c and load currents i_a, i_b, i_c are transformed into an α–β (two-phase) coordinate system:

$$\begin{vmatrix} e_\alpha \\ e_\beta \end{vmatrix} = \sqrt{\frac{2}{3}} \begin{vmatrix} 1 & -\dfrac{1}{2} & -\dfrac{1}{2} \\ 0 & \dfrac{\sqrt{3}}{2} & -\dfrac{\sqrt{3}}{2} \end{vmatrix} \begin{vmatrix} e_a \\ e_b \\ e_c \end{vmatrix} \tag{8.55}$$

and

$$\begin{vmatrix} i_\alpha \\ i_\beta \end{vmatrix} = \sqrt{\frac{2}{3}} \begin{vmatrix} 1 & -\dfrac{1}{2} & -\dfrac{1}{2} \\ 0 & \dfrac{\sqrt{3}}{2} & -\dfrac{\sqrt{3}}{2} \end{vmatrix} \begin{vmatrix} i_a \\ i_b \\ i_c \end{vmatrix} \tag{8.56}$$

The instantaneous real power p and the instantaneous imaginary power q are defined as

$$\begin{vmatrix} p \\ q \end{vmatrix} = \begin{vmatrix} e_\alpha & e_\beta \\ -e_\beta & e_\alpha \end{vmatrix} \begin{vmatrix} i_\alpha \\ i_\beta \end{vmatrix} \tag{8.57}$$

Here, p and q are not conventional watts and vars. The p and q are defined by the instantaneous voltage in one phase and the instantaneous current in the other phase:

$$p = e_\alpha i_\alpha + e_\beta i_\beta = e_a i_a + e_b i_b + e_c i_c \tag{8.58}$$

To define IRP, the space vector of imaginary power is defined as

$$q = e_\alpha i_\beta + e_\beta i_\alpha$$

$$= \frac{1}{\sqrt{3}} \left[i_a(e_c - e_b) + i_b(e_a - e_c) + i_c(e_b - e_a) \right] \tag{8.59}$$

Equation 8.57 can be written as

$$\begin{vmatrix} i_\alpha \\ i_\beta \end{vmatrix} = \begin{vmatrix} e_\alpha & e_\beta \\ -e_\beta & e_\alpha \end{vmatrix}^{-1} \begin{vmatrix} p \\ q \end{vmatrix} \tag{8.60}$$

These are divided into two kinds of currents:

$$\begin{vmatrix} i_\alpha \\ i_\beta \end{vmatrix} = \begin{vmatrix} e_\alpha & e_\beta \\ -e_\beta & e_\alpha \end{vmatrix}^{-1} \begin{vmatrix} p \\ 0 \end{vmatrix} + \begin{vmatrix} e_\alpha & e_\beta \\ -e_\beta & e_\alpha \end{vmatrix}^{-1} \begin{vmatrix} 0 \\ q \end{vmatrix} \tag{8.61}$$

This can be written as

$$\begin{vmatrix} i_\alpha \\ i_\beta \end{vmatrix} = \begin{vmatrix} i_{\alpha p} \\ i_{\beta p} \end{vmatrix} + \begin{vmatrix} i_{\alpha q} \\ i_{\beta q} \end{vmatrix} \tag{8.62}$$

where $i_{\alpha p}$ is the α-axis instantaneous active current:

$$i_{\alpha p} = \frac{e_\alpha}{e_\alpha^2 + e_\beta^2} p \tag{8.63}$$

$i\alpha_q$ is the α-axis instantaneous reactive current:

$$i_{\alpha q} = \frac{-e_\beta}{e_\alpha^2 + e_\beta^2} q \tag{8.64}$$

$i\beta_p$ is the β-axis instantaneous active current:

$$i_{\beta p} = \frac{e_\alpha}{e_\alpha^2 + e_\beta^2} p \tag{8.65}$$

and $i_{\beta q}$ is the β-axis instantaneous reactive current:

$$i_{\beta q} = \frac{e_\alpha}{e_\alpha^2 + e_\beta^2} q \tag{8.66}$$

The following equations exist as follows:

$$p = e_\alpha i_{\alpha P} + e_\beta i_{\beta P} \equiv P_{\alpha P} + P_{\beta P}$$
$$0 = e_\alpha i_{\alpha q} + e_\beta i_{\beta q} \equiv P_{\alpha q} + P_{\beta q} \tag{8.67}$$

where the α-axis instantaneous active and reactive powers are as follows:

$$P_{\alpha p} = \frac{e_\alpha^2}{e_\alpha^2 + e_\beta^2} p \quad P_{\alpha q} = \frac{-e_\alpha e_\beta}{e_\alpha^2 + e_\beta^2} q \tag{8.68}$$

The β-axis instantaneous active and reactive power is

$$P_{\beta q} = \frac{e_\beta^2}{e_\alpha^2 + e_\beta^2} p \quad P_{\beta q} = \frac{e_\alpha e_\beta}{e_\alpha^2 + e_\beta^2} q \tag{8.69}$$

The sum of the instantaneous active powers in two axes coincides with the instantaneous real power in the three-phase circuit. The IRPs $P\alpha_q$ and $P\beta_q$ cancel each other and make no contribution to the instantaneous power flow from the source to the load.

Consider instantaneous power flow in a three-phase cycloconverter. The IRP on the source side is the IRP circulating between source and cycloconverter, while the IRP on the output side is the IRP between the cycloconverter and the load. Therefore, there is no relationship between the IRPs on the input and output sides, and the instantaneous imaginary power on the input side is not equal to the instantaneous imaginary power on the output side. However, assuming zero active power loss in the converter, the instantaneous real power on the input side is equal to the real output power.

8.18 Harmonic Mitigation at Source

The harmonic mitigation at source, without filters, has attracted the attention of industry and researchers. This covers a wide field, spanning industrial applications to transmission systems and HVDC. We will briefly look at some systems from the point of view of harmonic mitigation.

8.18.1 Phase Multiplication

The principle of harmonic elimination by phase multiplication is discussed in Chapter 1. Figure 8.25 shows a 2300 V medium-voltage drive system, where each motor phase is driven by three PWM cells. Each group of power cells is wye connected with a floating neutral and is powered by an isolated secondary winding of the drive input transformer. A greatly improved voltage waveform is obtained due to phase displacements in the transformer secondary windings, and the harmonic distortion meets IEEE limits [1] without filters. Another advantage is that the common mode voltages are eliminated.

8.18.2 Parallel Connected 12-Pulse Converters with Interphase Reactor

Figure 8.26a shows the circuit of a conventional 12-pulse thyristor converter with interphase reactor and phase shift obtained through delta–delta and delta–wye input transformers. Figure 8.26b is a conventional stepped waveform of the 12-pulse converter. This can be rendered close to a sinusoid by superposition of a triangular current as shown in Figure 8.26c and the system has a better waveform than that of a 36-pulse thyristor converter [23].

8.18.3 Active Current Shaping

By using proper control systems, the input current of converters can be forced to follow a sinusoid in phase with voltage, addressing the need for reactive power compensation as well as harmonic elimination [19]. The load current can be written as

$$I_L(t) = Kv(t) + i_q(t) \tag{8.70}$$

where $Kv(t)$ is the active component of the load current, K is a coefficient that can be calculated in the control circuit, and $i_q(t)$ is the nonactive component of the current. The nonactive current must be compensated to have maximum power factor and harmonic rejection. The desired reference current in the active filter is

$$I_q(t) = I_L(t) - Kv(t) \tag{8.71}$$

8.18.4 Input Reactors to the PWM ASDs

By adding a choke (reactor) rated at 3% of the drive system kVA, the ASD current distortion is considerably reduced. Figure 8.27 is adapted from IEEE guide [24] and shows the reduction in the current distortion, the transformer impedance is assumed 5%.

8.19 Multilevel Inverters

Multilevel inverters are a new breed [25], which overcome some of the disadvantages of PWM inverters, namely as follows:

The carrier frequency is high, usually between 2 and 20 kHz. In a normal PWM waveform, the pulse height is the DC link voltage, dv/dt is high and causes a strong electromagnetic interference (EMI). This introduces a number of harmonics and possible ringing. PWM needs rigorous switching conditions and resultant switching losses. The inverter control circuitry is complex.

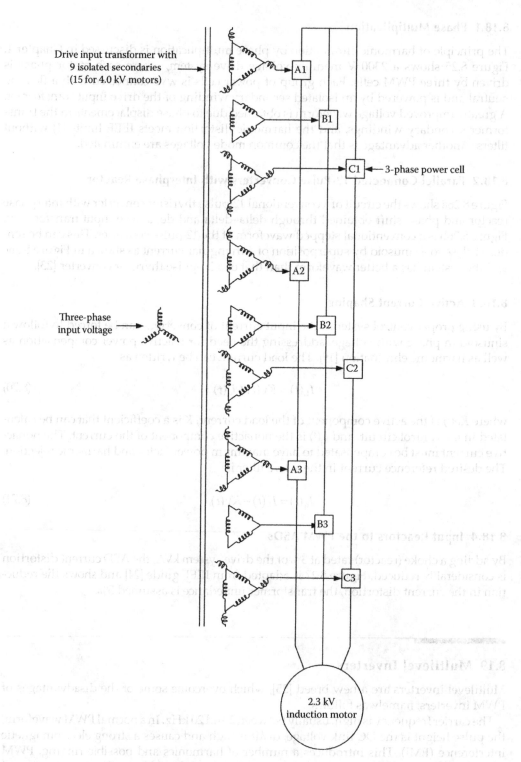

Drive input transformer with
9 isolated secondaries
(15 for 4.0 kV motors)

Three-phase
input voltage

A1

B1

C1 — 3-phase power cell

A2

B2

C2

A3

B3

C3

2.3 kV
induction motor

FIGURE 8.25
An ASD with secondary-phase multiplication and low harmonic distortion.

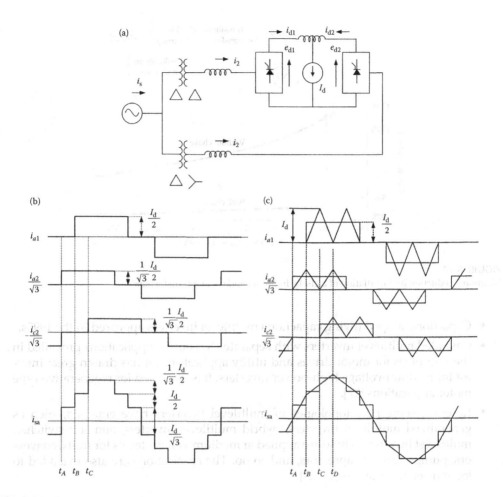

FIGURE 8.26
(a) A 12-pulse converter circuit with interphase reactor, (b) stepped voltage generation of a 12-pulse circuit, and (c) improvement in voltage waveform with triangular-wave superimposition.

The multilevel inverters emerged as a solution to high-power applications. The switching frequencies are low, equal to, or only a few times the output frequency. The pulse heights are low. For an *m*-level inverter, the pulse height is V_m/m, where V_m is the output voltage amplitude. This results in much smaller dv/dt and EMI as compared to PWM inverters. The harmonics and THD is further reduced. Smooth switching conditions are obtained with much lower switching power losses.

Multilevel converters have been applied to HVDC, large motor drives, railway traction applications, UPFC, STATCOM, and SVC. The quality of output voltage increases as the number of voltage levels increases. The applications have been extended to active power filters, voltage sag compensators, and photovoltaic systems.

The various types of multilevel converters are as follows:

- Diode-clamped multilevel inverter was proposed by Nabae in 1981 [26]. It is also called the neutral point-clamped (NPC) inverter, because the NPC inverter effectively doubles the device voltage without precise voltage matching.

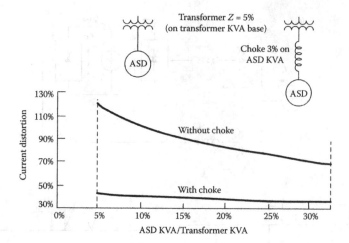

FIGURE 8.27
Harmonic reduction by installation of a choke in the secondary of a transformer.

- Capacitor-clamped (flying capacitors) multilevel inverters appeared in the 1990s.
- Cascaded multilevel inverters with separate DC sources applications prevailed in the mid-1990s for motor drives and utility applications. It has drawn great interest for medium-voltage high-power inverters. It is also used for regenerative-type motor applications [27].
- Recently, some new topologies of multilevel inverters have emerged, such as generalized multilevel inverters, hybrid multilevel inverters, and soft-switched multilevel inverters. These are applied at medium-voltage levels for mills, conveyors, pumps, fans, compressors, and so on. The applications are also extended to low-power applications [28, 29].

Figure 8.28a shows a single-phase, five-level diode clamp circuit with four DC bus capacitors, C_1, C_2, C_3, and C_4. The staircase voltage wave is synthesized from several levels of DC capacitor voltages. An m-level diode clamp converter consists of $m - 1$ capacitors on the DC bus and produces m levels of the phase voltage by appropriate switching. The voltage across each capacitor is $V_{dc}/4$. The staircase voltage shown in Figure 8.28b is generated by the five switch combinations shown in Figure 8.28a and the switching matrix shown below:

$$
\begin{array}{c c c c c c c c c}
      & S_{a1} & S_{a2} & S_{a3} & S_{a4} & S'_{a1} & S'_{a2} & S'_{a3} & S'_{a4} \\
V_4 & 1 & 1 & 1 & 1 & 0 & 0 & 0 & 0 \\
V_3 & 0 & 1 & 1 & 1 & 1 & 0 & 0 & 0 \\
V_2 & 0 & 0 & 1 & 1 & 1 & 1 & 0 & 0 \\
V_1 & 0 & 0 & 0 & 1 & 1 & 1 & 1 & 0 \\
V_0 & 0 & 0 & 0 & 0 & 1 & 1 & 1 & 1
\end{array}
\tag{8.72}
$$

With high switching levels, the harmonic content is low enough and filters are not needed. The disadvantages are large clamping diodes, unequal switching ratings, and real power control. The clamping diodes can be replaced with capacitors called flying capacitor-based

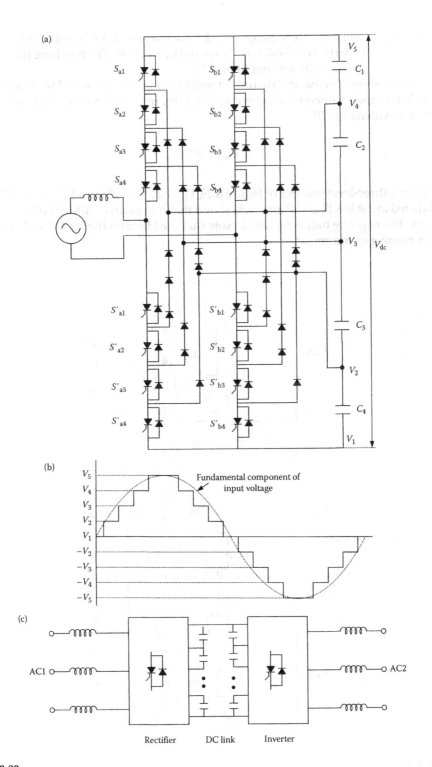

FIGURE 8.28
(a) Single-phase, full-bridge, five-level diode-clamp converter circuit, (b) stepped voltage generation resulting in low harmonic distortion, and (c) two-diode clamp multilevel converters for back-to-back intertie system.

control or multilevel converters using cascade converters with DC sources. An application in a back-to-back intertie connection is shown in Figure 8.28c. The resulting harmonic distortion is within IEEE limits without filters [25].

Figure 8.29 shows the basic structure of multilevel inverters using H-bridge (HB) converters. This shows a one-phase leg with three HBs. In binary hybrid multilevel inverter, the DC link voltage of HB is

$$V_{DCi} = 2^{i-1}E \qquad (8.73)$$

That is, for a three-level inverter of Figure 8.29, $V_{DC1}=E$, $V_{DC2}=2E$, and $V_{DC3}=4E$. The operation is listed in Table 8.14 and Figure 8.30 shows that the positive half-output waveform has 15 levels. The negative half is identical. Note that the HB with higher DC link voltage has a lower number of commutations.

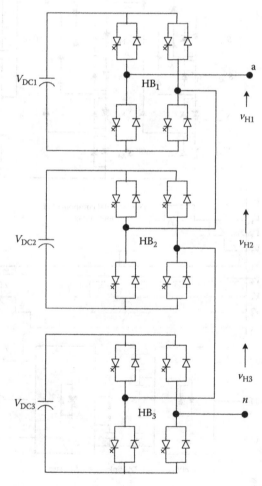

FIGURE 8.29
Multilevel inverter based upon series connections of HBs.

TABLE 8.14

Operation of Multilevel Inverter, Figure 8.29

E	v_{H1}	v_{H2}	v_{H3}
0	0	0	0
+1E	E	0	0
+2E	0	2E	0
+3E	E	2E	0
+4E	0	0	4E
+5E	E	0	4E
+6E	0	2E	4E
+7E	E	2E	4E
−E	−E	0	0
−2E	0	−2E	0
−3E	−E	−2E	0
−4E	0	0	−4E
−5E	−E	0	−4E
−6E	0	−2E	−4E
−7E	−E	−2E	−4E

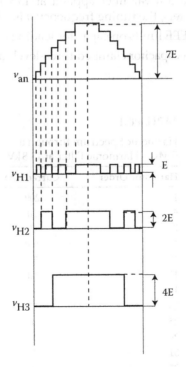

FIGURE 8.30
Positive half of waveform of binary hybrid mutilevel inverter, 15 levels.

In a trinary hybrid inverter (THMI), the DC line voltages are as follows:

$$V_{DCi} = 3^{i-1}E \tag{8.74}$$

That is for three-HB one-phase leg, $V_{DC1}=E$, $V_{DC2}=3E$, and $V_{DC3}=9E$. The output voltage, as before, is

$$V_{an} = \sum_{i=1}^{h} v_{Hi} = \sum_{i=1}^{h} F_i V_{DCi} \tag{8.75}$$

where F_i is a switching function. It can be shown that THMI has greatest number of output voltage levels and by proper switching angles, the odd and even harmonics can practically be eliminated (see Reference [29]).

The above techniques form an introduction, and references are provided for the interested reader to probe further.

Problems

8.1 The current through a 5 Mvar filter applied at 13.8 kV is given in Table P8.1. Calculate filter parameters; the tuning frequency n is 10.65.

8.2 In Problem 2, are the IEEE limits on capacitor loadings exceeded?

8.3 Calculate the THD in a capacitor clamped three-level inverter.

TABLE P8.1

Harmonic Spectrum through a 5 Mvar Harmonic Filter at 13.8 kV

Harmonic Order	Current (A)
1	209
5	20
7	45
11	100
13	50
17	10
19	8
23	12
25	4
29	2
31	2
35	1
37	1

References

1. IEEE Standard 519. IEEE recommended practice and requirements for harmonic control in electrical systems, 1992.
2. IEC. Electromagnetic Compatibility—Part 3: Limits-Section 2: Limits for harmonic current emission (equipment input current #16A per phase), 1995. Standard 61000-3-2.
3. E.W. Kimbark. *Direct Current Transmission*. Chapter 8. John Wiley, New York, 1971.
4. J.C. Das. *Transients in Electrical Systems—Recognition Analysis and Mitigation*. McGraw Hill, New York, 2010.
5. IEEE Standard C37.99. IEEE guide for protection of the capacitor banks, 2000.
6. IEEE. IEEE guide for application and specifications of harmonic filters. Standard 1531. 2003.
7. M.M. Swamy, S.L. Rossiter, M.C. Spencer, M. Richardson. Case studies on mitigating harmonics in ASD systems to meet IEEE-519-1992 standards. *IEEE-IAS Conference Record*, 1, 685–692, 1994.
8. H.M. Zubi, R.W. Dunn, F.V.P. Robinson, M.H. El-werfelli. Passive filter design using genetic algorithms for adjustable speed drives. *IEEE Power and Energy Society General Meeting*, Minneapolis, July 2010.
9. J.D. Anisworth. Filters, damping circuits and reactive voltamperes in HVDC converters. In *High Voltage Direct Current Converters and Systems*, ed. B.J. Cory. Macdonald, London, 1965, pp. 137–174.
10. J. Arrillaga, D.A. Bradley, P.S. Bodger. *Power System Harmonics*. John Wiley, New York, 1985.
11. M.A. Zamani, M. Moghaddasian, M. Joorabian, S. Gh Seifossadat, A. Yazdani. C-Type filter design based on power-factor correction for 12-pulse HVDC converters. *IEEE Industrial Electronics, IECON, 34th Annual Conference*, pp. 3039–3044, November 2008.
12. CIGRE Working Group 14.03. AC harmonic filters and reactive power compensation for HVDC. CIGRE Report, June 1990.
13. C.O. Gercek, M. Ermis, A. Ertas, K.N. Kose, O. Unsar. Design implementation and operation of a new C-type 2nd order harmonic filter for electric arc and ladle furnaces. *IEEE Trans Ind Appl*, 47(4), 1545–1557, 2011.
14. J.C. Das. Design and application of a second-order high-pass damped filter for 8000-hp ID fan drives—A case study. *IEEE Trans Ind Appl*, 51(2), 1417–1426, 2015.
15. J.C. Das. Analysis and control of harmonic currents using passive filters. *TAPPI Proceedings, Atlanta Conference*, Atlanta, pp. 1075–1089, 1999.
16. J.C. Das. Passive filters—potentialities and limitations. *IEEE Trans Ind Appl*, 40(1), 232–241, 2004.
17. H. Akagi. Trends in active power line conditioners. *IEEE Trans Power Electron*, 9, 263–268, 1994.
18. W.M. Grady, M.J. Samotyi, A.H. Noyola. Survey of active line conditioning methodologies. *IEEE Trans Power Deliv*, 5, 1536–1541, 1990.
19. H. Akagi. New trends in active filters for power conditioning. *IEEE Trans Ind Appl*, 32, 1312–1322, 1996.
20. A. Cavallini, G.C. Montanarion. Compensation strategies for shunt active filter control. *IEEE Trans Power Electron*, 9, 587–593, 1994.
21. C.V. Nunez-Noriega, G.G. Karady. Five step-low frequency switching active filter for network harmonic compensation in substations. *IEEE Trans Power Deliv*, 14, 1298–1303, 1999.
22. H. Akagi, A. Nabe. The p–q theory in three-phase systems under non-sinusoidal conditions. *ETEP*, 3, 27–30, 1993.
23. T. Tanaka, N. Koshio, H. Akagi, A. Nabae. Reducing supply current harmonics. *IEEE Ind Appl Manage*, 4, 31–35, 1998.
24. IEEE P519.1. Draft guide for applying harmonic limits on power systems, 2004.
25. J.S. Lai, F.Z. Peng. Multilevel converters—a new breed of power converters. *IEEE Trans Ind Appl*, 32, 509–517, 1996.

26. A. Nabae, I. Takahashi, H. Akagi. A neutral point clamped PWM inverter. *IEEE Trans Ind Appl,* 17, 518–523, 1981.
27. P.W. Hammond. New approach to enhance power quality for medium voltage AC drives. *IEEE Trans Ind Appl,* 33, 202–208, 1997.
28. A.M. Trzymadlowski. *Introduction to Modern Power Electronics.* John Wiley, New York 1998.
29. F.L. Luo, H. Ye. *Power Electronics—Advanced Conversion Technologies.* CRC Press, Boca Raton, FL, 2010.

9

Harmonic Analysis in Solar and Wind Generation

Wind and solar generation are important renewable (green) energy sources and their fundamental aspects are covered in Chapter 3, Volume 1. There are stringent requirements of interconnections with the utility systems with respect to reactive power compensation, voltage dip ride-through capability, and stability. These aspects will not be repeated here. A reader may familiarize with this chapter before proceeding with the harmonic generation and control.

9.1 Solar Inverters

The harmonic emissions from a large solar plant are modeled in this study. There has been an attempt to obtain larger currents and increased voltage from interconnections of solar arrays and innovations in the design of inverters. Table 9.1 shows the operating characteristics of a solar inverter. Its output power is approximately 1.98 Mvar. There is another trend in the design of solar inverters: raising the voltage of the solar inverters.

Figure 9.1 shows a 2.5/2.8 MVA 13.8–0.8 kV three-phase transformer connected to two solar inverters. Each inverter is rated for an output of 900 A. Thus, the maximum output from the inverters is approximately 2.5 MW.

The inverters will generate some harmonics. For a harmonic study, it is necessary that the spectrum of the harmonics is obtained from the manufacturer as it may vary considerably. The harmonic emission is shown in Table 9.2. This shows that the harmonic emission is at low levels. The vendors are applying harmonic filters or using multilevel inverter technologies to reduce the harmonic injection, see Chapter 8.

9.1.1 Configuration of the Solar Generating Plant

From Figure 9.1, solar arrays in series and parallel configuration give an output of 900 A maximum current for a three-phase inverter output at 800 V. See Chapter 3 of Volume 1 for the arrangements of solar arrays, using PV cells.

Figure 9.2 shows that 10 such 13.8–0.8 kV transformers, primary windings in delta connection and the secondary windings in wye connection, solidly grounded form "one chain" served from a 1200 A 13.8 kV circuit breaker. Thus, the maximum possible output from one chain is approximately 75 MW. Some power will be consumed in the auxiliary loads as shown in Figure 9.1. The cable connections between each 2.5/2.8 MVA transformer are 650 long consisting of 2/C 1000 kcmil shielded cables, 15 kV grade per phase. Each cable length is modeled with equivalent Π model. The capacitance of cables is important for load flow and also for harmonic analysis.

Three such chains are served from 3000, A 13.8 kV bus 1, see Figure 9.2. Thus, each 13.8 kV bus has approximately a generation capability of 75 MW. In any analysis, it must be

TABLE 9.1

Operating Particulars of a Large Commercially Available Solar Inverter

Parameter	Specification
Rated voltage	415 V, three-phase
Voltage variation	+10%, –12%
Frequency	60 Hz or 50 Hz
Frequency variation	+0.5 to –0.7 Hz
Power factor range	0.91 lead/lag
Maximum current	2760 A
Maximum efficiency	98%
DC voltage operating range	605–950 V
Protection dc side	Ground fault, overvoltage, dc reverse current, overcurrent
Protection ac side	Over/undervoltage, over/under frequency, overcurrent, anti-islanding
Fault ride through capability	Can be provided

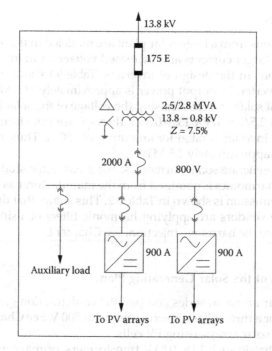

FIGURE 9.1
A substation module with solar inverters.

considered that solar generation can vary over large limits. A reactive power compensation device is also shown, which is discussed further in the study analysis.

Figure 9.3 illustrates that three such 13.8 kV buses (bus 1, bus 2, and bus 3) are each connected to 55/92 MVA, 13.8–230 kV step-up transformer. Thus, the total installed capacity of 9 chains, each chain consisting of 10, 2.5 MVA transformers (a total of 90 transformers), is approximately 225 MW of power generation at peak hours.

TABLE 9.2

Harmonic Emission, Solar Power Plant

Harmonic Order Odd Harmonics	% of Fundamental	Harmonic Order Even Harmonics	% of Fundamental
3	0.14	2	0.24
5	0.18	4	0.30
7	0.09	6	0.05
9	0.06	8	0.11
11	0.09	10	0.13
13	0.04	12	0.05
15	0.06	14	0.12
17	0.12	16	0.15
19	0.30	18	0.06
21	0.12	20	0.16
23	0.25	22	0.20
25	0.10	24	0.12
27	0.07	26	0.12
29	0.11	28	0.13
31	0.24	30	0.11
33	0.11	32	0.19
35	0.11	34	0.22
37	0.12	36	0.09
39	0.07	38	0.14
		40	0.14

The objective of the study is to ascertain the following:

- If there are harmonic distortion problems due to 180 solar inverters; two inverters on the secondary of each transformer. Will the harmonic distortion meet the IEEE 519 limits?
- Ascertain through load flow if any reactive power compensation is required for power factor improvement or bus voltage support.
- Study the impact on harmonic emission and distortions, with and without reactive power compensating devices.
- Considering impact of switching transients if shunt capacitor banks are used for reactive power compensation, recommend alternate reactive power compensating devices.

13.8 kV buses 1, 2, and 3 are declared as PCC (PCC-1) and 230 kV bus is declared as PCC-2.

9.1.2 Load Flow Study

First, a load flow study is conducted and its summary results are shown in Table 9.3. The load flow is carried out without any reactive power compensation shown on buses 1, 2, and 3, and with reactive power compensation.

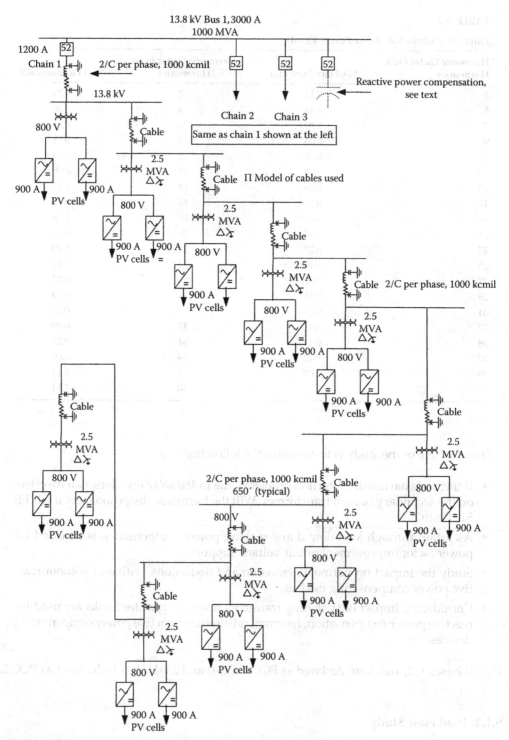

FIGURE 9.2
Configuration of a solar PV generating plant, details shown for distribution from one 55/92 MVA transformer.

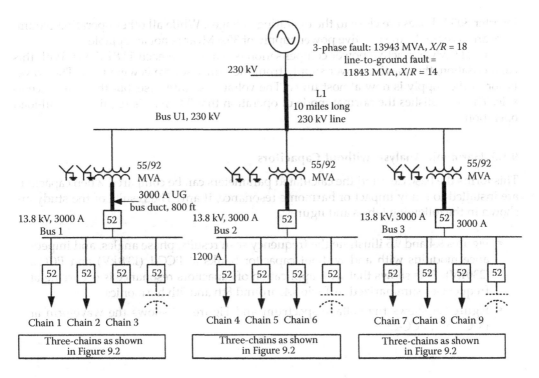

FIGURE 9.3
Overall configuration of a solar generating plant.

TABLE 9.3

Load Flow

Description	Without 3 × 11 Mvar Capacitors	With 3 × 11 Mvar Capacitors
Operation of solar converters (together)	2.5 MW, 120 kvar output	2.5 MW, 120 kvar output
Power supplied into the utility system at 230 kV	201.5 MW, −35.6 Mvar	201 MW, 0.4 Mvar
Losses	4.33 MW, 41.22 Mvar	4.13 MW, 39.22 Mvar
Voltage at secondary of 2.5/2.8 Mvar Transformer	100.59–100.11%	102.87–102.41%
Voltage at PCC-1	99.49%	101.79%
Voltage at PCC-2	100.23%	100.78%

Notes: All transformer taps are at rated voltage; Negative sign signifies that the reactive power is supplied by the utility system, which is not acceptable.

The load flow without reactive power compensation shows that 201.5 MW of active power is supplied into utility 230 kV system; however, the utility must supply 35.6 Mvar. This is not acceptable. The operating voltages in the system are acceptable; the voltages at PCC-1 and PCC-2 are 99.49% and 100.23% of the rated voltages, respectively. The inverters output 2.65 MW and 120 kvar at each 800 V substation. The operating voltages at the inverter 800 V buses are close to the operating voltages. While all other operating parameters are acceptable, the reactive power supply of 35.6 Mvar is not acceptable.

Thus, 11.0 Mvar reactive power compensation is required at each 13.8 kV bus. With this compensation, the reactive power supply from the utility source is wiped out. The power factor of the supply is now almost unity. The voltages slightly rise, but these are acceptable. This establishes the normal mode of operation that 33 Mvar is required at full-load operation.

9.1.3 Harmonic Analysis without Capacitors

This forms the base case and the calculated parameters can be compared when capacitors are installed to study impact of harmonic resonance, if any. The results of the study are shown in the following tables and figures:

- Figures 9.4 and 9.5 illustrate the frequency scan results, phase angles, and impedance modulus with and without capacitor banks at PCC-1 (13.8 kV) and PCC-2 (230 kV). This shows that with application of capacitors resonance is occurring at frequencies summarized in Table 9.4, around 8th and 9th harmonics.

- Figure 9.6 shows the voltage spectrum and Figure 9.7 shows the waveform at PCC-1 and PCC-2.

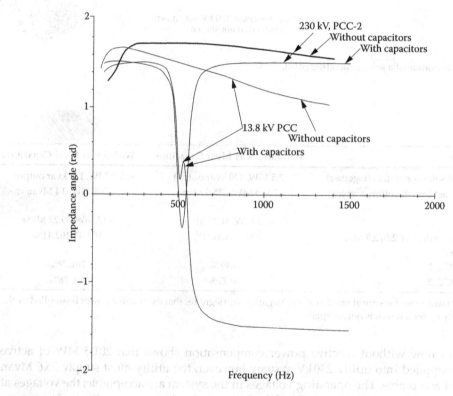

FIGURE 9.4
Impedance angle versus frequency plot with and without capacitors at PCC-1 and PCC-2.

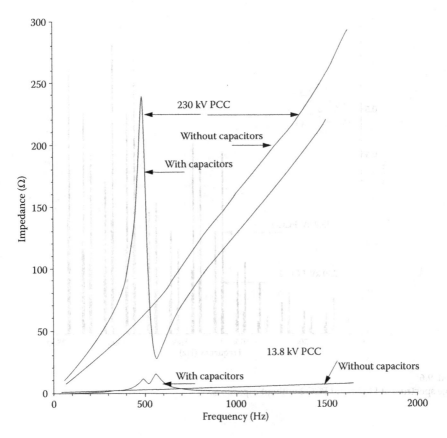

FIGURE 9.5
Impedance modulus versus frequency plot with and without capacitors at PCC-1 and PCC-2.

TABLE 9.4

Frequency Scan with 9 Mvar Capacitor Bank

Bus ID	Impedance Modulus	Harmonic	Frequency (Hz)
PCC-1 (13.8 kV)	11.44	8.20	492.00
	15.58	9.33	560.00
PCC-2 (230 kV)	238.91	8.17	490.00

- Figure 9.8 shows the current spectrum at 230 kV PCC-2 and Figure 9.9 depicts its waveform.
- Table 9.5 shows the calculation of TDD and the comparison of results with IEEE 519 (2014) limitations. The magnitude at harmonics 38 and 40 exceeds IEEE limits. The total TDD is 0.37 at 13.8 kV and 0.42 at 230 kV, and the corresponding IEEE limits are 5% and 1.5%, respectively.
- Table 9.6 shows the calculation of voltage distortion and the comparison of results with IEEE limitations. The THD voltage is 1.68% at 13.8 kV and 0.64% at 230 kV, and the corresponding IEEE limits are 5.0% and 1.5%, respectively.

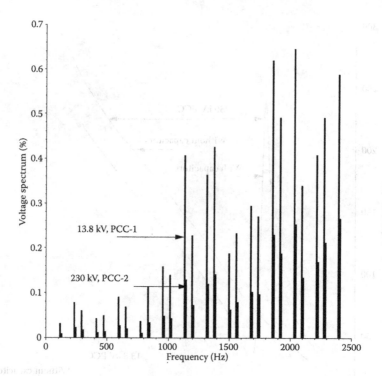

FIGURE 9.6
Voltage spectrum at PCC-1 and PCC-2 without capacitors.

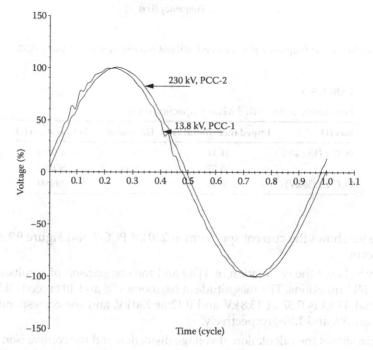

FIGURE 9.7
Voltage waveform at PCC-1 and PCC-2 without capacitors.

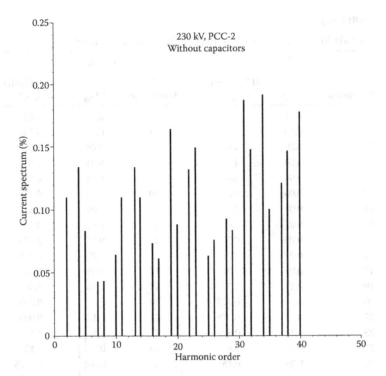

FIGURE 9.8
Current spectrum at PPC-2 (230 kV) without capacitors.

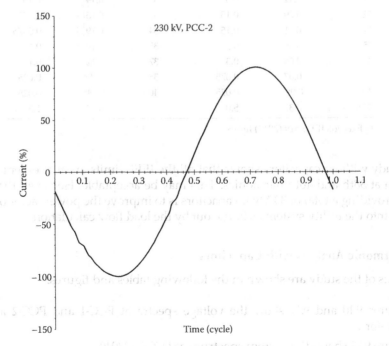

FIGURE 9.9
Current waveform at PCC-2 (230 kV) without capacitors.

TABLE 9.5

Calculation of TDD according to IEEE 519—No Capacitors

PCC-1			PCC-2		
Harmonic	Mag	IEEE Limits $I_s/I_r < 20$	Harmonic	Mag	IEEE Limits $I_s/I_r < 25$
2	0.11	1.0	2	0.11	0.25
4	0.13	1.0	4	0.13	0.25
5	0.08	4.0	5	0.08	1.0
7	0.04	4.0	7	0.04	1.0
8	0.04	1.0	8	0.04	0.25
10	0.06	1.0	10	0.06	0.25
11	0.04	2.0	11	0.04	0.5
13	0.02	2.0	13	0.02	0.5
14	0.05	0.5	14	0.06	0.125
16	0.07	0.5	16	0.07	0.125
17	0.05	1.5	17	0.06	0.38
19	0.14	1.5	19	0.16	0.38
20	0.07	0.375	20	0.09	0.095
22	0.11	0.375	22	0.13	0.095
23	0.12	0.6	23	0.15	0.15
25	0.05	0.6	25	0.06	0.15
26	0.06	0.15	26	0.08	0.0375
28	0.07	0.15	28	0.09	0.0375
29	0.06	0.6	29	0.08	0.15
31	0.12	0.6	31	0.19	0.15
32	0.09	0.15	32	0.15	0.075
34	0.11	0.15	34	0.19	0.0375
35	0.06	0.3	35	0.10	0.1
37	0.06	0.3	37	0.12	0.1
38	0.07	0.075	38	0.15[a]	0.025
40	0.08[a]	0.075	40	0.18[a]	0.025
TDD	0.37	5.0		0.42	1.5

[a] Exceeds IEEE 519 (2014) limits.

The study without capacitors shows that all the IEEE limits are met except the current distortion at 38th and 40th harmonics. This may be acceptable. However, the main purpose of providing a total of 33 Mvar capacitors is to improve the power factor of the power supplied into the utility system as borne out by the load flow calculation.

9.1.4 Harmonic Analysis with Capacitors

The results of the study are shown in the following tables and figures:

- Figures 9.10 and 9.11 show the voltage spectra at PCC-1 and PCC-2 and their waveforms.
- Figure 9.12 shows the current spectrum at PCC-2 (230 kV).

TABLE 9.6

Calculation of THD_V according to IEEE 519—No Capacitors

	PCC-1			PCC-2	
Harmonic	Mag	IEEE Limits	Harmonic	Mag	IEEE Limits
2	0.03	3.0	2	0.01	1.0
4	0.08	3.0	4	0.02	1.0
5	0.06	3.0	5	0.02	1.0
7	0.04	3.0	7	0.01	1.0
8	0.05	3.0	8	0.02	1.0
10	0.09	3.0	10	0.03	1.0
11	0.07	3.0	11	0.02	1.0
13	0.04	3.0	13	0.01	1.0
14	0.11	3.0	14	0.03	1.0
16	0.16	3.0	16	0.05	1.0
17	0.14	3.0	17	0.04	1.0
19	0.41	3.0	19	0.13	1.0
20	0.23	3.0	20	0.07	1.0
22	0.37	3.0	22	0.12	1.0
23	0.43	3.0	23	0.14	1.0
25	0.19	3.0	25	0.06	1.0
26	0.23	3.0	26	0.08	1.0
28	0.30	3.0	28	0.10	1.0
29	0.27	3.0	29	0.10	1.0
31	0.62	3.0	31	0.23	1.0
32	0.50	3.0	32	0.19	1.0
34	0.65	3.0	34	0.25	1.0
35	0.34	3.0	35	0.14	1.0
37	0.41	3.0	37	0.17	1.0
38	0.50	3.0	38	0.21	1.0
40	0.59	3.0	40	0.27	1.0
TDD	1.68	5.0		0.64	1.5

- Figure 9.13 shows the waveform of the current through 11 Mvar capacitor bank.
- Tables 9.7 and 9.8 show the calculation results of TDD and THDv. The IEEE limits are met.

A review of the spectra with and without capacitor bank shows that the higher order harmonics are much attenuated, while harmonics around the 8th increase. *The capacitor is acting like high-pass filters, see Chapter 8.* A shunt capacitor is a first-order high-pass filter. There is no need to turn any of the capacitors into filters.

A consideration of importance is that the solar generation can vary over large limits, and with it the reactive power compensation will vary. Practically, fixed capacitors will not be a proper choice and some form of variable reactive power source such as a TCR, TSC, or STATCOM should be provided, See Volume 2.

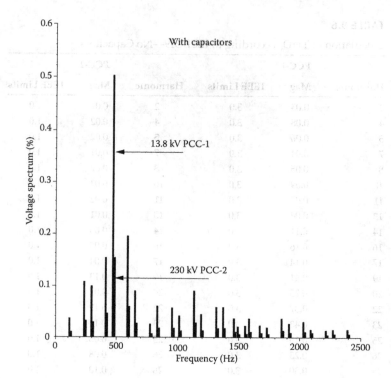

FIGURE 9.10
Voltage spectrum at PCC-1 and PCC-2 with capacitors.

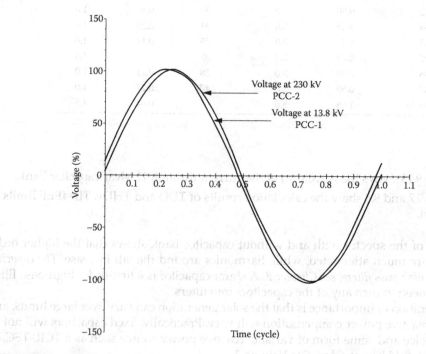

FIGURE 9.11
Voltage waveform at PCC-1 and PCC-2 with capacitors.

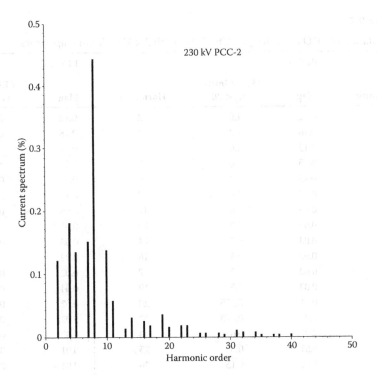

FIGURE 9.12
Current spectrum at PPC-2 (230 kV) with capacitors.

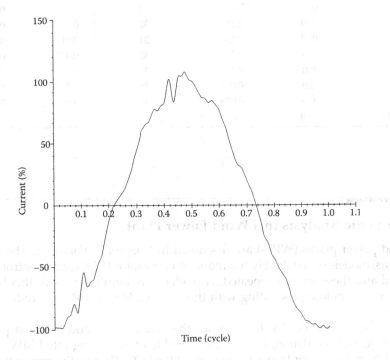

FIGURE 9.13
Current waveform through the capacitors.

TABLE 9.7

Calculation of TDD according to IEEE 519—with 3 × 11.00 Mvar Capacitors

PCC-1			PCC-2		
Harmonic	Mag	IEEE Limits $I_s/I_r < 20$	Harmonic	Mag	IEEE Limits $I_s/I_r < 25$
2	0.12	1.0	2	0.12	0.25
4	0.18	1.0	4	0.18	0.25
5	0.13	4.0	5	0.14	1.0
7	0.15	4.0	7	0.15	1.0
8	0.43	1.0	8	0.44	0.25
10	0.13	1.0	10	0.14	0.25
11	0.06	2.0	11	0.06	0.5
13	0.01	2.0	13	0.01	0.5
14	0.03	0.5	14	0.03	0.125
16	0.02	0.5	16	0.03	0.125
17	0.02	1.5	17	0.02	0.38
19	0.03	1.5	19	0.04	0.38
20	0.01	0.375	20	0.02	0.095
22	0.02	0.375	22	0.02	0.095
23	0.02	0.6	23	0.02	0.15
25	0.01	0.6	25	0.01	0.15
26	0.01	0.15	26	0.01	0.0375
28	0.01	0.15	28	0.01	0.0375
29	0.0	0.6	29	0.01	0.15
31	0.01	0.6	31	0.01	0.15
32	0.01	0.15	32	0.01	0.075
34	0.01	0.15	34	0.01	0.0375
35	0.0	0.3	35	0.0	0.1
37	0.0	0.3	37	0.0	0.1
38	0.0	0.075	38	0.0	0.025
40	0.0	0.075	40	0.0	0.025
TDD	0.68	5.0		0.55	1.5

9.2 Harmonic Analysis in a Wind Power Plant

The wind power plants (WPPs) are discussed in Chapter 3, Volume 1. The system configurations, models of the doubly fed induction generator (DFIG), and control systems are discussed and these are not repeated. A reader may familiarize with this fundamental conceptual data before proceeding with this section. Figure 9.14 is reproduced for quick reference.

Wind turbines are not able to maintain the voltage level and required power factor. According to one regulation, it should be capable of supplying rated MW at any point between 0.95 power factor lagging to leading at the PCC. The reactive power limits defined at rated MW at lagging power factor will apply at all active power output levels above 20%

TABLE 9.8

Calculation of THD_V according to IEEE 519—With 3×11 Mvar Capacitors

PCC-1			PCC-2		
Harmonic	Mag	IEEE Limits	Harmonic	Mag	IEEE Limits
2	0.03	3.0	2	0.01	1.0
4	0.10	3.0	4	0.03	1.0
5	0.10	3.0	5	0.03	1.0
7	0.15	3.0	7	0.05	1.0
8	0.49	3.0	8	0.15	1.0
10	0.19	3.0	10	0.06	1.0
11	0.09	3.0	11	0.03	1.0
13	0.02	3.0	13	0.01	1.0
14	0.06	3.0	14	0.02	1.0
16	0.05	3.0	16	0.02	1.0
17	0.04	3.0	17	0.01	1.0
19	0.09	3.0	19	0.01	1.0
20	0.04	3.0	20	0.01	1.0
22	0.05	3.0	22	0.02	1.0
23	0.06	3.0	23	0.02	1.0
25	0.02	3.0	25	0.01	1.0
26	0.02	3.0	26	0.01	1.0
28	0.02	3.0	28	0.01	1.0
29	0.02	3.0	29	0.01	1.0
31	0.04	3.0	31	0.01	1.0
32	0.03	3.0	32	0.01	1.0
34	0.03	3.0	34	0.01	1.0
35	0.01	3.0	35	0.01	1.0
37	0.01	3.0	37	0.01	1.0
38	0.02	3.0	38	0.01	1.0
40	0.02	3.0	40	0.01	1.0
TDD	0.60	5.0		0.18	1.5

of the rated MW output. Also the reactive power limits defined at rated MW at leading power factor will apply at all active power output levels above 50% of the rated MW output. See Figure 9.14 for further details; this figure is for interconnection at the grid, PCC, *and not for individual operating units*. Thus, the reactive power compensation, fault levels and short-circuit analysis, the variations in the active power due to wind speed in the particular area over the course of a day, month-to-month, peak and lowest raw electricity that will be generated are the first set of studies performed in the planning stage. These allow fundamental equipment ratings to be selected and protection systems designed. The considerations such as cables versus overhead lines for connection of collector buses to grid step-up transformer also arise. Further need for dynamic studies, voltage profiles, fault clearance times, studies documenting the grid connection requirements also arise. In fact, wind power generation and utility interconnections required extensive studies apart from harmonic considerations.

Due to the stochastic nature of wind turbine harmonics, probability concepts have been applied. Autoregressive moving average model is the statistical analysis of time series and

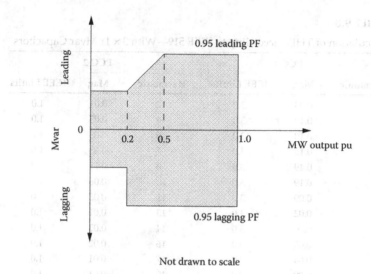

FIGURE 9.14
Operating power factor of a WPP with MW output for connection to utility grid.

provides parsimonious description of a stationary stochastic process in terms of two polynomials: one for auto regression and the other for moving average.

9.2.1 Configuration of WPP for the Study

Figure 9.15 shows a 100 MW wind farm generation for harmonic analysis (1000 MW wind generation in a single location is being planned). The length and sizes of cables are indicated and also the source impedance and transformer impedances. Each DFIG is rated at 2.5 MW output and is connected through a step-up transformer of 3.0/3.36 MVA.

Table 9.9 shows the planning level of harmonics recommended by the Energy Network Association, UK. The harmonic current spectra are usually supplied by the manufacturer; Table 9.10 shows the harmonic emissions from a WTG (Wind Turbine Generator).

For harmonic power flow analysis, the harmonic currents are modeled in parallel with the asynchronous machine model. A resistive load in parallel with the generator and current source is placed to model turbine auxiliary loads.

Considerable lengths of cables are involved and modeling of cable capacitance is important. Consider the following:

- Capacitance of the turbine capacitor bank, if provided
- Capacitance of the collector cables
- Capacitance of the substation capacitors

Some simplifying assumptions can be made; i.e., triplen harmonics will be trapped by transformer windings, and the harmonic currents produced by all turbines can be assumed to have same phase angles. The even harmonics arise due to slight mismatch between firing angles in the phases, and some phase asymmetry. The modeling as per emission data provided by the manufacturer is important.

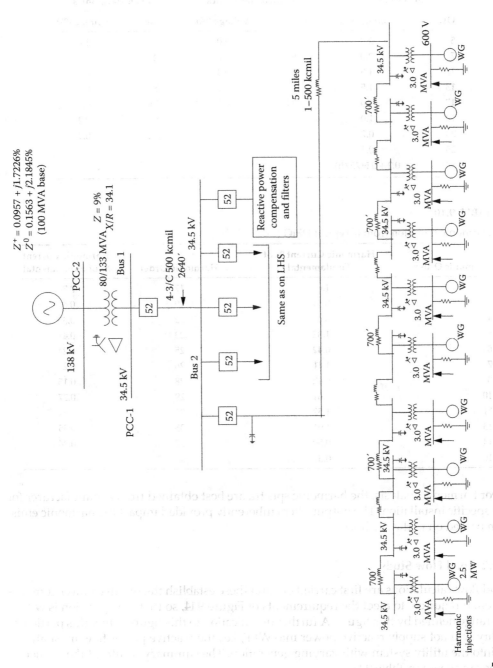

FIGURE 9.15
Configuration of a WPP for harmonic analysis.

TABLE 9.9

Planning Levels for Harmonic Voltages in Systems >20 kV and <145 kV

Odd Harmonics		Triplen Harmonics		Even Harmonics	
Order	Voltage (%)	Order	Voltage (%)	Order	Voltage (%)
5	2.0	3	2.0	2	1.0
7	2.0	9	1.0	4	0.8
11	1.5	15	0.3	6	0.5
13	1.5	21	0.2	8	0.4
17	1.0	>21	0.2	10	0.4
19	1.0			12	0.2
23	0.7			>12	0.2
25	0.7				
>25	$0.2 + 0.5(25/h)$				

TABLE 9.10

Harmonic Emission from a Typical DFIG

Harmonic Order	Harmonic Current % of Fundamental	Harmonic Order	Harmonic Current % of Fundamental
2	1.0	17	0.76
3	0.51	19	0.42
4	0.43	22	0.33
5	1.32	23	0.41
6	0.42	25	0.24
7	1.11	26	0.2
8	0.42	28	0.15
10	0.61	29	0.27
11	1.52	31	0.24
13	1.91	35	0.35
14	0.50	37	0.26
16	0.37		

For harmonic analysis, the harmonic spectra are best obtained from a manufacturer for the specific installation. The output filters inherently provided impact the harmonic emission passed on to the AC lines.

9.2.2 Load Flow Study

Load flow calculations are first carried out and these establish the reactive power compensation, if required, to meet the requirements of Figure 9.14, so that the operation is within the range defined by this figure. A further qualification to this figure is that this particular utility will not supply reactive power into WPP; i.e., the reactive power flow must always be into the utility system with varying generation. The summary results of these calculations are shown in Table 9.11.

The load flow shows that variable reactive power compensation is required; this can be obtained with STATCOM, TCR, and TSC. The comparative analyses of these devices have been shown in Volume 2.

TABLE 9.11

Summary of Load Flow Results in WPP

Description	Active Power into Utility Source (MW)	Reactive Power in-or-from Utility Source (Mvar)	Power Factor	System Losses		Voltage PCC-1 (34.5 kV) (%)	Voltage DFIG (600 V) (%)
				MW	Mvar		
Full generation operation, no reactive power compensation	89.90	−24.89	96.37 Not acceptable	2.82	20.17	98.03	98.53
Full generation operation, 30 Mvar capacitive compensation	89.56	7.36	99.66 Acceptable	2.61	18.64	101.62	102.21
Operation at 75% generation, 30 Mvar capacitive compensation	66.80	15.29	97.48 Acceptable	1.86	12.31	102.21	102.73
Operation at 48.8% generation, 30 Mvar capacitive compensation	44.02	21.87	89.55 Not acceptable	1.20	7.33	102.72	103.20
Operation at 48.8% generation, 15 Mvar capacitive compensation	44.16	5.83	99.14 Acceptable	1.22	7.00	100.92	101.38
Operation at 22% generation, 7.5 Mvar capacitive compensation	21.47	3.23	98.88 Acceptable	0.76	3.80	100.58	100.90

9.2.3 Harmonic Analysis

For this analysis, the 34.5 kV bus on the secondary of the transformer and the 138 kV bus on the primary of the transformer are declared as PCC-1 and PCC-2, respectively. It is more difficult to control the emissions at 138 kV PCC-2 as the harmonic emission limits are much lower.

Step 1

As a first step, the harmonic spectrum and penetration without any additional reactive power compensation should be ascertained. The results of HA (Harmonic Analysis) are shown in Table 9.12. It is seen that

- Even without any reactive power compensation, the total current distortions at PCC-2 exceed IEEE limits, 3.35% versus permissible limit of 2.5%.
- A resonance occurs around the 8th harmonic, due to cable and system capacitances. Cables are modeled as equivalent Π models. The harmonic resonance at even harmonics can occur.
- Figure 9.16 shows the voltage spectrum at PCC-1. The harmonics at 5th, 7th, 8th, 10th, 11th, 13th, and higher order are present. Thus, a wide spectrum of harmonics needs to be controlled.

Step 2

Looking at the characteristics of variable reactive power compensation devices, TCR, TSC, and STATCOM, the *STATCOM gives the least harmonic pollution*. Also, its other advantages are described in Volume 2.

TABLE 9.12

Harmonic Analysis, No Reactive Power Compensation

	Frequency Scan		Harmonic Analysis			
			Voltage Distortion (%)		Current Distortion (%)	
PCC	Impedance Modulus (Ω)	Harmonic	Calculated	IEEE Limits	Calculated	IEEE Limits
PCC-1	64.91	8.5	3.33	5.0	3.35	5.0
PCC-2	34.61	8.3	0.43	2.5	3.35	2.5

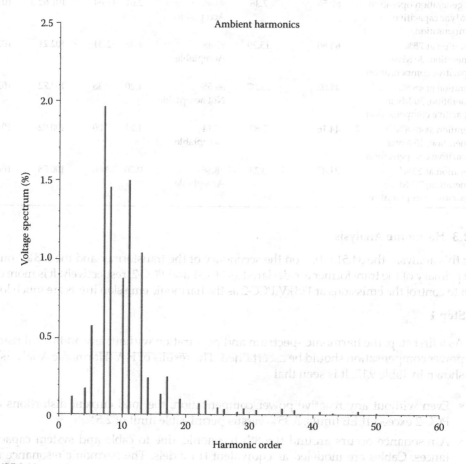

FIGURE 9.16
Ambient harmonics without any reactive power compensating device.

Thus, a harmonic analysis study is conducted with 30 Mvar STATCOM supplying capacitive reactive power into the system. The harmonic emission from the STATCOM as specified by a specific vendor is shown in Table 9.13. Though these emissions may be small, these should be modeled. The results of the study are shown in Table 9.14.

TABLE 9.13

Typical Harmonic Emission from the STATCOM (Based on a Manufacturer's Data)

Harmonic Order	% of Var Output	Harmonic Order	% of Var Output
3	1.84	19	0.20
5	3.89	41	0.20
6	0.20	42	0.20
7	2.05	46	0.20
9	0.2	48	0.20
11	0.41	49	0.20
13	0.20		
17	0.20		

TABLE 9.14

Harmonic Analysis, 30 Mvar STATCOM

	Frequency Scan		Harmonic Analysis			
			Voltage Distortion (%)		Current Distortion (%)	
PCC	Impedance Modulus (Ω)	Harmonic	Calculated	IEEE Limits	Calculated	IEEE Limits
PCC-1	72.30	9.97	4.30	5.0	5.96	5.0
PCC-2	41.82	9.77	0.55	2.5	6.0	2.5

It is seen that

- The resonant frequency has shifted to 9.97 and 9.77 at PCC-1 and PCC-2, respectively.
- The current harmonic distortions increase compared to Step 1, as expected due to additional harmonic emission from the STATCOM.
- Figure 9.17 shows the voltage spectrum. Compare it with Figure 9.16 without STATCOM. The amplitude of harmonics increases, as expected. Figure 9.18 shows the distorted waveform at PCC-1.

Step 3

It has been established that we need to use STATCOM with harmonic filters. Various types of filters have been used in WPP:

- Single-tuned filters, single or parallel combination. These are most effective at the notch frequency at which these are tuned to.
- Type C filter.
- Double-tuned filters.
- Damped filters.
- Single-tuned filters in combination with damped high-pass filters.

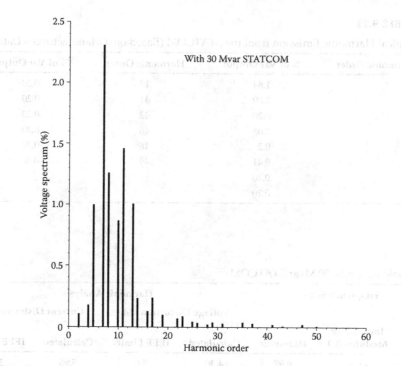

FIGURE 9.17
Harmonic spectrum with 30 Mvar STATCOM added, the harmonics increase over the spectrum depicted in Figure 9.16.

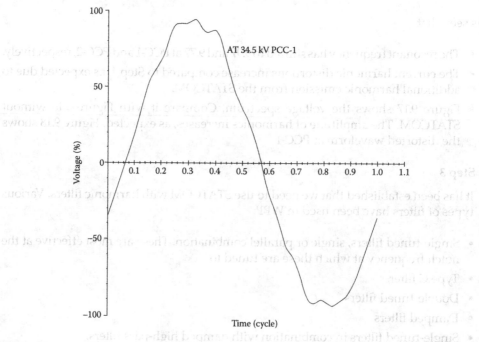

FIGURE 9.18
Voltage waveform at 34.5 kV PCC-1.

TABLE 9.15

Harmonic Filters and STATCOM Ratings

Harmonic Filter	Three-Phase Mvar Rating	Tuning Frequency Times Fundamental
5	5	4.75
7	5	7.50
11	5	10.76
13	5	12.80
STATCOM	15 Mvar capacitive to 7.5 Mvar reactive	

See Chapter 8 for their characteristics.

Step 4: Harmonic Analysis Generation at the Maximum

The design of filters is an iterative process—a number of trials must be made to optimize the filter size and its type, see Chapter 8. These trial runs are not documented. A number of filter types as stated earlier were tried. The final configuration of the filters is shown in Table 9.15.

The operation is as follows:

- At full generation, all filters are in service, and the STATCOM supplies 15 Mvar capacitive.
- At 50% generation, all filters are in service and the STATCOM just floats on the bus.
- At 20% generation, all filters are in service and the STATCOM supplies 7.5 Mvar reactive power into the system.

Fundamentally, all the filters are kept in service, and the STATCOM operates based on the power factor control at 138 kV PCC-2. In fact, a desired constant power factor of operation can be obtained.

The effectiveness of the filters is documented in the following figures and tables.

Figures 9.19 and 9.20 show the impedance angle and impedance modulus scan at PCC-1 and PCC-2. An ST filter does not eliminate the resonance, but the resonance frequency is shifted below the tuned frequency. A resonance can occur at these shifted frequencies. A number of switching conditions must be studied with varying generation, to ensure that this does not happen.

Figures 9.21 and 9.22 show the voltage spectrum and voltage waveforms at PCC-1 and PCC-2. The voltage waveform looks almost sinusoidal. Compare it with Figure 9.17.

Figures 9.23 and 9.24 show the current spectrum and its waveform at PCC. Again, this looks almost sinusoidal.

Figure 9.25 illustrates the effectiveness of filters. Each filter shunts away considerable amount of harmonic to which it is tuned.

Tables 9.16 and 9.17 illustrate the voltage and current distortions. The calculated values at each harmonic are compared with IEEE 519 (2014) limits. It is seen that the harmonic distortion at each harmonic and also the total harmonic distortion current and voltage are much below IEEE limits. The total voltage distortion at PCC-1 and PCC-2 is 1.01% and 0.15% versus permissible limits of 5% and 2.5%, respectively. The total current distortion at PCC-1 and PCC-2 is 1.52% and 1.52% versus permissible limits of 5% and 2.5%, respectively. The

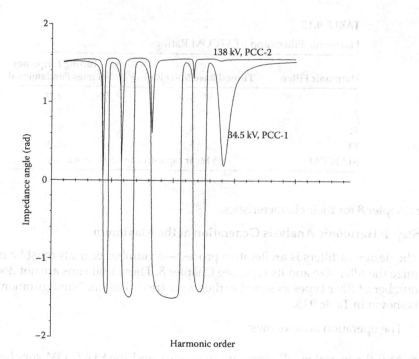

FIGURE 9.19
Impedance angle versus frequency plot with application of filters at PCC-1 and PCC-2.

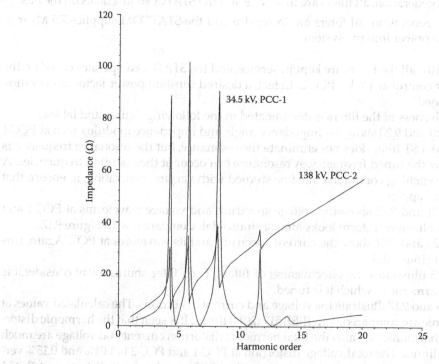

FIGURE 9.20
Impedance modulus versus frequency plot with application of filters at PCC-1 and PCC-2.

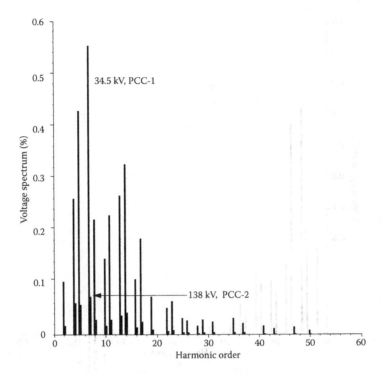

FIGURE 9.21
Voltage spectrum at PCC-1 and PCC-2 full generation.

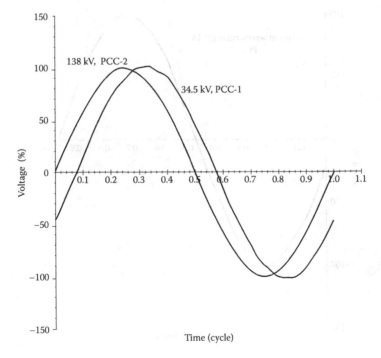

FIGURE 9.22
Voltage waveform at PCC-1 and PCC-2 full generation.

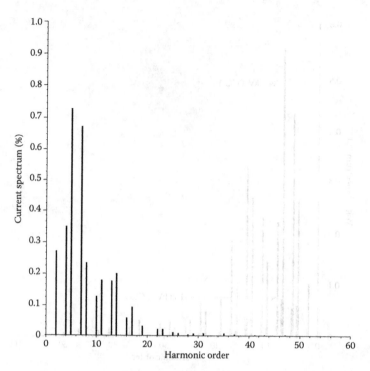

FIGURE 9.23
Current spectrum at PCC-2 full generation.

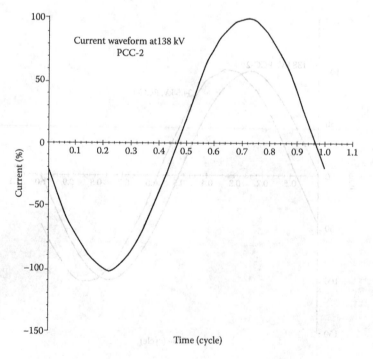

FIGURE 9.24
Current waveform at PCC-2 full generation.

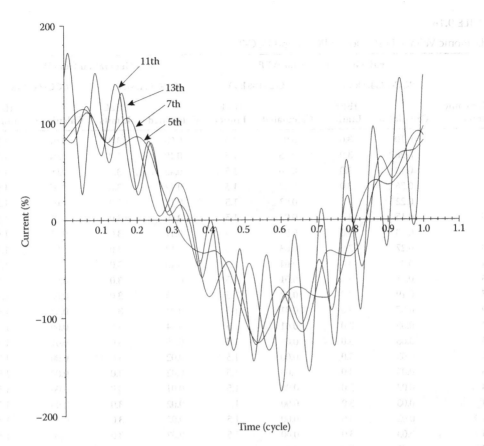

FIGURE 9.25
Current waveforms through the filters.

current distortion is the same as it is plotted in percentage of the fundamental frequency current on the secondary and primary sides of the utility interconnecting 80/133 Mvar transformer.

Step 5: Harmonic Analysis at 50% Generation

The results of the calculations at 50% generation are also shown in Tables 9.16 and 9.17. The calculated values at each harmonic are compared with IEEE 519 (2014) limits. It is seen that the harmonic distortion at each harmonic and also the total harmonic distortion current and voltage are much below IEEE limits. The total voltage distortion at PCC-1 and PCC-2 is 0.80% and 0.09% versus permissible limits of 5% and 2.5%, respectively. The total current distortion at PCC-1 and PCC-2 is 2.46% and 2.46% versus permissible limits of 5% and 2.5%, respectively. The current distortion is the same as it is plotted in percentage of the fundamental frequency current on the secondary and primary sides of the utility interconnecting 80/133 Mvar transformer.

Note that the harmonics may not reduce directly in proportion to generation and loads. The relative magnitude of current and voltage distortions depends upon the system impedances, which changes with respect to the loads.

TABLE 9.16

Harmonic Voltage Distortion at PCC-1 and PCC-2

| Harmonic Order | Full Generation from WPP | | | | 50% Generation from WPP | | | |
| | PCC-1 (34.5 kV) | | PCC-2 (138 kV) | | PCC-1 (34.5 kV) | | PCC-2 (138 kV) | |
	Calculated	IEEE Limits	Calculated	IEEE Limits	Calculated	IEEE Limits	Calculated	IEEE Limits
2	0.10	3.0	0.02	1.5	0.08	3.0	0.01	1.5
4	0.26	3.0	0.06	1.5	0.19	3.0	0.04	1.5
5	0.43	3.0	0.06	1.5	0.45	3.0	0.06	1.5
7	0.56	3.0	0.07	1.5	0.42	3.0	0.05	1.5
8	0.22	3.0	0.03	1.5	0.10	3.0	0.01	1.5
10	0.15	3.0	0.02	1.5	0.09	3.0	0.01	1.5
11	0.23	3.0	0.03	1.5	0.11	3.0	0.01	1.5
13	0.27	3.0	0.03	1.5	0.12	3.0	0.02	1.5
14	0.33	3.0	0.04	1.5	0.40	3.0	0.05	1.5
16	0.11	3.0	0.01	1.5	0.09	3.0	0.01	1.5
17	0.19	3.0	0.02	1.5	0.15	3.0	0.02	1.5
19	0.07	3.0	0.01	1.5	0.04	3.0	0.01	1.5
22	0.05	3.0	0.01	1.5	0.04	3.0	0.00	1.5
23	0.06	3.0	0.01	1.5	0.05	3.0	0.01	1.5
25	0.03	3.0	0.00	1.5	0.02	3.0	0.00	1.5
26	0.03	3.0	0.00	1.5	0.02	3.0	0.00	1.5
28	0.02	3.0	0.00	1.5	0.01	3.0	0.00	1.5
29	0.03	3.0	0.00	1.5	0.02	3.0	0.00	1.5
31	0.02	3.0	0.00	1.5	0.02	3.0	0.00	1.5
35	0.03	3.0	0.00	1.5	0.02	3.0	0.00	1.5
37	0.02	3.0	0.00	1.5	0.01	3.0	0.00	1.5
41	0.02	3.0	0.00	1.5	0.01	3.0	0.00	1.5
43	0.01	3.0	0.00	1.5	0.01	3.0	0.00	1.5
47	0.01	3.0	0.00	1.5	0.01	3.0	0.00	1.5
50	0.01	3.0	0.00	1.5	0.01	3.0	0.00	1.5
THD$_V$	1.01	5.0	0.12	2.5	0.80	5.0	0.09	2.5

Note: All values shown are in % of the rated voltages.

Step 5: Harmonic Analysis at 50% Generation.

The result of the calculation of 50% generation are also shown in Table 9.16 and 9.17. The calculated voltage or is harmonics are compared with IEEE 519 (2014) limits. It is seen that the harmonic distortion at each harmonic and also the total harmonic distortion current and voltage are much below IEEE limits. The total voltage distortion at PCC-1 and PCC-2 is 0.80% and 0.09% versus permissible limits of 5% and 2.5% respectively. The total current distortion at PCC-1 and PCC-2 is 2.49% and 2.35% versus permissible limits of 5% and 2.35% respectively. The current distortion is the same as that plotted in percentage of the fundamental frequency component from the secondary and primary sides of the utility interconnecting 230/138 MVA transformer.

Note that the harmonics may not reduce directly in proportion to generation as the relative magnitude of current and voltage distortions depends upon the system impedances, which changes with respect to the loads.

TABLE 9.17

Harmonic Current Distortion at PCC-1 and PCC-2

	Full Generation from WPP				50% Generation from WPP			
	PCC-1 (34.5 kV)		PCC-2 (138 kV)		PCC-1 (34.5 kV)		PCC-2 (138 kV)	
Harmonic Order	Calculated	IEEE Limits	Calculated	IEEE Limits	Calculated	IEEE Limits	Calculated	IEEE Limits
2	0.28	1.0	0.28	0.5	0.38	1.0	0.38	0.5
4	0.35	1.0	0.35	0.5	0.45	1.0	0.45	0.5
5	0.72	4.0	0.72	2.0	1.48	4.0	1.48	2.0
7	0.67	4.0	0.67	2.0	1.01	4.0	1.01	2.0
8	0.23	1.0	0.23	0.5	0.2	1.0	0.2	0.5
10	0.12	1.0	0.12	0.5	0.15	1.0	0.15	0.5
11	0.17	2.0	0.17	1.0	0.16	2.0	0.16	1.0
13	0.17	2.0	0.17	1.0	0.15	2.0	0.15	1.0
14	0.20	0.5	0.20	0.25	0.21	0.5	0.21	0.25
16	0.05	0.5	0.05	0.25	0.09	0.5	0.09	0.25
17	0.09	1.5	0.09	0.75	0.15	1.5	0.15	0.75
19	0.03	1.5	0.03	0.75	0.04	1.5	0.04	0.75
22	0.02	0.375	0.02	0.1875	0.03	0.375	0.03	0.1875
23	0.02	0.6	0.02	0.3	0.03	0.6	0.03	0.3
25	0.01	0.6	0.01	0.3	0.01	0.6	0.01	0.3
26	0.01	0.15	0.01	0.075	0.01	0.15	0.01	0.075
28	0.00	0.15	0.00	0.075	0.01	0.15	0.01	0.075
29	0.01	0.6	0.01	0.3	0.01	0.6	0.01	0.3
31	0.01	0.6	0.01	0.3	0.01	0.6	0.01	0.3
35	0.01	0.3	0.01	0.15	0.01	0.3	0.01	0.15
37	0.00	0.3	0.00	0.15	0.01	0.3	0.01	0.15
41	0.00	0.3	0.00	0.15	0.00	0.3	0.00	0.15
43	0.00	0.3	0.00	0.15	0.00	0.3	0.00	0.15
47	0.00	0.3	0.00	0.15	0.00	0.3	0.00	0.15
50	0.00	0.075	0.00	0.0375	0.00	0.075	0.00	0.0375
TDD	1.52	5.0	1.52	2.5	2.46	5.0	2.46	2.5

Note: All values shown are in % of the fundamental frequency currents.

References

1. ANSI/IEEE Std. 928. IEEE recommended criteria for terrestrial photovoltaic power systems, 1986.
2. IEEE Std. 1262. IEEE recommended practice for qualifications of photovoltaic (PV) Modules, 1995.
3. IEEE Std. 1547.2. IEEE application guide for IEEE Std. 1547, IEEE standard for interconnecting distributed resources with electrical power systems, 2008.
4. IEEE Std. 1021. IEEE recommended practice for utility interconnection of small wind energy conversion systems, 1989.
5. IEEE std. 1547. IEEE standard for interconnecting distributed resources with electric power systems, 2003.
6. J.C. Das. *Transients in Electrical Systems*. McGraw-Hill, New York, 2010.
7. J.A. Fleeman, R. Gutman, M. Heyeck, M. Baharman, B. Normark. EHV and HVDC transmission working together to integrate renewable power. *CIGRE and IEEE PES Joint Symposium*, Calgary. Canada, 2009.
8. J.D. McDonald. The next generation grid energy infrastructure of the future, *IEEE Power Energy Mag 7*, (2), 2009.
9. S. Heier. *Grid Integration of Wind Energy Conversion Systems*, 2nd ed. John Wiley & Sons, Chichester, 2009.
10. T. Burton, D. Sharpe, N. Jenkins, E. Bossanyi. *Wind Power Handbook*. John Wiley & Sons, New York, 2012.
11. Western Electricity Coordinating Council disturbance monitoring reports.
12. IEEE Standard 1001. Guide for interfacing dispersed storage and generation facilities with electric facility systems, 1988
13. P. Tenca, A.A. Rockhill, T.A. Lipo. Wind turbine current source converter providing reactive power control and reduced harmonics, *IEEE Trans Ind Appl 43*, 1050–1060, 2007.
14. R. Strzelecki, G. Benysek (Eds). *Power Electronics in Smart Electrical Energy Networks*. Springer, London, 2008.
15. Dynamic models for wind farms for power system studies. www.energy.sintef.no/wind/iea.asp
16. IEC-61400-21. Wind turbine generator systems. Part 21: Measurement and assessment of power quality characteristics of grid connected wind turbines, 2001.
17. E.A. DeMeo, W. Grant, M.R. Milligan, M.J. Schuerger. Wind power integration, *IEEE Power Energy Mag 3*, (6), 38–46, 2005.
18. T. Ackermann (Ed.). *Wind Power in Power Systems*. Wiley-Inter-science, New York, 2005.
19. IEEE Std. 1094. IEEE recommended practice for electrical design and operation of windfarm generating stations, 1991.
20. International Energy Agency. Solar energy perspectives: Executive summary, www.iea.org/.
21. Canadian Renewable Energy Network. Solar energy technologies and applications, www.canren.gc.ca/.
22. J.C. Das. Harmonic distortion control and reactive power compensation in a large wind mill generation plant, CIGRE US National Committee, *2015 Grid of the Future Symposium*, pp. 1–21, 2016.

Appendix A: *Fourier Analysis*

A.1 Periodic Functions

A function is said to be periodic if it is defined for all real values of t, and if there is a positive number T such that

$$f(t) = f(t+T) = f(t+2T) = f(t+nT) \qquad \text{(A.1)}$$

then T is called the period of the function.

If k is any integer and $f(t + kT) = f(t)$ for all values of t and if two functions $f_1(t)$ and $f_2(t)$ have the same period T, then the function $f_3(t) = af_1(t) + bf_2(t)$, where a and b are constants, also has the same period T. Figure A.1 shows a periodic function.

A.2 Orthogonal Functions

Two functions $f_1(t)$ and $f_2(t)$ are orthogonal over the interval (T_1, T_2) if

$$\int_{T_1}^{T_2} f_1(t) f_2(t) = 0 \qquad \text{(A.2)}$$

Figure A.2 shows two orthogonal functions over the period T.

A.3 Fourier Series and Coefficients

A periodic function can be expanded in a Fourier series. The series has the following expression:

$$f(t) = a_0 + \sum_{n=1}^{\infty} \left(a_n \cos\left(\frac{2\pi nt}{T}\right) + b_n \sin\left(\frac{2\pi nt}{T}\right) \right) \qquad \text{(A.3)}$$

where a_0 is the average value of function $f(t)$. It is also called the dc component, and a_n and b_n are called the coefficients of the series. A series such as Equation A.3 is called a trigonometric

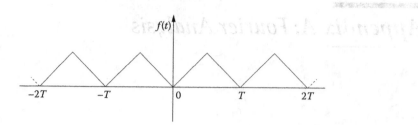

FIGURE A.1
A periodic function.

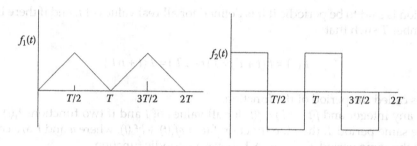

FIGURE A.2
Orthogonal functions.

Fourier series. The Fourier series of a periodic function is the sum of sinusoidal components of different frequencies. The term $2\pi/T$ can be written as ω. The nth term $n\omega$ is then called the nth harmonic and $n = 1$ gives the fundamental; a_0, a_n, and b_n are calculated as follows:

$$a_0 = \frac{1}{T} \int_{-T/2}^{T/2} f(t)\,dt \tag{A.4}$$

$$a_n = \frac{2}{T} \int_{-T/2}^{T/2} \cos\left(\frac{2\pi nt}{T}\right) dt \quad \text{for } n = 1, 2, \ldots, \infty \tag{A.5}$$

$$b_n = \frac{2}{T} \int_{-T/2}^{T/2} \sin\left(\frac{2\pi nt}{T}\right) dt \quad \text{for } 1, 2, \ldots, \infty \tag{A.6}$$

These equations can be written in terms of angular frequency:

$$a_0 = \frac{1}{2\pi} \int_{-\pi}^{\pi} f(x)\omega t\,d\omega t \tag{A.7}$$

$$a_n = \frac{1}{\pi} \int_{-\pi}^{\pi} f(x)\omega t\,\cos(n\omega t)\,d\omega t \tag{A.8}$$

$$b_n = \frac{1}{\pi} \int_{-\pi}^{\pi} f(x) \omega t \, \sin(n\omega t) \, d\omega t \tag{A.9}$$

This gives

$$x(t) = a_0 + \sum_{n=1}^{\infty} [a_n \cos(n\omega t) + b_n \sin(n\omega t)] \tag{A.10}$$

We can write

$$a_n \cos n\omega t + b_n \sin \omega t = \left[a_n^2 + b_n^2\right]^{1/2} [\sin \phi_n \cos n\omega t + \cos \phi_n \sin n\omega t]$$

$$= \left[a_n^2 + b_n^2\right]^{1/2} \sin(n\omega t + \phi_n) \tag{A.11}$$

where

$$\phi_n = \tan^{-1} \frac{a_n}{b_n}$$

The coefficients can be written in terms of two separate integrals:

$$a_n = \frac{2}{T} \int_0^{T/2} x(t) \cos\left(\frac{2\pi n t}{T}\right) dt + \frac{2}{T} \int_{-T/2}^{0} x(t) \cos\left(\frac{2\pi n t}{T}\right) dt$$

$$b_n = \frac{2}{T} \int_0^{T/2} x(t) \sin\left(\frac{2\pi n t}{T}\right) dt + \frac{2}{T} \int_{-T/2}^{0} x(t) \sin\left(\frac{2\pi n t}{T}\right) dt \tag{A.12}$$

Example A.1

Find the Fourier series of a function defined by

$$\begin{matrix} x+\pi & 0 \le x \le \pi \\ -x-\pi & -\pi \le x < 0 \end{matrix}$$

and

$$f(x+2\pi) = f(x)$$

Find a_0, which is given by

$$a_0 = \frac{1}{\pi} \int_{-\pi}^{0} (-x-\pi) \, dx + \frac{1}{\pi} \int_{0}^{\pi} (x+\pi) \, dx = \pi$$

and a_n is given by

$$a_n = \frac{1}{\pi} \int_{-\pi}^{0} (-x-\pi) \cos nx \, dx + \frac{1}{\pi} \int_{0}^{\pi} (x+\pi) \cos nx \, dx$$

$$= -\frac{4}{n^2 \pi} \text{ if } n \text{ is odd}$$

$$= 0 \text{ if } n \text{ is even}$$

b_n is given by

$$b_n = \frac{1}{\pi} \int_{-\pi}^{0} (-x - \pi) \sin nx \, dx + \frac{1}{\pi} \int_{0}^{\pi} (x + \pi) \sin nx \, dx$$

$$= \frac{4}{n} \text{ if } n \text{ is odd}$$

$$= 0 \text{ if } n \text{ is even}$$

Thus, the Fourier series is

$$f(x) = \frac{\pi}{2} - \frac{4}{\pi} \left(\frac{\cos x}{1^2} + \frac{\cos 3x}{3^2} + \cdots \right) + 4 \left(\frac{\sin x}{1} + \frac{\sin 3x}{3} + \cdots \right)$$

A.4 Odd Symmetry

A function $f(x)$ is said to be an odd or skew symmetric function if

$$f(-x) = -f(x) \tag{A.13}$$

The area under the curve from $-T/2$ to $T/2$ is zero. This implies that

$$a_0 = 0, \, a_n = 0 \tag{A.14}$$

$$b_n = \frac{4}{T} \int_{0}^{T/2} f(t) \sin\left(\frac{2\pi nt}{T} \right) dt \tag{A.15}$$

Figure A.3a shows a triangular function, having odd symmetry. The Fourier series contains only sine terms.

A.5 Even Symmetry

A function $f(x)$ is even symmetric if

$$f(-x) = f(x) \tag{A.16}$$

The graph of such a function is symmetric with respect to the y-axis. The y-axis is a mirror of the reflection of the curve:

$$a_0 = 0, b_n = 0 \tag{A.17}$$

(a)

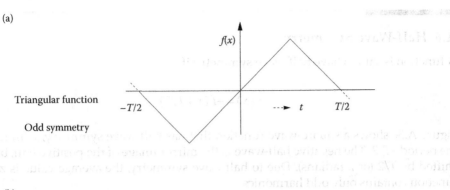

Triangular function

Odd symmetry

(b)

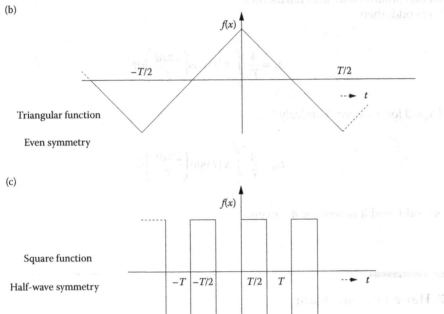

Triangular function

Even symmetry

(c)

Square function

Half-wave symmetry

FIGURE A.3
(a) Triangular function with odd symmetry; (b) triangular function with even symmetry; and (c) square function with half-wave symmetry.

$$a_n = \frac{4}{T} \int_0^{T/2} f(t) \cos\left(\frac{2\pi nt}{T}\right) dt \qquad (A.18)$$

Figure A.3b shows a triangular function with odd symmetry. The Fourier series contains only cosine terms. Note that the odd and even symmetry has been obtained with the triangular function, by shifting the origin.

A.6 Half-Wave Symmetry

A function is said to have half-wave symmetry if

$$f(x) = -f(x + T/2) \tag{A.19}$$

Figure A.3c shows a square-wave function that has half-wave symmetry, with respect to the period $-T/2$. The negative half-wave is the mirror image of the positive half, but phase shifted by $T/2$ (or π radians). Due to half-wave symmetry, the average value is zero. The function contains only odd harmonics.

If n is odd, then

$$a_n = \frac{4}{T} \int_0^{T/2} x(t) \cos\left(\frac{2\pi nt}{T}\right) dt \tag{A.20}$$

and $a_n = 0$ for n = even. Similarly

$$b_n = \frac{4}{T} \int_0^{T/2} x(t) \sin\left(\frac{2\pi nt}{T}\right) dt \tag{A.21}$$

for n = odd, and it is zero for n = even.

A.7 Harmonic Spectrum

The Fourier series of a square-wave function is

$$f(t) = \frac{4k}{\pi}\left(\frac{\sin \omega t}{1} + \frac{\sin 3\omega t}{3} + \frac{\sin 5\omega t}{5} + \cdots\right) \tag{A.22}$$

where k is the amplitude of the function. The magnitude of the nth harmonic is $1/n$, when the fundamental is expressed as 1 pu. The construction of a square wave from the component harmonics is shown in Figure A.4a and the plotting of harmonics as a percentage of the magnitude of the fundamental gives the harmonic spectrum of Figure A.4b. A harmonic spectrum indicates the relative magnitude of the harmonics with respect to the fundamental and is not indicative of the sign (positive or negative) of the harmonic, nor its phase angle.

A.7.1 Constructing Fourier Series from Graphs and Tables

When the function is given as a graph or table of values, a Fourier series can be constructed as follows:

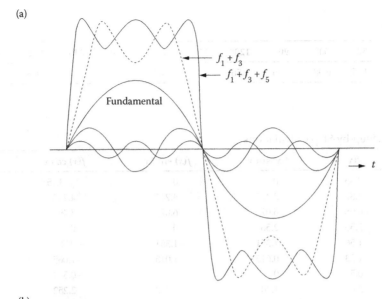

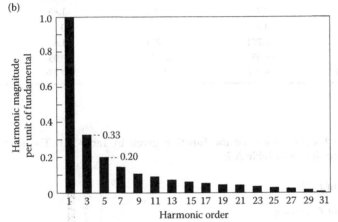

FIGURE A.4
(a) Construction of a square wave from its harmonic components and (b) harmonic spectrum of a square wave.

$$a_0 = \frac{1}{\pi} \int_0^{2\pi} f(x)\,dx = \frac{2}{2\pi - 0} \int_0^{2\pi} f(x)\,dx$$

$$= \text{twice mean value of } f(x),\ 0 \text{ to } 2\pi$$

(A.23)

Similarly

$$a_n = \text{twice mean value of } f(x)\cos nx,\ 0 \text{ to } 2\pi \tag{A.24}$$

$$b_n = \text{twice mean value of } f(x)\sin nx,\ 0 \text{ to } 2\pi \tag{A.25}$$

TABLE A.1

Example A.2

x	0°	30°	60°	90°	120°	150°	180°	210°	240°	270°	300°	330°
f(x)	3.45	4.87	6.98	2.56	1.56	1.23	0.5	2.6	3.7	4.8	3.2	1.1

TABLE A.2

Example A.2: Step-by-Step Calculations

χ (°)	$f(x)$	$f(x) \sin x$	$f(x) \sin 2x$	$f(x) \cos x$	$f(x) \cos 2x$
0	3.45	0	0	3.45	3.45
30	4.87	2.435	4.217	4.217	2.435
60	6.98	6.07	6.07	3.49	−3.49
90	2.56	2.56	0	0	−2.56
120	1.56	1.351	−1.351	−0.78	−0.78
150	1.23	0.615	−1.015	−1.065	0.615
180	0.5	0	0	−0.5	0.5
210	2.6	−1.30	2.252	−2.252	1.3
240	3.7	−3.204	3.204	−1.85	−1.85
270	4.8	−4.8	0	0	−4.8
300	3.2	−2.771	−2.771	1.6	−1.6
330	1.1	−0.55	−0.957	0.957	0.55
Total	36.55	0.406	9.649	7.276	−6.23

Example A.2

Construct the Fourier series for the function given in Table A.1. The step-by-step calculations are shown in Table A.2:

$$a_0 = 2 \times \text{mean of } f(x)$$

$$a_1 = 2 \times \text{mean of } f(x) \cos x$$

$$b_1 = 2 \times \text{mean of } f(x) \sin x$$

$$a_0 = 2 \times \text{mean of } f(x) = (2)(36.55) / 12 = 6.0917$$

From the summation of functions in Table A.2

$$f(x) = a_0 + a_1 \cos x + a_2 \cos 2x + \cdots b_1 \sin x + b_2 \sin 2x + \cdots$$

$$= 6.092 + 1.212 \cos x - 1.038 \cos 2x + \cdots + 0.068 \sin x + 1.608 \sin 2x + \cdots$$

A.8 Complex Form of Fourier Series

A vector with amplitude A and phase angle θ with respect to a reference can be resolved into two oppositely rotating vectors of half the magnitude so that

$$|A|\cos\theta = |A/2|e^{j\theta} + |A/2|e^{-j\theta} \tag{A.26}$$

Thus

$$a_n \cos n\omega t + b_n \sin n\omega t \tag{A.27}$$

can be substituted by

$$\cos(n\omega t) = \frac{e^{jn\omega t} + e^{-jn\omega t}}{2} \tag{A.28}$$

$$\sin(n\omega t) = \frac{e^{jn\omega t} - e^{-jn\omega t}}{2j} \tag{A.29}$$

Thus

$$x(t) = \frac{a_0}{2} + \frac{1}{2}\sum_{n=1}^{n=\infty}(a_n - jb_n)e^{jn\omega t} + \frac{1}{2}\sum_{n=1}^{n=\infty}(a_n - jb_n)e^{-jn\omega t} \tag{A.30}$$

We introduce negative values of n in the coefficients, i.e.,

$$a_{-n} = \frac{2}{T}\int_{-T/2}^{T/2} x(t)\cos(-n\omega t)\,dt = \frac{2}{T}\int_{-T/2}^{T/2} x(t)\cos(n\omega t)\,dt = a_n \quad n = 1, 2, 3, \ldots \tag{A.31}$$

$$a_{-n} = \frac{2}{T}\int_{-T/2}^{T/2} x(t)\sin(-n\omega t)\,dt = -\frac{2}{T}\int_{-T/2}^{T/2} x(t)\sin(n\omega t)\,dt = -b_n \quad n = 1, 2, 3, \ldots \tag{A.32}$$

Hence

$$\sum_{n=1}^{\infty} a_n e^{-jn\omega t} = \sum_{n=-1}^{\infty} a_n e^{jn\omega t} \tag{A.33}$$

and

$$\sum_{n=1}^{\infty} jb_n e^{-jn\omega t} = \sum_{n=-1}^{\infty} jb_n e^{jn\omega t} \tag{A.34}$$

Therefore, substituting in Equation A.30, we obtain

$$x(t) = \frac{a_0}{2} + \frac{1}{2}\sum_{n=-\infty}^{\infty}(a_n - jb_n)e^{jn\omega t} = \sum_{n=-\infty}^{\infty} c_n e^{jn\omega t} \tag{A.35}$$

This is the expression for a Fourier series expressed in exponential form, which is the preferred approach for analysis. The coefficient c_n is complex, which is given by

$$c_n = \frac{1}{2}(a_n - jb_n) = \frac{1}{T}\int_{-T/2}^{T/2} x(t)e^{-jn\omega t}\,dt \quad n = 0, \pm 1, \pm 2, \ldots \tag{A.36}$$

A.8.1 Convolution

When two harmonic phasors of different frequencies are convoluted, the result will be harmonic phasors at sum and difference frequencies:

$$|a_k| \sin(k\omega t + \theta_k) |b_m| \sin(m\omega t + \theta_m)$$

$$= \frac{1}{2} |a_k||b_m| \left[\sin\left((k-m)\omega t + \theta_k - \theta_m + \frac{\pi}{2}\right) - \sin\left((k+m)\omega t + \theta_k + \theta_m + \frac{\pi}{2}\right) \right] \quad \text{(A.37)}$$

Convolution is a process of correlating one time series with another time series that has been reversed in time.

A.9 Fourier Transform

Fourier analysis of a continuous periodic signal in the time domain gives a series of discrete frequency components in the frequency domain. The Fourier integral is defined by the following expression:

$$X(f) = \int_{\infty}^{-\infty} x(t) e^{-j2\pi ft} \, dt \quad \text{(A.38)}$$

If the integral exists for every value of parameter f (frequency), then this equation describes the Fourier transform. The Fourier transform is a complex quantity:

$$X(f) = RX(f) + jIX(f) \quad \text{(A.39)}$$

where $RX(f)$ is the real part of the Fourier transform and $IX(f)$ is the imaginary part of the Fourier transform. The amplitude or *Fourier spectrum of x(t)* is given by

$$|X(f)| = \sqrt{R^2(f) + I^2(f)} \quad \text{(A.40)}$$

$\phi(f)$ is the phase angle of the Fourier transform which is given by

$$\phi(f) = \tan^{-1}\left[\frac{\operatorname{Im} X(f)}{\operatorname{Re} X(f)}\right] \quad \text{(A.41)}$$

The inverse Fourier transform or the backward Fourier transform is defined as follows:

$$x(t) = \int_{-\infty}^{\infty} X(f) e^{j2\pi ft} \, df \quad \text{(A.42)}$$

Inverse transformation allows determination of a function of time from its Fourier transform. Equations E.38 and E.42 are a Fourier transform pair and the relationship can be indicated by

$$x(t) \leftrightarrow X(f) \tag{A.43}$$

Example A.3

Consider a function, which is defined as follows:

$$x(t) = \beta e^{-\alpha t} \, t > 0$$

$$= \beta e^{-\alpha t} \, t > 0 \tag{A.44}$$

It is required to write its forward Fourier transform.
From Equation A.38

$$X(f) = \int_0^\infty \beta e^{-\alpha t} e^{-j2\pi f t} \, dt$$

$$= \frac{-\beta}{\alpha + j2\pi f} e^{-(\alpha + j2\pi f)t} \Big|_0^\infty$$

$$= \frac{\beta}{\alpha + j2\pi f} = \frac{\beta \alpha}{\alpha^2 + (2\pi f)^2} - j \frac{2\pi f \beta}{\alpha^2 + (2\pi f)^2}$$

This is equal to

$$\frac{\beta}{\sqrt{\alpha^2 + (2\pi f)^2}} e^{j \tan^{-1}[-2\pi f / \alpha]} \tag{A.45}$$

Example A.4

Convert the function arrived at in Example A.3 to $x(t)$.
The inverse Fourier transform is

$$x(t) = \int_{-\infty}^\infty X(f) e^{j2\pi f t} \, df$$

$$= \int_{-\infty}^\infty \left[\frac{\beta \alpha}{\alpha^2 + (2\pi f)^2} - j \frac{2\pi f \beta}{\alpha^2 + (2\pi f)^2} \right] e^{j2\pi f t} \, df$$

$$= \int_{-\infty}^\infty \left[\frac{\beta \alpha \cos(2\pi f t)}{\alpha^2 + (2\pi f)^2} + \frac{2\pi f \beta \sin(2\pi f t)}{\alpha^2 + (2\pi f)^2} \right] df$$

$$+ j \int_{-\infty}^\infty \left[\frac{\beta \alpha \sin(2\pi f t)}{\alpha^2 + (2\pi f)^2} + \frac{2\pi f \beta \cos(2\pi f t)}{\alpha^2 + (2\pi f)^2} \right] df$$

The imaginary term is zero, as it is an odd function.
This can be written as follows:

$$x(t) = \frac{\beta\alpha}{(2\pi)^2} \int_{-\infty}^{\infty} \frac{\cos(2\pi tf)}{(\alpha/2\pi)^2 + f^2} \, df + \frac{2\pi\beta}{(2\pi)^2} \int_{-\infty}^{\infty} \frac{f\sin(2\pi tf)}{(\alpha/2\pi)^2 + f^2} \, df$$

As

$$\int_{-\infty}^{\infty} \frac{\cos\alpha x}{b^2 + x^2} \, dx = \frac{\pi}{b} e^{-ab}$$

and

$$\int_{-\infty}^{\infty} \frac{x\sin ax}{b^2 + x^2} \, dx = \pi e^{-ab}$$

$x(t)$ becomes

$$x(t) = \frac{\beta\alpha}{(2\pi)^2}\left[\frac{\pi}{\alpha/2\pi} e^{-(2\pi t)(\alpha \div \pi)}\right] + \frac{2\pi\beta}{(2\pi)^2}\left[\pi e^{-(2\pi t)(\alpha \div \pi)}\right]$$

$$= \frac{\beta}{2} e^{-\alpha t} + \frac{\beta}{2} \div^{-\alpha t} = \beta e^{-\alpha t} \quad t > 0$$

i.e.,

$$\beta e^{-\alpha t} t > 0 \leftrightarrow \frac{\beta}{\alpha + j2\pi f} \tag{A.46}$$

Example A.5

Consider a rectangular function defined by

$$x(t) = K \quad |t| \leq T/2$$

$$= 0 \quad |t| > T/2 \tag{A.47}$$

The Fourier transform is

$$X(f) = \int_{-T/2}^{T/2} K e^{-j2\pi ft} \, dt = KT\left[\frac{\sin(\pi fT)}{\pi fT}\right] \tag{A.48}$$

This is shown in Figure A.5. The term in parentheses in Equation A.48 is called the *sinc function*. The function has zero value at points $f = n/T$.

A.10 Sampled Waveform: Discrete Fourier Transform

The sampling theorem states that if the Fourier transform of a function $x(t)$ is zero for all frequencies greater than a certain frequency f_c, then the continuous function $x(t)$ can be uniquely determined by knowledge of the sampled values. The constraint is that $x(t)$ is

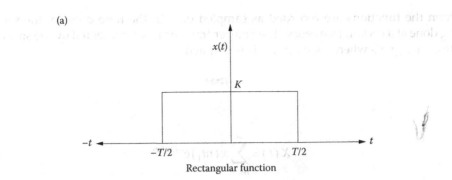

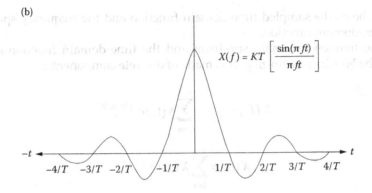

FIGURE A.5
(a) Bandwidth-limited rectangular function with even symmetry, amplitude K and (b) the sinc function, showing side lobes.

zero for frequencies greater than f_c, i.e., the function is band limited at frequency f_c. The second constraint is that the sampling spacing must be chosen so that

$$T = 1/(2f_c)$$ (A.49)

The frequency $1/T = 2f_c$ is known as the *Nyquist sampling rate*.

Aliasing means that the high-frequency components of a time function can impersonate a low frequency if the sampling rate is low. Figure A.6 shows a high frequency as well as a low frequency that share identical sampling points. Here, a high frequency is impersonating a low frequency for the same sampling points. The sampling rate must be high enough for the highest frequency to be sampled at least twice per cycle, $T = 1/(2f_c)$.

FIGURE A.6
High-frequency impersonating a low frequency—to illustrate aliasing.

Often the functions are recorded as sampled data in the time domain, the sampling being done at a certain frequency. The Fourier transform is represented by the summation of discrete signals where each sample is multiplied by

$$e^{-j2\pi fnt_1} \tag{A.50}$$

i.e.,

$$X(f) = \sum_{n=-\infty}^{\infty} x(nt_1)e^{-j2\pi fnt_1} \tag{A.51}$$

Figure A.7 shows the sampled time-domain function and the frequency spectrum for a discrete time-domain function.

Where the frequency-domain spectrum and the time-domain function are sampled functions, the Fourier transform pair is made of discrete components:

$$X(f_k) = \frac{1}{N} \sum_{n=0}^{N-1} x(t_n)e^{-j2\pi kn/N} \tag{A.52}$$

$$X(t_n) = \sum_{k=0}^{N-1} X(f_k)e^{j2\pi kn/N} \tag{A.53}$$

Figure A.8 shows the discrete time and frequency functions. *The discrete Fourier transform approximates the continuous Fourier transform.* However, errors can occur in the approximations involved. Consider a cosine function $x(t)$ and its continuous Fourier transform $X(f)$, which consists of two impulse functions that are symmetric about zero frequency, see Figure A.9a.

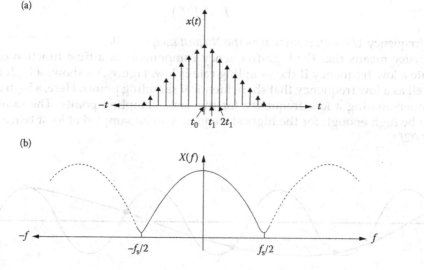

(a)

(b)

FIGURE A.7
(a) Sampled time-domain function and (b) frequency spectrum for the discrete time-domain function.

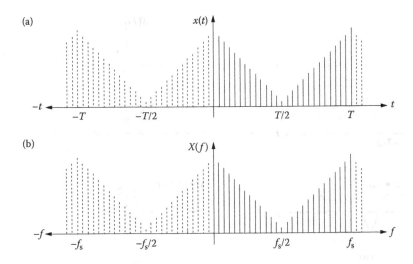

FIGURE A.8
Discrete (a) time-domain and (b) frequency-domain functions.

The finite portion of $x(t)$, which can be viewed through a unity amplitude window $w(t)$, and its Fourier transform $W(f)$, which has side lobes, are shown in Figure A.9b.

Figure A.9c shows that the corresponding convolution of two frequency signals results in blurring of $X(f)$ into two sine x/x shaped pulses. Thus, the estimate of $X(f)$ is fairly corrupted.

The sampling of $x(t)$ is performed by multiplying with $c(t)$, see Figure A.9d; the resulting frequency-domain function is shown in Figure A.9e.

The continuous frequency-domain function shown in Figure A.9e can be made discrete if the time function is treated as one period of a periodic function. This forces both the time- and frequency-domain functions to be infinite in extent, periodic and discrete, see Figure A.9e. *The discrete Fourier transform is reversible mapping of N terms of the time func-*tion into N terms of the frequency function. Some problems are outlined as follows.

A.10.1 Leakage

Leakage is inherent in the Fourier analysis of any finite record of data. The function may not be localized on the frequency axis and has side lobes (Figure A.5). The objective is to localize the contribution of a given frequency by reducing the leakage through these side lobes. The usual approach is to apply a data window in the time domain, which has lower side lobes in the frequency domain, compared to a rectangular data window. An extended cosine bell data window is shown in Figure A.10. A raised cosine wave is applied to the first and last 10% of the data and a weight of unity is applied for the middle 90% of the data. A number of other types of windows which give more rapidly decreasing side lobes have been described in the literature [1].

A.10.2 Picket Fence Effect

The picket fence effect can reduce the amplitude of the signal in the spectral windows, when the signal being analyzed falls in between the orthogonal frequencies, say between the 3rd and 4th harmonics. The signal will be experienced by both the 3rd and 4th harmonic

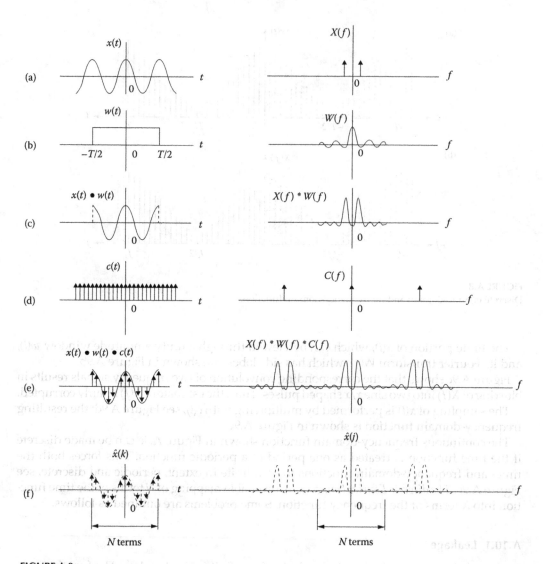

FIGURE A.9
Fourier coefficients of the discrete Fourier transform viewed as corrupted estimate of the continuous Fourier transform. (Bergland, G.D., *IEEE Spectrum* © 1969 IEEE.)

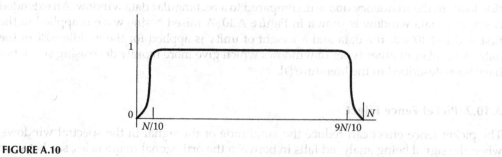

FIGURE A.10
An extended data window.

spectral windows, and in the worst case halfway between the computed harmonics. By analyzing the data with a set of samples that are identically zero, the fast Fourier transform (FFT) algorithm (Section A.11) can compute a set of coefficients with terms lying in between the original harmonics.

A.11 Fast Fourier Transform

The FFT is simply an algorithm that can compute the discrete Fourier transform more rapidly than any other available algorithm.

Define

$$W = e^{-j2\pi/N} \tag{A.54}$$

The frequency-domain representation of the waveform is

$$X(f_k) = \frac{1}{N} \sum_{n=0}^{N=1} x(t_n) W^{kn} \tag{A.55}$$

The equation can be written in a matrix form

$$
\begin{vmatrix} X(f_0) \\ X(f_1) \\ . \\ X(f_k) \\ . \\ X(f_{N-1}) \end{vmatrix}
= \frac{1}{N}
\begin{vmatrix}
1 & 1 & . & 1 & . & 1 \\
1 & W & . & W^k & . & W^{N-1} \\
. & . & . & . & . & . \\
1 & W^k & . & W^{k2} & . & W^{k(N-1)} \\
. & . & . & . & . & . \\
1 & W^{N-1} & . & W^{(N-1)k} & . & W^{(N-1)^2}
\end{vmatrix}
\begin{vmatrix} x(t_0) \\ x(t_1) \\ . \\ x(t_n) \\ . \\ x(t_{N-1}) \end{vmatrix}
\tag{A.56}
$$

or in a condensed form

$$\left[\overline{X}(f_k) \right] = \frac{1}{N} \left[\overline{W}^{kn} \right] \left[\overline{x}(t_n) \right] \tag{A.57}$$

where $\left[\overline{X}(f_k) \right]$ is a vector representing N components of the function in the frequency domain and $\left[\overline{x}(t_n) \right]$ is a vector representing N samples in the time domain. Calculation of N frequency components from N time samples, therefore, requires a total of $N \times TN$ multiplications.

For $N = 4$

$$
\begin{vmatrix} X(0) \\ X(1) \\ X(2) \\ X(3) \end{vmatrix}
=
\begin{vmatrix}
1 & 1 & 1 & 1 \\
1 & W^1 & W^2 & W^3 \\
1 & W^2 & W^4 & W^6 \\
1 & W^3 & W^6 & W^9
\end{vmatrix}
\begin{vmatrix} x(0) \\ x(1) \\ x(2) \\ x(3) \end{vmatrix}
\tag{A.58}
$$

However, each element in matrix $\left[\overline{W}^{kn}\right]$ represents a unit vector with clockwise rotation of $2n/N$ ($n = 0, 1, 2,..., N-1$). Thus, for $N = 4$ (i.e., four sample points), $2\pi/N = 90°$. Thus

$$W^0 = 1 \tag{A.59}$$

$$W^1 = \cos \pi/2 - j \sin \pi/2 = -j \tag{A.60}$$

$$W^2 = \cos \pi - j \sin \pi = -1 \tag{A.61}$$

$$W^3 = \cos 3\pi/2 - j \sin 3\pi/2 = j \tag{A.62}$$

$$W^4 = W^0 \tag{A.63}$$

$$W^6 = W^2 \tag{A.64}$$

Hence, the matrix can be written in the following form:

$$
\begin{vmatrix} X(0) \\ X(1) \\ X(2) \\ X(3) \end{vmatrix} =
\begin{vmatrix} 1 & 1 & 1 & 1 \\ 1 & W^1 & W^2 & W^3 \\ 1 & W^2 & W^0 & W^2 \\ 1 & W^3 & W^2 & W^1 \end{vmatrix}
\begin{vmatrix} x(0) \\ x(1) \\ x(2) \\ x(3) \end{vmatrix} \tag{A.65}
$$

This can be factorized into

$$
\begin{vmatrix} X(0) \\ X(2) \\ X(1) \\ X(3) \end{vmatrix} =
\begin{vmatrix} 1 & W^0 & 0 & 0 \\ 1 & W^2 & 0 & 0 \\ 0 & 0 & 1 & W^1 \\ 0 & 0 & 1 & W^2 \end{vmatrix}
\begin{vmatrix} 1 & 0 & W^0 & 0 \\ 0 & 1 & 0 & W^0 \\ 1 & 0 & W^2 & 0 \\ 0 & 1 & 0 & W^2 \end{vmatrix}
\begin{vmatrix} x(0) \\ x(1) \\ x(2) \\ x(3) \end{vmatrix} \tag{A.66}
$$

Equation A.65 yields a square matrix in Equation A.66 except that rows 1 and 2 have been interchanged.

First let

$$
\begin{vmatrix} x_1(0) \\ x_1(1) \\ x_1(2) \\ x_1(3) \end{vmatrix} =
\begin{vmatrix} 1 & 0 & W^0 & 0 \\ 0 & 1 & 0 & W^0 \\ 1 & 0 & W^2 & 0 \\ 0 & 1 & 0 & W^2 \end{vmatrix}
\begin{vmatrix} x_0(0) \\ x_0(1) \\ x_0(2) \\ x_0(3) \end{vmatrix} \tag{A.67}
$$

The column vector on the left is equal to the product of second matrix and last column vector in Equation A.66.

Element $x_1(0)$ is computed with one complex multiplication and one complex addition:

$$x_1(0) = x_0(0) + W^0 x_0(2) \tag{A.68}$$

Element $x_1(1)$ is also calculated by one complex multiplication and addition. One complex addition is required to calculate $x_1(2)$:

$$x_1(2) = x_0(0) + W^2 x_0(2) = x_0(0) - W^0 x_0(2) \tag{A.69}$$

because
$W^0 = -W^2$ and $W^0 x_0(2)$ is already computed.
Then, Equation A.66 is

$$
\begin{vmatrix} X(0) \\ X(2) \\ X(1) \\ X(3) \end{vmatrix} = \begin{vmatrix} x_2(0) \\ x_2(1) \\ x_2(2) \\ x_2(3) \end{vmatrix} = \begin{vmatrix} 1 & 0 & W^0 & 0 \\ 0 & 1 & 0 & W^0 \\ 1 & 0 & W^2 & 0 \\ 0 & 1 & 0 & W^2 \end{vmatrix} \begin{vmatrix} x_1(0) \\ x_1(1) \\ x_1(2) \\ x_1(3) \end{vmatrix} \tag{A.70}
$$

The term $x_2(0)$ is determined by one complex multiplication and addition:

$$x_2(0) = x_1(0) + W^0 x_1(1) \tag{A.71}$$

$x_1(3)$ is computed by one complex addition and no multiplication.

Computation requires four complex multiplications and eight complex additions. Computation of Equation A.58 requires 16 complex multiplications and 12 complex additions. The computations are reduced. In general, the direct method requires N^2 multiplications and $N(N-1)$ complex additions.

For $N = 2^\gamma$, the FFT algorithm is simply factoring $N \times N$ matrix into γ matrices, each of dimensions $N \times N$. These have the properties of minimizing the number of complex multiplications and additions.

The matrix factoring does introduce the discrepancy that instead of

$$
X(n) = \begin{vmatrix} X(0) \\ X(1) \\ X(2) \\ X(3) \end{vmatrix} \tag{A.72}
$$

it yields

$$
\bar{X}(n) = \begin{vmatrix} X(0) \\ X(2) \\ X(1) \\ X(3) \end{vmatrix} \tag{A.73}
$$

This can be easily rectified in matrix manipulation by unscrambling.

The $\bar{X}(n)$ can be rewritten by replacing n with its binary equivalent:

$$\bar{X}(n) = \begin{vmatrix} X(0) \\ X(2) \\ X(1) \\ X(3) \end{vmatrix} = \begin{vmatrix} X(00) \\ X(10) \\ X(01) \\ X(11) \end{vmatrix} \qquad (A.74)$$

If the bits are flipped over, then

$$\bar{X}(n) = \begin{vmatrix} X(00) \\ X(10) \\ X(01) \\ X(11) \end{vmatrix} \text{ flips to } \begin{vmatrix} X(00) \\ X(01) \\ X(10) \\ X(11) \end{vmatrix} = X(n) \qquad (A.75)$$

The second vertical array of nodes is vector $x_1(k)$ and the next vector is $x_2(k)$. There will be γ computation arrays where $N = 2^\gamma$.

While this forms an overview, an interested reader may like to probe further. Sixty-two further references are listed in [1]. Other works [2–16] provide further reading.

References

1. G.D. Bergland. A guided tour of the fast Fourier transform. *IEEE Spectrum* 6, 41–52, 1969.
2. J.F. James. *A Student's Guide to Fourier Transforms*, 3rd ed., Cambridge University Press, Cambridge, 2012.
3. I.N. Sneddon. *Fourier Transforms*, Dover Publications Inc., New York, 1995.
4. H.F. Davis. *Fourier Series and Orthogonal Functions*, Dover Publications, New York, 1963.
5. R. Roswell. *Fourier Transform and Its Applications*, McGraw-Hill, New York, 1966.
6. R.N. Bracewell. *Fourier Transform and Its Applications*, 2nd ed., McGraw-Hill, New York, 1878.
7. E.M. Stein, R.R. Shakarchi. *Fourier Analysis* (Princeton Lectures in Analysis), Princeton University Press, Princeton, 2003.
8. B. Gold, C.M. Rader. *Digital Processing of Signals*, McGraw Hill, New York, 1969.
9. J.W. Cooley, P.A.W. Lewis, and P.D. Welch. Application of fast Fourier transform to computation of Fourier integral, Fourier series and convolution integrals, *IEEE Trans Audio Electroacoust* AU-15, 79–84, 1967.
10. H.D. Helms. Fast Fourier transform method for calculating difference equations and simulating filters, *IEEE Trans Audio Electroacoust* AU-15, 85–90, 1967.
11. G.D. Bergland. A fast Fourier transform algorithm using base 8 iterations, *Math Comput* 22, 275–279, 1968.
12. J.W. Cooley. Harmonic analysis complex Fourier series, SHARE Doc. 3425, February 7, 1966.
13. R.C. Singleton. On computing the fast Fourier transform, *Commun Assoc Comput Mach* 10, 647–654, 1967.
14. J. Arsac. *Fourier Transform*, Prentice Hall, Englewood Cliffs, NJ, 1966.
15. G.D. Bergland. A fast Fourier algorithm for real value series, *Numer Anal* 11, (10), 703–710, 1968.
16. F.F. Kuo. *Network Analysis and Synthesis*, John Wiley and Sons, New York, 1966.

Appendix B: Solution to the Problems

B.1 Solutions to the Problems: Chapter 1

B.1.1 Problem 1.1

The waveform is shown in Figure B.1. With Fourier analysis

$$a_n = \frac{1}{\pi}\left[\int_{\pi/6}^{5\pi/6} I_d \cos n\omega t \, d(\omega t) - \int_{7\pi/6}^{11\pi/6} I_d \cos n\omega t \, d(\omega t)\right] = 0$$

$$b_n = \frac{1}{\pi}\left[\int_{\pi/6}^{5\pi/6} I_d \sin n\omega t \, d(\omega t) - \int_{7\pi/6}^{11\pi/6} I_d \sin n\omega t \, d(\omega t)\right] = \frac{4I_d}{n\pi}\sin\frac{n\pi}{3} \text{ for } n = 1, 3, 5...$$

$$= 0, \text{ for } n = \text{even.}$$

Therefore, the series is

$$\sum_{n=1,2,...}^{\infty} b_n \sin n\omega t$$

$$= \frac{2\sqrt{3}}{\pi} I_d\left[\sin \omega t - \frac{1}{5}\sin 5\omega t + \frac{1}{7}\sin 7\omega t - \cdots\right]$$

Consider now an ANSI standard delta–wye transformer. The voltage applied to a converter with 30° shift lags by 30°:

$$= \frac{2\sqrt{3}}{\pi} I_d\left[\sin(\omega t - 30°) - \frac{1}{5}\sin(5\omega t - 150°) + \frac{1}{7}\sin(7\omega t - 210°)...\right]$$

The positive sequence voltages or currents undergo a shift of plus 30° while the negative sequence undergo a phase shift of –30° (see Section 2.5).
Then

$$= \frac{2\sqrt{3}}{\pi} I_d\left[\sin(\omega t - 30° + 30°) - \frac{1}{5}\sin(5\omega t - 150° - 30°) + \frac{1}{7}\sin(7\omega t - 210° + 30°)\cdots\right]$$

$$= \frac{2\sqrt{3}}{\pi} I_d\left[\sin(\omega t) + \frac{1}{5}\sin(5\omega t) - \frac{1}{7}\sin(7\omega t)\cdots\right]$$

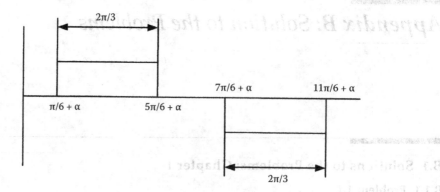

FIGURE B.1
Line current waveform, Problem 1.1.

B.1.2 Problem 1.2

$$I_L = \frac{2\sqrt{3}}{\pi} I_d \left(\sin(\omega t - \alpha) - \frac{1}{5} \sin 5(\omega t - \alpha) - \frac{1}{7} \sin(\omega t - \alpha) - \cdots \right)$$

$$I_L = 11.03 \left(\sin(\omega t - 45°) - 2.206 \sin 5(\omega t - 45°) - 1.575 \sin(\omega t - 45°) - \cdots \right)$$

The output voltage is

$$\frac{3\sqrt{3}E_m}{\pi} \cos\alpha + 2 \sum [e_n \cos m\omega t + e'_n \sin \omega t]$$

where

$$e_n = \frac{3\sqrt{3}E_m}{2\pi} (-1)^{n-1} \left[\frac{\cos(n-1)\alpha}{n-1} - \frac{\cos(n+1)}{n+1} \right]$$

$$e'_n = \frac{3\sqrt{3}E_m}{2\pi} (-1)^{n-1} \left[\frac{\sin(n-1)\alpha}{n-1} - \frac{\sin(n+1)}{n+1} \right]$$

The first term is

$$\frac{3\sqrt{3} \times 480 \times \sqrt{2}}{\pi \times \sqrt{3}} (0.707) = 459.6 \text{ V}$$

Therefore, the output voltage is

$$459.6 + 159.3(6\omega t + \alpha_6) + 79.63(12\omega t + \alpha_{12}) + \ldots$$

Thus, the ripple factor is

$$\frac{\sqrt{159.3^2 + 79.63^2}}{\sqrt{2} \times 459.6} = 0.274$$

B.1.3 Problem 1.3

This problem does not provide the data on the input of the converter in Problem 1.2. Assume a converter transformer of approximately 10 kVA, and percentage impedance = 2%. Then $(X_s + X_t)$ is 0.02 pu approximately.

The overlap angle can be calculated from Equation 1.66:

$$\mu = \cos^{-1}\left[\cos\alpha - (X_s + X_t)I_d\right] - \alpha = 1.6°$$

B.1.4 Problem 1.4

See expressions for the power factor and displacement factor and their definitions in the text.

The waveform of a full-wave single-phase bridge rectifier is shown in Figure 1.13b.

The average value of dc voltage is

$$V_{dc} = \int_{\alpha}^{\pi+\alpha} V_m \sin\omega t \, d(\omega t)$$

$$= \frac{2V_m}{\pi}\cos\alpha$$

It can be controlled by the change of conduction angle α.

From Figure B.2, the instantaneous input current can be expressed in Fourier series as follows:

$$I_{input} = I_{dc} + \sum_{n=1,2,\ldots}^{\infty} (a_n \cos n\omega t + b_n \sin n\omega t)$$

$$I_{dc} = \frac{1}{2\pi}\int_{\alpha}^{2\pi+\alpha} i(t)d(\omega t) = \frac{1}{\pi}\left[\int_{\alpha}^{\pi+\alpha} I_a d(\omega t) + \int_{\pi+\alpha}^{2\pi+\alpha} I_a d(\omega t)\right] = 0$$

Also

$$a_n = \frac{1}{\pi}\int_{\alpha}^{2\pi+\alpha} i(t)\cos n\omega t \, d(\omega t)$$

$$= -\frac{4I_a}{n\pi}\sin n\alpha \quad \text{for } n = 1, 3, 5$$

$$= 0 \quad \text{for } n = 2, 4, \ldots$$

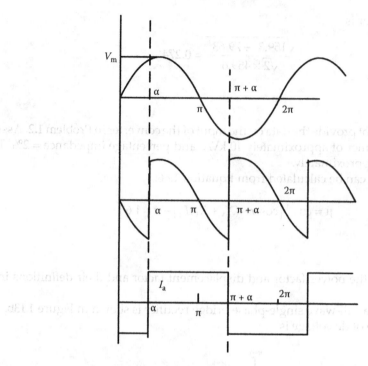

FIGURE B.2
Waveforms of a fully controlled single-phase bridge, Problem 1.4.

$$b_n = \frac{1}{\pi} \int_{\alpha}^{2\pi+\alpha} i(t) \sin n\omega t \; d(\omega t)$$

$$= \frac{4I_a}{n\pi} \cos n\alpha \quad \text{for } n = 1, 3, 5$$

$$= 0 \quad \text{for } n = 2, 4, \ldots$$

We can write the instantaneous input current as follows:

$$I_{input} = \sum_{n=1,2,\ldots}^{\infty} \sqrt{2} I_n \sin(\omega t + \varphi_n)$$

where

$$\varphi_n = \tan^{-1}\left(\frac{a_n}{b_n}\right) = -n\alpha$$

$\varphi_n = -n\alpha$ is the displacement angle of the nth harmonic current. The rms value of the nth harmonic input current is

$$I_n = \frac{1}{\sqrt{2}} (a_n^2 + b_n^2)^{1/2} = \frac{2\sqrt{2}}{n\pi} I_a$$

The rms value of the fundamental is

$$I_1 = \frac{2\sqrt{2}}{\pi} I_a$$

Thus, the rms value of the input current is

$$I_{rms} = \left(\sum_{n=1,2,\dots}^{\infty} I_n^2 \right)^{1/2}$$

The harmonic factor is

$$HF = \left[\left(\frac{I_{rms}}{I_1} \right)^2 - 1 \right]^{1/2} = 0.4834$$

The displacement factor is

$$DF = \cos \varphi_1 = \cos(-\alpha)$$

The power factor is

$$PF = \frac{V_{rms} I_1}{V_{rms} I_{rms}} \cos \varphi_1 = \frac{2\sqrt{2}}{\pi} \cos \alpha$$

B.1.5 Problem 1.5

The characteristic harmonics for a 12-pulse circuit are given by Equation 1.54, i.e., these are of the order of 11, 13, 23, 25, 35, 37, and so on. The theoretical magnitude of the harmonics is inversely proportional to the order of harmonic. On this basis, the maximum magnitude of the harmonics is 9.09, 7.69. 4.35. 4.0, 2.86, and 2.70, respectively. Note that noncharacteristic harmonics are not zero, and are always present to an extent, see Figure 1.39.

Tables 5.1 and 5.2 from the IEEE are applicable for six-pulse converters. For higher pulse numbers, the characteristic harmonics are increased by a factor of $\sqrt{P/6}$, provided that the amplitude of noncharacteristic harmonics is less than 25% of the limits specified in these tables, while evaluating the TDD.

Here, considering only the theoretical magnitude

$$HF = \sqrt{\frac{(0.09)^2 + (0.0769)^2 + (0.0435)^2 + (0.04)^2 + (0.0286)^2 + (0.0270)^2}{1}} = 13.8\%$$

Even after ignoring all the noncharacteristic harmonics, this is fairly high and harmonics up to 37th are considered. Practically, the characteristic harmonics will be reduced due to overlap angle.

Figure B.3 shows the spectrum.

B.1.6 Problem 1.6

See Table B.1; this is based on a 30° phase shift between the winding connections (see Figure 1.21).

B.1.7 Problem 1.7

The waveform pattern with firing angle α is shown in Figure B.4.

Then, as the wave is symmetrical, dc component is zero.

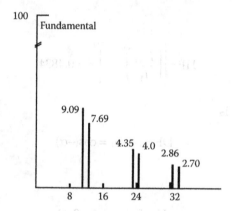

FIGURE B.3
Spectrum of harmonics, 12-pulse converter, Problem 1.5.

TABLE B.1

Transformer Winding Connections to Give Two-Pulse Operation

Two-Winding Transformers		Three-Winding Transformers	
Primary	**Secondary**	**Primary**	**Secondary**
Wye (0)	Delta (30° lag)	Wye(0)	Wye (0)
Wye (0)	Wye (0)		Delta (30° lag)
Wye (0)	Delta (30° lead)	Wye (0)	Wye (0)
Wye (0)	Wye (0)		Delta (30° lead)
Delta (0)	Delta (0)	Delta (0)	Delta (0)
Delta (0)	Wye (30° lag)		Wye (30° lag)
Delta (0)	Delta (0)	Delta (0)	Delta (0)
Delta (0)	Wye (30° lead)		Wye (30° lead)
Wye (0)	Wye (0)		
Wye (0)	Zigzag(30° lead)		
Wye (0)	Wye (0)		
Wye (0)	Zigzag (30° lag)		

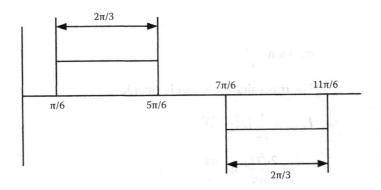

FIGURE B.4
Waveform for Problem 1.7.

The Fourier series of the input current is

$$\sum_{n=1}^{\infty} (a_n \cos n\omega t + b_n \sin n\omega t)$$

$$a_n = \frac{1}{\pi} \left[\int_{\pi/6+\alpha}^{5\pi/6+\alpha} I_d \cos n\omega t \, d(\omega t) - \int_{7\pi/6+\alpha}^{11\pi/6+\alpha} I_d \cos n\omega t \, d(\omega t) \right]$$

$$= -\frac{4I_d}{n\pi} \sin \frac{n\pi}{3} \sin n\alpha, \quad \text{for} \quad n = 1,3,5,\dots$$

$$= 0, \quad \text{for } n = 2,6,\dots$$

$$b_n = \frac{1}{\pi} \left[\int_{\pi/6+\alpha}^{5\pi/6+\alpha} I_d \sin n\omega t \, d(\omega t) - \int_{7\pi/6+\alpha}^{11\pi/6+\alpha} I_d \sin n\omega t \, d(\omega t) \right]$$

$$= \frac{4I_d}{n\pi} \sin \frac{n\pi}{3} \cos n\alpha \quad \text{for} \quad n = 1,3,5\dots$$

$$= 0, \text{for} \quad n = \text{even}$$

We can write the Fourier series as follows:

$$i = \sum_{n=1,2,\dots}^{\infty} \sqrt{2} I_n \sin(n\omega t + \varphi_n)$$

where

$$\varphi_n = \tan^{-1}\frac{a_n}{b_n} = -n\alpha.$$

The rms value of the nth harmonic is

$$I_{n,\text{rms}} = \frac{1}{\sqrt{2}}\left(a_n^2 + b_n^2\right)^{1/2}$$

$$= \frac{2\sqrt{2}I_d}{n\pi}\sin\frac{n\pi}{3}$$

The fundamental rms current is

$$I_1 = \frac{6}{\pi}I_d = 0.7797\,I_d$$

See Section 1.11.2.2 for the effect of source reactance. The overlap angle and the effect of source impedance are neglected in the above equations.

B.1.8 Problem 1.8

First calculate the overlap angle. Based on the data provided:
   pu (per unit) impedance of the transformer on converter base $X_t = 0.05457$ pu
   source reactance $= X_s = $ neglect.
   Then, from Equation 1.66, $\mu = 5.7°$.
   Neglecting overlap angle, for a six-pulse converter, the rms input current (including the effect of harmonics) is

$$I_{\text{input,rms}} = \left[\frac{2}{2\pi}\int_{\pi/6+\alpha}^{5\pi/6+\alpha} I_d^2 d(\omega t)\right]^{1/2} = I_d\sqrt{\frac{2}{3}} = 0.8165$$

$$HF = \left(\left[\frac{I_{\text{input,rms}}}{I_{\text{fund,rms}}}\right]^2 - 1\right)^{1/2} = 31.08\%$$

$$DF = \cos(-\alpha)$$

$$PF = \frac{3}{\pi}\cos\alpha = 0.9549\,DF$$

Substituting α, $DF = 0.866$ and $PF = 0.827$.
   Equation 1.63 considers overlap angle and ignores ripple content. Using this equation, Table B.2 is constructed.

TABLE B.2

Calculations of Harmonics

Harmonic Order	A	B	Harmonic (%)
5	0.049414	0.041025	25.43
7	0.058808	0.048439	21.54
11	0.047712	0.046840	11.95
13	0.046840	0.04582	9.70

A reader may refer to Chapter 7 for the following expression:

$$PF_{\text{disortion}} = \frac{1}{\sqrt{1+(THD_I/100)^2}}$$

$$\%THD_I = \sqrt{(0.2543)^2 +(0.2154)^2 +(0.1195)^2 +(0.0970)^2} = 0.364$$

Therefore, the distortion power factor is 0.9396. Note that only harmonics up to 13th have been included.

If the dc output current is given as 500 A, then the fundamental component of the ac input current is 389.85 A.

From the current distortion factor calculated above, the rms input current is 414.8 A.

The dc output voltage is 2400 V, and the ac voltage with $\alpha = 30°$ is 1185 rms V.

Then, the input volt-ampère is 1474.6 kVA.

Therefore, the total power factor is 0.8137. The difference can be attributed as we considered only harmonics up to 13th.

B.1.9 Problem 1.9

See Equation 1.80 in the text which gives the harmonic voltage and fundamental voltage expressions. If V_f is assumed as $V_{\max} = 1$ pu at $\sigma = 180°$, that is equal to $(\sqrt{2} / \pi)V_d = 1$ pu, a plot of the fundamental voltage and harmonic voltages can be created. This has been done for the fundamental, 3rd, and 5th harmonics only in Figure B.5. Similarly, other harmonics can be added to the plot. Complete plots are available in many texts.

B.1.10 Problem 1.10

In a three-phase fully controlled converter, neglecting overlap

$$HF = \left[\left(\frac{\pi}{3}\right)^2 -1\right]^{1/2} = 31.08\%$$

$$DF = \cos(-\alpha) = \cos 60° = 0.5$$

$$PF = \left(\frac{3}{\pi}\right)\cos \alpha = 0.478 \text{ (lagging)}$$

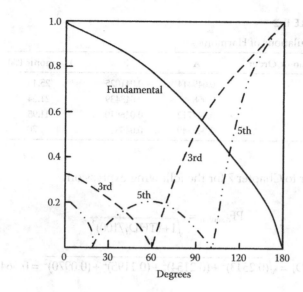

FIGURE B.5
Plots of fundamental, 3rd and 5th harmonics, Problem 1.9.

B.2 Solutions to the Problems: Chapter 2

B.2.1 Problem 2.1

The three-phase short-circuit MVA at the point of connection of furnace load is 478 MVA; the furnace load is 1 MW; then

$$SCVD = \frac{2MW_{furn}}{MVA_{sc}} = \frac{2}{478} = 0.0042$$

From the text of Chapter 2, an SCVD of 0.02–0.022 may be in the acceptable zone. Thus, flicker is not expected.

B.2.2 Problem 2.2

Harmonics due to 60 Hz source = $12 \times 60 = 720$ Hz, $2 \times 12 \times 60 = 1440$ Hz....
  Harmonics due to 43 Hz inverter = $12 \times 43 = 516$ Hz, $2 \times 12 \times 43 = 1032$ Hz....
  Harmonics of the order: $(12n\omega_1 + \omega_m)$,
  where ω_m is the mth harmonic frequency, which may be an integer harmonic of either of the two ac systems—call it a disturbing frequency.

$$12 \times 60 + 11 \times 60 = 1380 \text{ Hz},...$$

$$12 \times 60 + 11 \times 43 = 1193 \text{ Hz}$$

The third set of frequencies from Equation 2.10 will also appear in the dc current, so that $\omega_m = 12(n\omega_1 \pm p\omega_2)$. Therefore, this results in the following frequencies:

$$12n\omega_1 \pm 12(m\omega_1 \pm p\omega_2) = 12[(n \pm m)\omega_1 \pm p\omega_2)] = 12(k\omega_1 \pm p\omega_2)$$

where n, m, p, and k are integers.

$$12(60 \pm 43) = 1236 \text{ Hz}, \ 204 \text{ Hz}; \ldots$$

B.2.3 Problem 2.3

The subharmonic frequency is given by Equation 2.32 in the text:

$$f_r = 60\sqrt{0.3} = 5.4 \text{ Hz}$$

B.2.4 Problem 2.4

A reader may peruse Appendix C of Volume 1. Practically, the ferroresonance will occur on operation of one or two fuses on the HV side of a transformer protected with fuses. This forms a circuit of excitation of a nonlinear reactor (transformer windings) with a series capacitor (capacitance of the interconnecting cables). See Figure C.21 in Appendix C of Volume 1.

B.3 Solutions to the Problems: Chapter 3

B.3.1 Problem 3.1

All the parameters as calculated in Example 3.3 are valid for the 7th harmonic also.
    The 7th harmonic can be read from Figure 3.4b. Note that $r_c = 2.28$ falls beyond the x-axis of this figure. Projecting the curve for $\mu = 5°$, approximately the 7th harmonic is 8%–9%.

B.3.2 Problem 3.2

The analytical procedure is the same as for Example 3.4 in the text.
    Here

$$g_h = \frac{\sin 220.8°}{8} + \frac{\sin 165.6°}{6} - \frac{2(\sin 193.2° \sin 62.4°)}{7} = 0.01624$$

Then

$$\frac{I_h}{I_c} = \frac{2\sqrt{2}}{\pi}\left[\frac{\sin 210° \sin 33.6°}{2.0525} + \frac{2.28(0.0162)\cos 210°}{0.115}\right] = -0.16$$

Therefore, from Figure 3.8, the 7th harmonic in terms of fundamental frequency current is 8.9%.

B.4 Solutions to the Problems: Chapter 4

B.4.1 Problem 4.1

This is a straight application of Equation 4.57 in the text. For a 250 MVA short circuit, the resonant frequency is

$$60 \times \sqrt{\frac{250 \times 10^3}{1500}} = 12.92 \times 60 \text{ Hz}$$

Similarly, for 500, 750, and 1000 MVA, the resonant harmonics are 18.26, 22.36, and 25.82, respectively.

B.4.2 Problem 4.2

For series resonance

$$jX_L \times h = -jX_c / h$$

At fundamental frequency, the inductance is 37.7 Ω, and the capacitance is 942 Ω. Therefore

$$h = \sqrt{\frac{942}{37.7}} \approx 5$$

That is, the resonance will occur at 300 Hz. At this frequency

$$j\omega L = \frac{1}{j\omega C}$$

Therefore, the impedance is that of the series resistance = 0.01 Ω.
   In parallel resonance also the resonant frequency is 300 Hz, but the impedance is infinite:

$$\frac{j\omega L \times \dfrac{1}{j\omega C}}{j\omega L + \dfrac{1}{j\omega c}}$$

The denominator is zero.

B.4.3 Problem 4.3

At 600 Hz

$$X_c = X_L$$

$X_c = 300\ \Omega$ at fundamental frequency. Therefore, at 600 Hz, it is equal to 30 Ω.

Therefore, X_L at 600 Hz is equal to 30 Ω. At fundamental frequency, it is equal to 3 Ω. The Q factor (at fundamental frequency) is 50. Therefore, $R = 0.06\ \Omega$. At resonance, the impedance (R) is equal to 0.06 Ω. The plot of the impedance of the series circuit is, therefore, shown in Figure B.6.

B.4.4 Problem 4.4

Using the equations in the text for the parallel circuit, the calculated impedance plot is shown in Figure B.7.

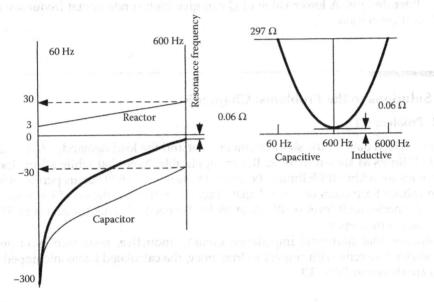

FIGURE B.6
Series resonance circuit calculation, Problem 4.3.

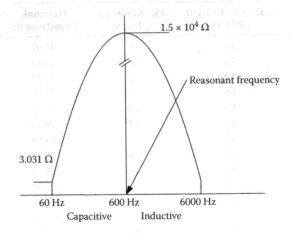

FIGURE B.7
Parallel resonance circuit calculations, Problem 4.4.

B.4.5 Problem 4.5

We considered that the specified Q is at fundamental frequency. Therefore, at the resonant frequency $Q_0 = 500$. Using Equation 4.16, the frequencies are not much different from the resonant frequency of 600 Hz.

The half-power frequencies determine the sharpness of tuning. Here, the resistance is low. *If we calculate for lower Q values,* with $Q = 30$ or higher, the frequencies are not much different from the resonant frequency. If we consider Q at resonance of say 10 only, then the frequencies can be calculated approximately 6.24% around the resonant frequency taken as unity.

This shows the impact on sharpness of tuning with change in resistance. Practically, the fundamental frequency loss on the selection of a certain Q becomes important in harmonic filter designs. A lower value of Q can give high fundamental frequency loss and much heat generation.

B.5 Solutions to the Problems: Chapter 5

B.5.1 Problem 5.1

From the given data on the short-circuit current and the load demand, I_s/I_r is < 20. Thus, the IEEE limits as shown in Table B.3 are applicable. Note that while all the individual harmonics are within IEEE limits, the total TDD is not. It is 6.19 versus permissible, =5%.

The voltage harmonics can be calculated based on the calculation of harmonic impedances. A short-circuit level of 10 kA at 69 kV is specified. This translates to 3.984 Ω at fundamental frequency.

Assuming that harmonic impedance (mainly inductive, resistance is neglected) is proportioned directly with respect to frequency, the calculated harmonic impedances in ohms are shown in Table B.3.

TABLE B.3

Problem 5.1

Harmonic Order	Harmonic Current at PCC (%)	IEEE Limits for I_s/I_r < 20	Harmonic Impedance (Ω)	Harmonic Voltage (kV)
5	3.8	4	19.92	0.787
7	3.6	4	27.89	0.808
11	1.9	2	43.82	0.865
13	1.8	2	51.79	0.969
17	1.3	1.5	67.73	0.915
19	1.2	1.5	75.70	0.944
23	0.5	0.6	91.60	0.476
25	0.5	0.6	99.60	0.518
29	0.4	0.6	115.54	0.480
31	0.4	0.6	123.50	0.513
33	0.3	0.3	131.47	0.410
35	0.2	0.3	139.44	0.290
37	0.1	0.3	147.41	0.153
	TDD = 6.19	TDD = 5%		THDv = 0.093%

Then, the harmonic voltages are

$$V_h = \sqrt{3}\omega I_h X_L$$

where X_L is at the fundamental frequency. Then, the calculated voltage distortion is 0.093%.

B.5.2 Problem 5.2

IEEE 519 gives the following expression for THD:

$$\text{THD}_{\text{Max}} = 0.074\sqrt{\frac{A_N}{\rho}}\%$$

where
ρ = the ratio of the total inductance to the common system inductance
A_N = notch area in volt-microsecond.
This expression is valid only for a 480 V line-to-line system.
We have already calculated $A_N = 51628$ µVs at point C:

$$\rho = \frac{0.64 + 34.4 + 30}{30} = 2.15$$

Substituting these values, $\text{THD}_{\text{Max}} = 11.47\%$, which is high.

B.6 Solutions to the Problems: Chapter 6

B.6.1 Problem 6.1

From Figure 6.1

$$K_{5,7} = 0.43$$

$$K_{11,13} = 0.39$$

$$K_{17,19} = 0.38$$

$$K_{23,25} = 0.38$$

Therefore, from Equation 6.5

$$I_{2\text{equiv}} = \begin{bmatrix} \sqrt{3}(0.43)(0.175+0.111)^2 + \sqrt{6}(0.39)(0.045+0.029)^2 + \sqrt{9}(0.38)(0.015+0.010)^2 \\ +\sqrt{12}(0.38)(0.009+0.008)^2 \end{bmatrix}$$

$$= 0.0672$$

Even the continuous negative sequence capability of 10% of the generator is not exceeded.

B.6.2 Problem 6.2

Construct Table B.4.
  From this table

$$F_{HL} = \frac{2.946}{1.047} = 2.8137$$

The copper loss based on the given data is

$$1.5\left[(0.2747)(209.2)^2 + (0.02698)(693.95)^2\right] = 37.52 \text{ kW}$$

Considering the loss of 45 kW

$$P_{TSL-R} = 45 - 37.52 = 7.48 \text{ kW}$$

The winding eddy loss is

$$P_{EC-R} = 0.67 \times 7480 = 5011.6$$

The transformer has a ratio less than 4:1, but the secondary current is less than 1000 A. $K_1 = 0.6$ multiplied by 4.0 pu:

$$P_{EC-Rmax} = \frac{2.4 \times 5011.6}{1.5(0.02698)(693.95)^2} = 0.617 \text{ pu}$$

Then

$$I_{max(pu)} = \sqrt{\frac{1.617}{1 + 2.8137 \times 0.617}} = 0.769 \text{ pu}$$

TABLE B.4

Calculations of Parameters for Problem 6.2

h	I_h/I	$\left(\dfrac{I_h}{I}\right)^2$	h^2	$\left(\dfrac{I_h}{I}\right)^2 h^2$	$h^{0.8}$	$\left(\dfrac{I_h}{I}\right)^2 h^{0.8}$
1	1	1	1	1	1	1
5	0.175	0.031	25	0.775	3.62	0.1122
7	0.111	0.0123	49	0.6027	4.74	0.0583
11	0.045	0.0020	121	0.2420	6.81	0.01362
13	0.029	0.00084	169	0.14196	7.78	0.00654
17	0.015	0.000225	289	0.0650	9.65	0.00217
19	0.010	0.00010	361	0.0361	10.54	0.00105
23	0.009	0.000081	529	0.0428	12.28	0.00099
25	0.008	0.000064	625	0.0400	13.13	0.00084
	$\sum$	1.047		2.946		1.1957

B.6.3 Problem 6.3

$$P_{\text{OSL-R}} = 7480 - 5011.6 = 2468.4 \text{ W}$$

The temperature rises are as follows:
  HV and LV windings average rise = 55°C
  Top oil rise = 55°C.
  Hottest spot conductor rise = 65°C.
  The losses are as follows:
  No load = 6000 W
  Copper = 37520 W
  Winding eddy = 5011.6 W
  Other stray loss = 2486 W
  Total loss = 51017.6 W.
  Considering 100% load factor, the correction for rms current is

$$P_{\text{LL}} = I^2_{\text{rms-pu}} L^2_{\text{f}} = \left(\sqrt{1.047}\right)^2 \times 1^2 = 1.047$$

Also, the harmonic loss factor for other stray loss from Table B.4 is (1.1957/1.047) = 1.142.

$$\text{Winding eddy} = 1.047 \times 2.8137 \times 5011.6 = 14665.2 \text{ W}$$

$$\text{Other stray loss} = 1.047 \times 2486 \times 1.142 = 2972.4 \text{ W}$$

$$\text{Copper loss} = 1.047 \times 37520 = 39283$$

New total losses = 62920.6 W.
  Then, the top oil temperature is

$$\theta_{\text{TO}} = 55 \times \left(\frac{62.92}{51.02}\right)^{0.8} = 65.04°\text{C}$$

Inner winding loss compensated for rms current is

$$1.047 \times 1.5(0.02698)(693.95)^2 = 20378.5 \text{ W}$$

The hottest spot conductor rise over top oil temperature is

$$\theta_{\text{g}} = (65-55) \times \left(\frac{20378.5 + 14665.2 \times 2.4}{19463.7 + 5011.6 \times 2.4}\right)^{0.8} = 15.76°\text{C}$$

Then, the hottest spot temperature is 65.04 + 15.76 = 80.8°C. The transformer on full-load factor will be soon damaged.

B.6.5 Problem 6.4

Assume that at the reduced load also, the same harmonic distribution in pu is applicable as shown in Table B.4. Then PU rms current = $\sqrt{1.047} = 1.0232$. The harmonic loss factor is 2.8137. The harmonic factor for other stray loss is

$$\frac{1.1957}{1.047} = 1.142$$

At 75% load factor

$$P_{\text{LL-pu}} = (1.0232)^2 (0.75)^2 = 0.59$$

The copper loss at 75% load factor is

$$1.5[(0.2747)(156.75)^2 + (0.02698)(520.46)^2] = 21087 \text{ W}$$

Table B.5 can be constructed.
  Then

$$\theta_{\text{TO}} = 55 \times \left(\frac{39452}{51017.6}\right)^{0.8} = 44.8^\circ$$

Inner winding losses are

$$1.5(0.02698)(693.95)^2 = 19489 \text{ W}$$

The losses corrected for specified load conditions are

$$19489(1.047)(0.75)^2 = 11498 \text{ W}$$

Then

$$\theta_g = (65 - 55) \times \left(\frac{(11498) + (2.4)(10293)}{19489 + (2.4)(5011.6)}\right)^{0.8} = 11.17^\circ$$

TABLE B.5

Transformer Losses, Load Factor 75%

Type of Loss	Rated Values (W)	At 75% Load Factor (W)	Harmonic Factor	Corrected Loss (W)
No load	6000	6000		6000
$I^2 R$	39283	21087		21087
Winding eddy	5011.6	3658	2.8137	10293
Other stray	2486	1815	1.142	2072
Total loss	51017.6	32560		39452

Then hottest spot over ambient is

$$44.8° + 11.17° = 56.5°$$

This is safe loading for the specified harmonics.

B.6.4 Problem 6.5

From Table B.4, the UL K factor is 2.946.

B.7 Solutions to the Problems: Chapter 7

B.7.1 Problem 7.1

Calculate the overlap angle

$$\mu = \cos^{-1}\left[\cos\alpha - (X_s + X_t)I_d\right] - \alpha$$

$$= \cos^{-1}(0.866 - (0.00342 + 0.055)0.75) - 60°$$

$$= 4.69°$$

Note that the converter is loaded to 75%.

Then, from the following equations in Chapter 1:

$$I_h = I_d\sqrt{\frac{6}{\pi}}\frac{\sqrt{A^2 + B^2 - 2AB\cos(2\alpha + \mu)}}{h[\cos\alpha - \cos(\alpha + \mu)]}$$

where

$$A = \frac{\sin\left[(h-1)\dfrac{\mu}{2}\right]}{h-1}$$

and

$$B = \frac{\sin\left[(h+1)\dfrac{\mu}{2}\right]}{h+1}$$

Substituting the values, for the 5th harmonic, $A = 0.04074$, $B = 0.04052$, and 5th harmonic = 12.83% of I_d.

Substituting the values, for the 7th harmonic, $A = 0.04051$, $B = 0.04020$, and 7th harmonic = 10.78%.

Similarly, other harmonic magnitudes can be calculated.

B.7.2 Problem 7.2

Figure 7.6 is applied for the adjustment of resistance with frequency, while saturation is neglected and reactance is proportionate with respect to frequency. Table B.6 shows the results.

B.7.3 Problem 7.3

Equations 7.45 and 7.46 are repeated here. The motor impedance at the hth harmonic is

$$R_1 + jhX_1 + \frac{jhx_m \left(\dfrac{r_2}{s_h} + jhx_2 \right)}{(r_2/s_h) + jh(x_m + x_2)}$$

$$s_h = 1 - \frac{n}{hn_s} \quad h = 3n + 1$$

$$s_h = 1 + \frac{n}{hn_s} \quad h = 3n - 1$$

From the given data, the motor full-load current is 547.18 A.

Therefore, the locked rotor current is 3283.1 A at a power factor of 20%.

$$(R_1 + r_2) + j(X_1 + x_2) = \frac{2.3 \times 10^3}{\sqrt{3}(3283.1)(0.2 - j0.9798)} = 0.081 + j0.396 \ \Omega$$

Assume that $R_1 = r_2 = 0.0405 \ \Omega$
$X_1 = x_2 = 0.1980 \ \Omega$

TABLE B.6

Calculation of Resistance and Reactance with Frequency

Harmonic Order	% Increase in Transformer Resistance due to Frequency, Figure 19.6	Resistance per Unit 100 MVA Base	Reactance per unit 100 MVA Base Directly Proportional to Frequency, Ignoring Saturation
Fundamental	100	0.099504	0.99504
5th	105	0.104479	4.9752
7th	107	0.106469	6.9653
11th	112	0.111444	10.9454
13th	114	0.113435	12.9355
17th	123	0.122389	16.9157
19th	126	0.125375	18.9058
21st	135	0.134330	20.8958
25th	145	0.144281	24.7526

For the induction motor x_m can be assumed to be 12 Ω approximately. Also assume the full-load slip to be 3%.

Then, the impedance at fundamental frequency from Equation 19.38 is

$$0.0405 + j0.1980 + \frac{j12.0\left(\dfrac{0.0405}{0.03} + j0.1980\right)}{\dfrac{0.0405}{0.03} + j(12 + 0.1980)}$$

$$= 1.3365 + j0.4340$$

The 5th harmonic is negative. Then, $S_h = 1.20$ (consider $n_s = n$). Again using the same equation, impedance at the 5th harmonic is

$$0.0405 + j4.90 + \frac{j60.0\left(\dfrac{0.0405}{1.2} + j4.90\right)}{\dfrac{0.0405}{1.2} + j(60 + 4.90)}$$

$$= 0.0695 + j9.43 \ \Omega$$

Similarly, the impedance at other frequencies can be calculated. Motor windings have infinite impedance for triplen harmonics.

B.7.4 Problem 7.4

If we neglect the overlap angle, then

$$DF = \cos(-\alpha) = 0.866$$

The rms value of the fundamental current:

$$I_1 = \frac{\sqrt{6}}{\pi} I_d$$

The rms input current (including effect of harmonics):

$$I_s = \sqrt{\frac{2}{3}} I_d$$

$$HF = \left[\left(\frac{I_s}{I_1}\right)^2 - 1\right]^{1/2} = 0.3108 = 31.08\%$$

$$PF = \frac{3}{\pi} \cos\alpha = 0.9549$$

$$DF = 0.827$$

Alternatively, the power factor can also be calculated as follows.

The converter voltage is not given in Problem 6.1. Consider a voltage of 480 V. Then

$$V_{\mathrm{m}} = \sqrt{\frac{2}{3}} \times 480 = 391.9 \text{ V}$$

The maximum dc output voltage is

$$V_{\mathrm{dm}} = \frac{3\sqrt{3}}{\pi} V_{\mathrm{m}} = 648 \text{ V}$$

At a firing angle of 30°, the average voltage is 561 V.
The rms value of the output voltage is given by the following expression:

$$V_{\mathrm{rms}} = \sqrt{6} V_{\mathrm{m}} \left(\frac{1}{4} + \frac{3\sqrt{3}}{8\pi} \cos 2\alpha \right)^{1/2}$$

$$= 570.65 \text{ V}$$

The converter operates at 75% of the rating of transformer which is equal to 2250 kW:

$$P_0 = I_{\mathrm{rms}}^2 R = \left(\frac{V_{\mathrm{rms}}}{R} \right)^2 R = 2250 \times 10^3$$

This gives the rms output current as follows:

$$I_{\mathrm{rms}} = 3.94 \text{ kA}$$

The rms input line current is

$$\sqrt{\frac{2}{3}} \times 3.94 \times 10^3 = 3.217 \text{ kA}$$

Thus, the input volt-ampères is

$$3 \times 277 \times 3.217 = 2674.6 \text{ kVA}$$

Then, the PF is equal to 0.841.

B.7.5 Problem 7.5

See Section 7.6 and Figure 7.13, for the consequence of inadequate modeling. A sensitivity analysis is required. See Sections 7.6.4 and 7.2.1 for the transmission line models and Section 7.2.4 for the transformer models.

B.7.6 Problem 7.6

The line-to-line voltage is 44 kV, and the line-to-neutral voltage is 25.40 kV. Referring to Table 7.2, and using capacitors of 13.28 kV rated voltage, two series groups are required. This gives a voltage of 26.56 kV. But the overvoltage factor is of only 1.045. Generally, it is desirable to have a factor of 1.1. Move to the next rated voltage of 13.8 kV. This gives an overvoltage factor of 1.087. It will be desirable to use 14.4 kV rated capacitors. The banks size is 15 Mvar. This means 5 Mvar per phase. As expulsion fuses are to be used a single wye-connected bank cannot be designed. Thus, it has to be double-wye connection. Select 400 kvar capacitor cans. The voltage is now 28.8 kV (two series groups). To meet the requirements of 2.5 Mvar per phase for each Y-group, we must have

$$(2.5)\left(\frac{28.8}{25.40}\right)^2 = 3.214\,\text{Mvar}$$

As 400 kvar units are being used, 8.03 units per phase will be required. If we use 8 units, the effective Mvar at operating voltage of 44 kV will be 14.93, instead of desired 15 Mvar. Consider this acceptable. Check that the bank thus formed meets the requirements in Table 7.3 for the number of units in double Y and also that expulsion fuses can be used.

The formation of the bank is shown in Figure B.8.

B.7.7 Problem 7.7

The voltage and current at the sending and receiving ends of a transmission line are not identical.

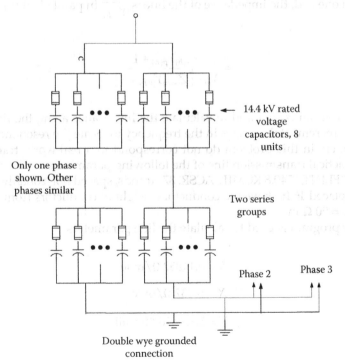

FIGURE B.8
Formation of a double-wye capacitor grounded bank, Problem 7.6.

B.7.8 Problem 7.8

Example 10.6 uses a nominal π model. *The Equivalent π model is derived as follows:*

$$\left| \begin{array}{c} I_s \\ I_r \end{array} \right| = \frac{1}{B} \left| \begin{array}{cc} D & CB-DA \\ 1 & -A \end{array} \right| \left| \begin{array}{c} V_s \\ V_r \end{array} \right|$$

$$= \frac{1}{Z_0 \sin h\gamma l} \left| \begin{array}{cc} \cos h\gamma l & -1 \\ 1 & -\cos h\gamma l \end{array} \right| \left| \begin{array}{c} V_s \\ V_r \end{array} \right|$$

$$= \left| \begin{array}{cc} \dfrac{1}{Z_s}+Y_p & -\dfrac{1}{Z_s} \\ \dfrac{1}{Z_s} & -\dfrac{1}{Z_s}-Y_p \end{array} \right| \left| \begin{array}{c} V_s \\ V_r \end{array} \right|$$

Note that

$$CB - DA = \sin h^2(\gamma l) - \cos h^2(\gamma l) = -1$$

Looking from one end, the impedance of the line is $\dfrac{1}{Y_{p(h)}}$ in parallel with $\left(Z_{s(h)} + \dfrac{1}{Y_{p(h)}} \right)$

$$= \frac{Z_{s(h)}Y_{p(h)}+1}{Y_{p(h)}\left[Z_{s(h)}Y_{p(h)}+2 \right]}$$

A computer program is required to calculate the impedance along the line using above Equation with incremental changes in the frequency to locate the resonance points.

The parameters in this problem do not correspond to a real-world transmission line. Consider a practical transmission line of the following parameters:

Conductor "FLINT," 740.8 KCMIL, ACSR, 37 strands, spaced horizontally 25 ft apart, one ground wire placed 16 ft above the conductors, height of conductors from ground = 60 ft, soil resistivity = 90 Ω-m.

A computer program is used to calculate the line parameters:

$$R = 0.14852 \ \Omega/\text{mile}$$

$$X = 0.837 \ \Omega/\text{mile}$$

$$Y = 5.15 \times 10^{-6} \ \text{S/mile}$$

Then

$L = 0.00222 \text{ H/mile}$

$C = 0.01366 \times 10^{-6} \text{ F/mile}$

$Z_0 = 403 \ \Omega$

$\gamma = \sqrt{zy} = [(0.14852 + j0.837)(j5.15 \times 10^{-6})]^{1/2} = (0.184 + j2.084) \times 10^{-3}/\text{miles}$

$\lambda = \dfrac{2\pi}{\beta} = 3014 \text{ miles}$

$v = f\lambda = 1.808 \times 10^{5} \text{ miles/s}$

$f_{\text{osc}} = \dfrac{1}{\sqrt{LC}} = 3016 \text{ Hz}(300 \text{ miles-line})$

Also

$$Z_s = Z_c \sin h(\gamma l) = \sqrt{\frac{0.14852 + j0.837}{j5.15 \times 10^{-6}}} \sin h(0.0552 + j0.6252)$$

$$= (405 - j36) \sin h(0.0552 + j0.6252)$$

$$\sin h(0.0552 + j0.6252) = \sin h(0.0552)\cos(0.6252) + j\cos h(0.0552)\sin(0.6252)$$

$$= (0.0552)(0.8108) + j(1.0)(0.5853)$$

$$= 0.045 + j0.5853$$

Then,

$$Z_s = (405 - j36)(0.045 + j0.5833)$$

$$= 39.224 + j234.6$$

$$Y_p = \tanh(\gamma l/2)/Z_c$$

$$\tan h(\gamma l/2) = \frac{\sin h(\gamma l/2)}{\cos h(\gamma l/2)} =$$

$\sin h(\gamma l/2) = \sin h(0.0275 + j0.326) = (0.0275)(0.947) + j(1)(0.32) = 0.026 + j0.32$

$\cos h(\gamma l/2) = \cos h(0.275)(0.947) + j(0.0275)(0.32) = 0.947 + j0.0088$

$\tan h(\gamma l/2) = 0.031 + j0.338$

$Y_p = \dfrac{0.031 + j0.338}{405 - j36} = 2.341 \times 10^{-6} + j8.348 \times 10^{-4} \approx j8.348 \times 10^{-4}$

Then, the line impedance at fundamental frequency is

$$\frac{Z_{s(h)}Y_{p(h)}+1}{Y_{p(h)}\left[Z_{s(h)}Y_{p(h)}+2\right]}$$

$$=\frac{(405-j36)(j8.348\times10^{-4})+1}{(j8.348\times10^{-4})[(405-j36)(j8.348\times10^{-4})+2]}$$

$$=\frac{1.03+j0.338}{-2.822\times10^{-4}+1.695\times10^{-3})}$$

$$=98.46-j600.36$$

This shows the procedure: a computer program will run these calculations at close increment of frequency to capture the series and shunt resonance frequencies; plots as demonstrated in Figure 7.5.

B.8 Solutions to the Problems: Chapter 8

B.8.1 Problem 8.1

$$\text{THD}_i=\frac{\sqrt{\displaystyle\sum_{h=2}I_h^2}}{I_1}=59.1\%$$

$$I_{\text{rms}}=\sqrt{\sum_{h=1}I_h^2}=242.78\text{ A}$$

$$I_{\text{rms}}=I_1\sqrt{1+\text{THD}_i^2}=1.162$$

As the filter is 5 Mvar

$$X_c=\frac{kV^2}{\text{Mvar}}=\frac{13.8^2}{5}=38.09\ \Omega$$

As the tuning frequency of the filter is $n=10.65$:

$$X_L=\frac{X_c}{n^2}=0.336\ \Omega$$

The output of the filter is

$$\frac{n^2}{n^2-1} \times 5 = 5.044 \text{ Mvar}$$

B.8.2 Problem 8.2

As per IEEE capacitor bank ratings, I_{rms} is no more than 1.35 times the fundamental current. This limit is met as calculated in Problem 8.1.

Capacitor rms voltage is within 1.1 pu limit.

The capacitor voltage is calculated in Table B.7 using the following expression:

$$V_{ch} = \sqrt{3} I_h X_c / h$$

We have already calculated X_c in Problem 8.1

$$V_{c,rms} = \sqrt{\sum_{h=1} V_{ch}^2} = 13.944 \text{ kV}$$

This is 1.0104 times the rated voltage and, therefore, is acceptable.

The peak capacitor voltage should be within 1.2 pu limit as per IEEE.

$$\sum_{h=1} V_{ch} = 15.583$$

This is 1.129 times the rated and, therefore, is acceptable

TABLE B.7

Problems 8.1 and 8.2

Harmonic Order	Harmonic Current (A)	Harmonic Voltage (kV)	I_h (pu)
5	20	0.264	0.095
7	45	0.424	0.214
11	100	0.600	0.476
13	50	0.254	0.238
17	10	0.038	0.047
19	8	0.027	0.038
23	12	0.034	0.057
25	4	0.010	0.019
29	2	0.004	0.009
31	2	0.004	0.009
35	1	0.002	0.005
37	1	0.002	0.005

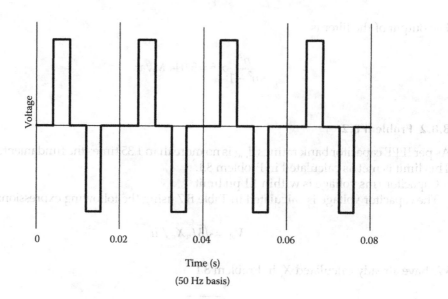

FIGURE B.9
Output waveform of a three-level inverter, Problem 8.3.

Finally, the kvar limit is 1.365 pu. This is calculated as follows:

$$Q_{C,\mathrm{pu}} = \sum_{h=1} \left(\frac{I_{h,\mathrm{pu}}^2}{h} \right) = 1.03 \text{ pu}$$

(See Table B.7.)
Thus, none of the limits are exceeded.

B.8.3 Problem 8.3

The operation time in each level is 90° (Figure B.9). The switching angles in a cycle are

$$\alpha_1 = 45°$$

$$\alpha_2 = 135°$$

$$\alpha_3 = 225°$$

$$\alpha_4 = 315°$$

Thus, the harmonics are

$$h_3 = \frac{\sin(3 \times 90^\circ / 2)}{3 \sin(90^\circ / 2)} = \frac{1}{3}$$

$$h_5 = \frac{\sin(5 \times 90^\circ / 2)}{5 \sin(90^\circ / 2)} = \frac{1}{5}$$

$$h_7 = \frac{1}{7}$$

Considering harmonics only up to 7th the THD = 43.7%.

Index

Note: Page numbers followed by "*f*" and "*t*" refer to figures and tables, respectively.